D0847963

REFRIGERATION:
HOME and COMMERCIAL

by Edwin P. Anderson
and
Roland E. Palmquist

THEODORE AUDEL & CO.
a division of

HOWARD W. SAMS & CO., INC.
4300 West 62nd Street
Indianapolis, Indiana 46268

FIRST EDITION

Fourth Printing—1981

Copyright © 1977 by Howard W. Sams & Co., Inc., Indianapolis, Indiana 46268. Printed in the United States of America.

All rights reserved. Reproduction or use, without express permission, of editorial or pictorial content, in any manner, is prohibited. No patent liability is assumed with respect to the use of the information contained herein. While every precaution has been taken in the preparation of this book, the publisher assumes no responsibility for errors or omissions. Neither is any liability assumed for damages resulting from the use of the information contained herein.

International Standard Book Number: 0–672–23286–3
Library of Congress Catalog Card Number: 77–71583

Foreword

This book has been prepared to assist personnel that are involved with the installation, servicing, and operating of refrigeration equipment. Fundamentally, the principles and physics of refrigeration have not changed to any degree during the many years that we have relied on refrigeration as a means of keeping foods from spoiling.

Household as well as commercial types of refrigeration will be covered. Some of the items in this book will of necessity involve both household and commercial refrigeration, as the principles are basically the same.

It is not the intent of this book to cover large industrial installations, as this is a separate subject, involving refrigerating machines of very large capacity.

The personnel involved in installing, servicing and operating refrigeration equipment must have a firm foundation and knowledge of the principles of refrigeration, including the fundamentals of refrigeration and the laws of physics that govern these fundamentals.

Trouble charts will be included as necessary, so that one may refer to them, for troubles that present themselves in the servicing and operating of refrigeration equipment.

It is most important to have a thorough understanding of the working parts of both electrically driven and absorption-type refrigeration units. This will include the properties of the various refrigerants used, including some of the obsolete refrigerants, as it is possible that you may come across some of these obsolete refrigerants still in use. Once you have the theory of refrigeration well organized in your mind, troubles in the systems will be much easier to locate.

My most sincere thanks goes to Mr. Carl W. McPhee, Editor, of the ASHRAE HANDBOOKS AND PRODUCT DIRECTORY. ASHRAE, is the American Society of Heating, Refrigeration and Air-Conditioning Engineers, Inc. This Society is the foundation source of information on Refrigeration, Heating, and Air-Conditioning and has issued many books on these subjects. Credit will be properly given to them for any portions used from their Handbooks, as we proceed.

Contents

control by suction pressure—servicing valves—summary—review questions

Units of electricity—the electrical circuit—direct current—single phase current—frequency and slip—three-phase current—transformer connections—energy—motor efficiency—pulley speed and sizes—calculation of driver diameter—calculation of the driven diameter—calculation of driver rpm—alternating-current motors—single-phase motors—motor-compressor controls—oil-pressure controls—temperature controls—defrost controls—electrical system maintenance—compressor motor replacement—electrical system trouble chart—summary—review questions

Effect of pressure on refrigerants—essential parts of a refrigeration plant—condensers, fittings, and accessories—cooling water requirements—cooling system components—refrigeration system operation—compressor trouble chart—summary—review questions

Indirect refrigeration system—holdover tanks—congealing tanks—brine—brine system components—preparing new brine—summary—review questions

Can ice—plate ice—air piping—ice cans—inferior ice—freezing tanks—can dumps—plate ice systems—automatic ice makers—dry ice—ice making calculations—ice storage—summary—review questions

Display case arrangement—open display cases—display case lighting—radiant heat effect—closed meat display cases— multiple connection—display case maintenance—installation—mechanical centers—combined systems—display case defrosting—drainage requirements—walk-in coolers—meat preparation—summary—review questions

Locker plant construction—design considerations—insulation requirements—equipment and controls—summary—review questions

Water coolers—ice cream refrigeration—bottled and canned beverage coolers—milk coolers—servicing and repairs—milk pasteurization—summary—review questions

CHAPTER 1

Fundamentals of Refrigeration

Refrigeration, regardless of the means by which it is obtained, may be defined as a process of the removal of heat from a substance or from a space. Thus, if heat is removed from any substance or space, that substance or space becomes cool. In making the statements, warm or cool, one must remember that these are relative statements. What is warm or what is cool is relative to the surrounding temperatures. Thus, we must consider temperatures as measured by means of thermometers, to arrive at the relationship between heat and cool.

From the foregoing statement, it follows that the problem of refrigeration concerns itself with heat removal from a substance or a space. Heat is a form of *energy*, like electrical, chemical, and mechanical energy. It must be stated that energy can neither be created or destroyed, but can be converted from one form to another, with temperature indicting its relative intensity.

The first method of refrigeration in recent history, was by means of cutting and storing ice in cabinets. Later, considerable progress was made in the commercial manufacture of artificial ice.

Probably the most widely used form of refrigeration, was the manufacture of ice by means of a electrically driven compressor, with ammonia used as the refrigerant from which 100 lb. cakes of ice were produced commercially. The development of household refrigeration followed, and since 1923 this type of refrigeration machine has been manufactured in large quantities and has proven a commercial success. Today, this type of machine is no longer considered a luxury, but is considered as a necessity by all of us. From this effort, we further expanded into the commercial refrigeration and the frozen foods field.

APPLICATIONS OF REFRIGERATION

Food preservation is the largest application of mechanical refrigeration today. Without modern refrigerating machinery the packing house industry, the transportation of perishable foods, and the preparation of many other edibles would not only be difficult, but in many cases almost impossible. Refrigeration has not only saved quantities of meat, fish, eggs, milk, and cream from spoilage, it has also played an important part in the diet revision of the world. No longer are the inhabitants of one hemisphere, country or locality dependent upon local foods, but many draw upon the entire world as a source. This has made possible great developments in agriculture and livestock raising in countries far distant from potential markets. It has permitted the utilization of the great productivity of the tropics in supplying fruits and foods for other parts of the world.

Nearly all of the industries involved in the preparation of foods and drinks make extensive use of artificial ice or mechanical refrigeration. The dairy industry, for example finds that for precooling milk and cream, and in the manufacture of butter and ice cream, refrigeration is indispensable. In many industries, the cooling and conditioning of air is an important phase of the manufacturing process. Among these processes may be mentioned the manufacture of photographic films, explosives, the machine production of cigars and cigarettes, candy and chewing gum, and rayon. They are important users of refrigeration, not only because of their size, but also because of their dependence on conditioned air for successful operation.

The cooling of liquids plays an important part in the mechanical and chemical industries. Oil-tempering baths are kept at constant low temperatures through refrigeration, and in the jackets of nitrators and mixing machines for the celluloid, smokeless powder, and rubber industries, cooled brines are used to keep the mixtures at the correct temperatures. In the chemical industry, the use of refrigeration for drying air and gases and for the purification of solutions by means of crystallization at definite temperatures is steadily increasing.

PHYSICAL UNITS

In order to obtain a clear conception of the functioning of a mechanical refrigeration system, it is imperative to consider and understand the physical and thermal properties underlying the production of artificial cold.

Practical Dimensions

We are apparently in a transition period from our system of measurements to the Metric system. Thus, we must consider both systems in our discussions. You will notice that many items that you purchase in stores, give weights and measures both in the English and the Metric system. A comparison will now be made between the two different systems, but the book will primarily use our present system. From the following information given you may readily convert from one system to another.

Prefixes used in the Metric system:

Micro = one millionth
Milli = one thousandth
Centi = one hundredth
Deci = one tenth
Deca = ten
Hecto = one hundred
Kilo = one thousand
Mega = one million
10 millimeters = 1 centimeter
100 centimeters = 1 meter

1000 millimeters = 1 meter
25.4 millimeters = 1 inch
2.54 centimeters = 1 inch
30.48 centimeters = 1 foot
304.8 millimeters = 1 foot
3.28083 feet = 1 meter
39.36996 inches = 1 meter
0.62137 miles = 1 kilometer
1.60935 kilometers = 1 mile
0.9144 meters = 1 yard

Fig. 1. The relationship of centimeters to feet and meters to yards.

Square Measure

Common units of area, covering both our system and the Metric system are:

1 square foot = 144 sq. inches
1 square yard = 9 square feet
1 square foot = 929.03 square centimeters
1 square yard = 083612736 square meters
1 foot = 30.48 centimeters
1 yard = 0.9144 meters

Fig. 2. Illustrates the relationship between various Metric and English square units.

Cubic Measure or Measure of Solids

Common measurements of three dimensional spaces of volume, in both our system and the Metric system.

1 cubic foot = 1728 cubic inches
1 cubic yard = 27 cubic feet
1 cubic foot = 28,316.84659 cubic centimeters
1 cubic foot = 0.02832 cubic meters
1 cubic yard = 0.7646 cubic meters

Fig. 3. Relative amount of space occupied by a cubic foot, as compared a cubic yard.

Mass

The quantity of matter which a body contains is called its *mass*. The space a body occupies is called its *volume*. The relative quantity of matter contained in a given volume is called its *density*. The relation of volume to mass is expressed by the term density, and is written:

$$\text{density} = \frac{\text{mass}}{\text{volume}}$$

$$\text{mass} = \text{density} \times \text{volume}$$

$$\text{volume} = \frac{\text{mass}}{\text{density}}$$

The relative density of a substance is the ratio of its density to the density of pure water at a temperature of approximately 39.2° F (Fahrenheit) or 4° C (Celsius, formerly Centigrade)

Mass is a property of all matter. Gas, air, water, and metals have mass. The weight of a substance is due to the earth's attraction on a substance (gravity). The only condition in which a substance has no weight, is when it is falling in a perfect vacuum under the influence of gravity. At all other times it has weight, therefore, weight divided by acceleration due to gravity is mass.

13

g = acceleration due to gravity = 32.2 feet per second per second.

The unit of mass is the slug. The definition of a slug is: 32.74 pounds or 14.505 kilograms.

Our units of weight are the ounce, the pound, and the ton.

$$16 \text{ ounces } = 1 \text{ pound}$$
$$2000 \text{ pounds } = 1 \text{ ton}$$

Example: What is the mass of a 10 pound substance?

$$\text{Mass} = \frac{W}{g}$$

where,

W = weight of substance,
g = acceleration due to gravity (32.2 feet per second per second).

$$\text{Mass} = \frac{10}{32.2} = 3.22 \text{ slugs}$$

The metric units of weight and their conversion to our system:

$$1 \text{ grain} = 0.6480 \text{ grams}$$
$$1 \text{ ounce} = 28.3495 \text{ grams}$$
$$1 \text{ pound} = 0.45359 \text{ kilograms}$$
$$1 \text{ short ton} = 0.9718 \text{ metric tons}$$
$$1 \text{ long ton} = 1.01605 \text{ metric tons}$$
$$1 \text{ gram} = 15.4324 \text{ grains}$$
$$1 \text{ gram} = 0.03527 \text{ ounces}$$
$$1 \text{ kilogram} = 2.20462 \text{ pounds}$$
$$1 \text{ metric ton} = 1.10231 \text{ short tons}$$
$$1 \text{ metric ton} = 0.98421 \text{ long tons}$$

Before defining specific gravity, it should be explained that there must be a certain temperature used in arriving at specific gravity as liquids, solids, and gases will expand when heated and contract when cooled. As

14

you noticed in the above discussion about density of water, 39.2° F (4° C) was used.

Specific Gravity

Specific gravity is the ratio between the weight of a given volume of any substance and the weight of the same volume of some other substance taken as a standard. For solids and liquids, this standard is distilled water at a temperature of 39.2° F. Fig. 4 illustrates the volume and weight of

Fig. 4. Illustrating volume and weight of one gallon of water at 62° F.

one gallon of water at 62° F. In Fig. 5, the weight of one cubic inch of water is compared with one cubic foot. For gases and vapors, some substance must be taken that is itself a gas, and so the standard for air is *hydrogen*. For solids and liquids, therefore:

$$\text{specific gravity} = \frac{\text{weight of body in air}}{\text{weight of same volume of water}}$$

15

Fig. 5. Showing weight of one cubic inch of water and volume of one cubic foot and their metric equivalents.

$$\text{specific gravity} = \frac{\text{weight of body in air}}{\text{loss of weight in water}}$$

Fig. 6 shows how to determine the specific gravity of a piece of cast iron. It will be noted that its weight in air equals 15 pounds, and the

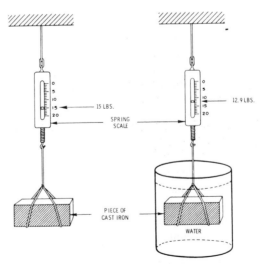

Fig. 6. Illustrating the specific gravity of a piece of cast iron in air and in water.

weight registered by the same scale when the pieces are totally submerged in water is 12.9 pounds. Substituting formula,

$$\text{specific gravity} = \frac{W}{W-w} = \frac{15}{15-12.9} = 7.14$$

ATMOSPHERIC, ABSOLUTE, AND GAUGE PRESSURES

It is of the utmost importance that the refrigeration student understand the meaning of the various kinds of pressure as related to refrigeration.

Atmospheric Pressure

Atmospheric pressure is that pressure which is exerted by the atmosphere in all directions, as indicated by a barometer. Standard atmospheric pressure is considered to be 14.695 pounds per square inch (usually written 14.7) which is equivalent to 29.92 inches of mercury.

When using 14.7 lbs. or 29.92 inches of Hg as atmospheric pressure, bear in mind that this is the perfect atmospheric pressure at sea level, and will vary plus or minus, with altitude and the local atmospheric pressures. 62° F is the temperature that is used for determining standard atmospheric pressure.

As the altitude increases above sea level, the barometric pressure decreases. This is mentioned since in working with refrigeration, you will find later that vacuum measurements must be made. Altitude also affects the boiling point of the refrigerants. To show the results, the following pressures are given in Table 1 for various altitudes.

Barometers should always be corrected to the sea level reading, but from the above figures you can find the actual barometric pressure for the altitude at which you are working. This is a standard reading, and does not take into account high and low atmospheric pressures, as occur with weather changes.

Absolute Pressure

Absolute pressure is the sum of gauge pressure and atmospheric pressure at any particular time. For example, if the pressure gauge at one particular time reads 53.7 pounds, the absolute pressure would be 53.7 + 14.7, or 68.4 pounds per square inch.

The aforementioned definitions may be written as follows:

$$\text{Abs. pressure} = \text{gauge pressure} + 14.7$$

Table 1. Effects of Altitude on Atmospheric Pressure

Altitude in Feet	Pounds Per Sq. In.	Inches of Mercury
Sea Level = 0	14.72	29.92
1000	14.17	28.80
2000	13.64	27.73
3000	13.13	26.69
4000	12.64	25.69
5000	12.17	24.74
6000	11.71	23.80
7000	11.27	22.91
8000	10.85	22.05
9000	10.45	21.24
10,000	10.06	20.45
11,000	9.69	19.70
12,000	9.33	18.96
13,000	8.98	18.25
14,000	8.64	17.56
15,000	8.32	16.91

where,

14.7 is the normal atmospheric pressure, then,

Gauge pressure = abs. pressure − 14.7

BOYLE'S LAW FOR PERFECT GASES

Boyle's law refers to the relationship between the pressure and volume of a gas and may be stated as follows: *With temperatures constant, the volume of a given weight of gas varies inversely as its absolute pressure.*

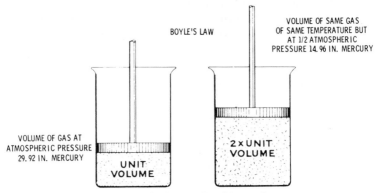

Fig. 7. Illustrating Boyle's law for a perfect gas.

This is illustrated if Fig. 7. Boyle's law is written mathematically as

$$P_1V_1 = P_2V_2$$

or,

$$\frac{V_1}{V_2} = \frac{P_2}{P_1}$$

where,

P_1 = absolute pressure of a quantity of perfect gas before a pressure change.

P_2 = absolute pressure after pressure change.

V_1 = volume of gas at pressure P_1.

V_2 = volume of gas at pressure P_2.

Since P_1V_1 for any given case is a definite constant quantity, it follows that the product of the absolute pressure and volume of a gas is constant,

or

$PV = C$ (when the temperature is kept constant)

In this connection it should be mentioned that any change in the pressure and volume of a gas at constant temperature is called an *isothermal change*.

CHARLES' LAW

Charles' law refers to the relationship between the pressure, volume, and temperature of a gas, and may be stated as follows: *At a constant pressure, the volume of a gas varies directly as the absolute temperature; at a constant volume, the pressure varies directly as the absolute temperature.* This is illustrated in Fig. 8.

When heat is added to a constant volume, the relation is written:

$$P_1T_2 = P_2T_1$$

or,

$$\frac{P_1}{P_2} = \frac{T_1}{T_2}$$

For the same temperature range at a constant pressure:

$$V_1T_2 = V_2T_1$$

or,

$$\frac{V_1}{V_2} = \frac{T_1}{T_2}$$

VOLUME OF GAS
AT 0⁰ C. OR 273⁰ ABS.

VOLUME OF SAME GAS
AT 273⁰ C. OR 546⁰ ABS.

PRESSURE UNCHANGED

Fig. 8. Illustrating Charles' law for a perfect gas.

Combined, these laws read:

$$P_1 V_1 T_2 = P_2 V_2 T_1$$

or,

$$\frac{P_1 V_1}{T_1} = \frac{P_2 V_2}{T_2}$$

Since volume is proportional to weight, the relationship between P, V and T for any weight of gas W is:

$$PV = WRT$$

where,

P = Absolute pressure of the gas in points.
V = volume of gas.
W = weight of gas.
R = constant depending on the gas under consideration.
T = absolute temperature in ° F.

Fig. 9 illustrates the four standard temperature scales, with a comparison of the Absolute Celsius and Celsius scales, showing the Absolute scale has its zero degrees at 273° below the freezing point of water, which

Fig. 9. The four standard temperature scales.

is 0° on the Celsius scale, so 0° Celsius would therefore be 273° C Absolute and 10° Celsius would be 283° C Absolute.

This figure also shows the Fahrenheit scale and Absolute Fahrenheit scale. Freezing of water on the Fahrenheit scale would be 32° and on the Absolute scale would be 492°.

The methods and scales used to measure temperature have been chosen by scientists, and the following standards have been established. The common American scales are the Fahrenheit scale and the Fahrenheit Absolute scale (Rankine scale), while the metric system embodies the Celsius and Celsius absolute scales (Kelvin scale).

The Fahrenheit scale is so fixed that it divides the heat level from the melting temperature of ice to the boiling point of water into 180 equal divisions and sets the melting point of ice at 32 divisions above the zero scale. Therefore, ice melts at 32° F and water boils at 212° F. (180° F + 32° F) assuming the standard atmospheric pressure of 14.7 pounds per square inch at sea level. The Fahrenheit absolute (F_a) scale is a scale using the same divisions as the Fahrenheit scale but setting zero of the scale at the temperature where molecular action of all substances ceases (absolute zero), where no more heat exists in the body and the temperature cannot be lowered any further. This temperature corresponds to $-460°$ F and water boils at 672° F_a assuming standard atmospheric pressure.

21

The Celsius scale has coarser divisions than the Fahrenheit scale and the zero (0°) of this scale is set at the melting temperature of ice. The boiling point of water is fixed 100 divisions above that point or 100° C assuming atmospheric pressure to be 14.7 pounds per square inch at sea level.

You will on occasion have to convert Celsius readings to Fahrenheit readings and also Fahrenheit readings to Celius readings.

To convert Celsius degrees into Fahrenheit degrees:

$$° F = 9/5 \ (° \ C \ + \ 32)$$

To convert Fahrenheit degrees into Celsius degrees:

$$° C = 5/9 \ (° \ F \ - \ 32)$$

To convert Fahrenheit degrees to Fahrenheit absolute:

$$° F_a = ° F + 460$$

To convert Celsius degrees into Celsius absolute:

$$° C_a = ° C + 273$$

POWER AND WORK

Power may be defined as the capability of performing mechanical work as measured by the rate at which it is or can be done. Stated in simple language, *power is the rate of doing work.* If a horse, as shown in Fig. 10, pulls a weight of 220 pounds against gravity at the rate of 2.5 feet per second, and if the friction in the pulling arrangement is neglected, the horse will develop one horsepower. The work done in a given time divided by the units of time gives *the average rate of doing work, or the power.*

It may be written:

$$\text{Power} = \frac{\text{work}}{\text{time}}$$

Fig. 10. Illustrating the meaning of one horsepower.

UNITS OF POWER

The most commonly used units of power are the foot-pounds per unit time, and horsepower. The horsepower is the rate of doing work at 550 foot-pounds per second or 33,000 foot pounds per minute. It is equivalent to the power which is used to raise a weight of 33,000 pounds against gravity at a rate of one foot per minute.

$$\text{H.P.} = \frac{\text{foot-pounds}}{33,000 \times \text{time in minutes}}$$

Another way of illustrating one horsepower is shown in Fig. 11. The motor is said to develop one horsepower when the 550-pound weight is lifted against gravity at a rate of one foot per second. If a weight of 3000 pounds is lifted through a distance of 35 feet in two minutes, the required number of horsepower would be:

$$\text{H.P.} = \frac{3000 \times 35}{33,000 \times 2} = 1.59$$

ENERGY

A body is said to possess energy when it can do work. Energy exists in various forms, such as:

23

Fig. 11. Another illustration of one horsepower.

1. Mechanical energy
2. Electrical energy
3. Heat energy (thermal)

Other well known forms are *kinetic* and *potential energy*.

Energy may be transmitted from one form to another by various processes, such as mechanical, thermal, chemical, or electrical, but the energy lost or gained is either kinetic or potential. If the process is mechanical, such as impact, compression, or the application of a mechanical force, work is done. If it is thermal, radiation, or conduction, then heat is added or withdrawn. There are very simple relations that exist between mechanical, electrical, and heat energy. These relationships may be written as follows:

$$778 \text{ foot-pounds} = 1 \text{ British thermal unit (Btu)}$$
$$2546 \text{ British thermal units (Btu)} = 1 \text{ horsepower-hour}$$
$$39,685 \text{ British thermal units (Btu)} = 1 \text{ kilogram-calorie}$$
$$746 \text{ watts} = 1 \text{ horsepower}$$

Other equivalents are:

> 1 Btu = 0.001036 lb. of water evaporated at 212° F.
> 0.0000688 lb. of carbon oxidized
> 0.000393 horsepower-hour
> 107.6 kilogram-meters
> 778 foot-pounds
> 1055 watt-seconds

From these equivalents it may immediately be apparent that heat is an entity, a real something, for it has a unit of measure, and each unit is convertible to other forms of energy at a constant rate of exchange. Since refrigeration is an art that is concerned with the heat problem, those who desire a thorough understanding of the refrigeration process must know the simple facts concerning heat.

THEORY OF HEAT

The technical nature of heat was not understood for a long time, and several different theories were advanced to explain it. The concept that is generally accepted is the so-called *molecular theory* because it is based on the theory that all matter is composed of innumerable, separate, and minute particles called *molecules*. The molecule is composed of two or more atoms and is so small that extremely powerful microscopes are required to view them.

Molecular Theory

The molecular theory is based on the supposition that the molecules of a substance are not attached to each other by any bond or cement, but are held together by a force known as cohesion or mutual attraction, a phenomenon somewhat similar to the attraction offered by a magnet for steel particles. The molecules, however, are not physically bound or in contact with each other, like the iron filings, but are actually separated to such an extent in some instances (and usually so at more elevated temperatures) that the space separating two adjoining molecules is larger than either particle. Furthermore, the molecules are not fixed or stationary, but revolve and vibrate within the orbits or limits of their allotted space.

Each substance on this earth is composed of different ingredients or

25

various combinations of molecules and therefore each particular kind and mixture, together with additional peculiarities in physical assembly, have a structure differing from other materials. It is the rapidity of the motion of the molecules composing a body or mass that determines the intensity of its heat. As a specific example to point out the connection between the theory of heat and the molecular theory, a bar of lead may be used for illustration. When cold, the molecules undergo comparatively slow motion. But if the bar is heated, the molecular activity becomes more rapid, until its temperature has been increased to a point where molecular motion becomes so rapid and the individual particles are so far separated from each other that, with further assimilation of heat, they become so weakly cohesive they no longer can hold the body in a rigid or solid mass, and it devolves into a liquid.

Further application of heat forces the molecules to greater separation and speeds up their motion to such an extent that the liquid becomes more mobile and volatilization finally results, so that a gas or vapor is produced. The vapor thus formed no longer has a definite volume, such as it had in either the solid or molten (liquid) form, but will expand and completely fill any space that is provided for it. The vapor, of course, contains the same number of molecules that were in the original solid, but with the difference that molecular action is extremely active in the vapor phase, comparatively slow in the instance of the liquid, and slowest in the solid form. The motion of the molecule is somewhat like a bicycle rider; at a slow speed a very small orbit can be adhered to, but with each advance in the rate of travel, larger orbits are required.

To one learning the theory of heat and matter for the first time, it may appear extremely complex. Whether the molecular theory herein described (which is the one accepted by leading scientists), or any other theory is used as a basis of explanation, the facts regarding heat as we know it are in agreement with the molecular theory, and from all tangible proof substantiate the theory in all respects. One of the simplest indications of value in support of the molecular theory is the age-old knowledge concerning expansion and contraction, caused respectively by heat and cold, or, as we look at it from the molecular theory, the rate of molecular activity.

Temperature

Heat and temperature are closely allied, but it must be remembered that heat is energy itself representing the kinetic activity of the molecules composing a substance, while temperature is but a measure of the condition of a mass or body, as it affects our ideas of warmth or cold. Heat is convertible into electrical, chemical, or mechanical energy.

Our ideas of cold and hot are merely relative ranges of temperature that affect our sense of feeling, which we term hot and cold. In a way, heat itself may be likened to water, for of its own accord, it will only flow downhill; that is, it will pass from the hotter (or higher) range of temperature to the colder (or lower) plane.

Two miscible liquids poured together will find their own temperature level, as will a solid and a liquid, or even two solids in intimate contact. For instance, if an ice cube is dropped into a tumbler of tepid water, it will be found that the colder substance will be able to assimilate a certain quantity of heat from the warmer substance. Thus, through this extraction, we find that the water has been cooled or reduced in temperature. Two substances of different temperatures in intimate contact tend to reach an equilibrium or balance by the dispersion of heat by one body and assimilation of heat by the colder. If a thermometer is placed in contact with the substance, the degree or level of its temperature, or the measure of the intensity of its heat, will be indicated.

THERMOMETERS

The instrument in common use for measuring temperature, known as the thermometer, operates on the principle of the expansion and contraction of liquids (and solids) under varying intensities of heat. The ordinary mercury thermometer operates with a fair degree of accuracy over a wide range. It becomes useless, however, where temperatures below $-38°$ F are to be indicated, because mercury freezes at that point. Some other liquid, such as alcohol (usually colored for easy observation) must be substituted. The upper range for mercurial thermometers is quite high, about $900°$ F, so it is apparent that for ordinary service and general use the mercury thermometer is usually applicable.

The operation of the thermometer depends on the effect of heat on the main body of mercury or alcohol contained in a bulb or reservoir. The liquid will expand or contract (rise or drop) in the capillary tube which is inscribed with the various increments of an arbitrary scale, as shown in Fig. 12. Several thermometer scales are in existence and are used in various countries. The English or Fahrenheit scale is commonly used in this country. Since the Celsuis scale is so widely used in scientific work in all countries, an illustration of the comparison of thermometers is presented so that any one scale may be converted to another.

The freezing on the Fahrenheit scale is fixed at 32°. On the Celsius scale, freezing is placed at zero degrees. On the Fahrenheit scale, the boiling point of pure water under the normal atmospheric pressure encountered at sea level is 212°; on the Celsius scale under the same conditions, it would have a boiling point of 100°.

Fig. 12. Illustrating the relation between a Fahrenheit and Celsuis temperature scale.

In Fig. 9, all four scales were shown. At this point in Fig. 12, just the Fahrenheit and Celsius scales are shown, as these are the two scales that you will be using most of the time.

As previously stated, temperature is a measurement of the intensity of the heat contained in a body, accomplished by means of a thermometer, just as a yardstick is used to measure the length of a body. A thermometer will not, however, indicate the quantity or amount of heat contained in a body.

ABSOLUTE ZERO

In the study of thermometer scales, the question of thermal limits is naturally considered. From the molecular theory we are led to believe that with the removal of heat, molecular action is slowed down accordingly. It must naturally follow that, at some point, all heat will be removed and molecular activity will cease entirely.

The calculation in Fig. 13 indicates that absolute zero is attained at a temperature of 460° below zero on the Fahrenheit scale. Temperatures as low as this have never been reached, although in some instances it has been approached within a few tenths of one degree. Bodies subjected to extremely low temperatures take on characteristics entirely different from those exhibited under normal conditions.

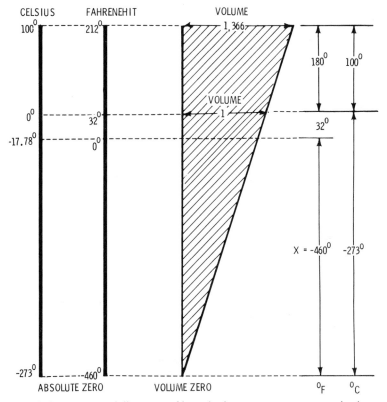

Fig. 13. Diagrammatical illustration of how absolute zero temperature may be determined.

This calculation is based upon the fact that atmospheric air expands 1.366 of its volume on being heated from 32° to 212° F. It can therefore be assumed inversely, if all possible heat were withdrawn, the volume of air would shrink to zero. With reference to our diagram, the volume of air at 32° F = 1. It follows that the volume of air at 212° F = 1.366. Constructing our triangles as shown, the following relations are obtained:

$$\frac{32 - X}{212 - X} = \frac{1}{1.366}$$

$$X = \frac{1.366 \times 32 - 212}{0.366} = -460° \text{ F.}$$

UNITS OF HEAT

The quantity of heat contained in a body is not measurable by thermometer which indicates only the temperature or intensity. For example, a gallon of water and a pint of water may have the same temperature, but we are certainly aware that the larger body must have more heat as energy within it, than the smaller.

The unit of heat measure employed in this country is called the British thermal unit, more commonly referred to as the Btu and it is that quantity of heat required to be added to one pound of pure water initially at a temperature of its greatest density (that is 39° F) to raise its temperature one degree on the Fahrenheit thermometer (in this case to 40° F). Roughly, a Btu may be said to be the quantity of heat required to raise the temperature of one pound of water one degree on the Fahrenheit scale. This is shown in Fig. 14.

Just as the thermometer is used as a measure of the intensity of the heat of a body, the Btu is used to represent the quantitative energy. For instance, one body at 50° F may contain twice as many Btu's as another body at the same temperature, because the bodies may be different in size or weight and, of great importance, may have different capacities for absorbing heat.

SPECIFIC HEAT

It could be said that each and every substance on the earth has a different capacity for absorbing heat. Some identical materials, especially natural formations, even give different values for samples secured in different localities. As an illustration of the difference in the heat capacities of materials, let us consider a pound of iron and a pound of water, both at 80° F. If the heat is removed from these bodies and the number of Btu's extracted in cooling each mass to the same temperature is recorded, it will be found that each will give up a different amount. If

Fig. 14. Showing the unit of heat or Btu.

each body had been at a temperature of 80° F, and was cooled to 60° (a reduction of 20° on the Fahrenheit scale), it would be found that 1 pound of water at 80° cooled to 60° would evolve 20 heat units or 20 Btu.

If the iron (of the same weight) is cooled over the same range, only 2.6 Btu will be extracted (approximately ⅛ of that taken from the same weight of water under identical conditions). We therefore come to the conclusion that all materials absorb heat in different capacities, and by comparing the heat absorbing qualities with a standard, we have a standard of measure, a gauge to compare all substances. This measure is the amount of heat expressed in Btu's required by one pound of a substance to change its temperature one degree Fahrenheit. Since water has a very large heat capacity, it has been taken as a standard, and since one pound of water requires one Btu to raise its temperature one degree, its rating on the specific-heat scale is 1.00. Iron has a lower specific heat, its average rating being 0.130; that of ice, 0.504; of air, 0.238; and of wood, 0.330. The more water an object contains, as in the case of fresh

31

food or air, the higher the specific heat. Materials usually stored in a refrigerator have a high specific heat, averaging about 0.80.

Table 2 lists the average specific heats of a number of common substances. With this table, their relative heat capacities can be compared with water. Observe that metals have limited heat storing powers as compared with water. This is one of the reasons why scalds from hot water burn so deeply, for the water contains so much heat energy that a considerable amount is released and causes a worse burn than molten metal at a much higher temperature.

Table 2.   Average Specific Heats

Water	1.000	Pine	0.650
Copper	0.900	Strong brine	0.700
Vinegar	0.920	Oak	0.570
Alcohol	0.659	Ice	0.504
Air	0.238	Glass	0.194
Mercury	0.033	Iron	0.130
Coal	0.241	Sulfur	0.202
Brass	0.094	Zinc	0.095

In the preceding table, various substances are listed with their average specific values, and by finding this factor, the amount of heat in Btu to be added or taken from a substance of known weight to bring about a change of one degree in its temperature may be calculated.

Specific heat problems may be calculated easily by use of the following formula:

British thermal units = Sp. heat $\times$ W $(t_2 - t_1)$.

where,

W = weight of the substance in pounds,
Sp. heat = specific heat of the substance to be heated,
$t_2 - t_1$ = temperature change in deg. F.

Where very accurate scientific work is done, certain allowances are made for the fact that the specific heat of a substance does not remain constant throughout the entire temperature range. Since the difference is not appreciable (except over a wide range), the refrigeration engineer usually regards the specific heat factor as a constant. Other materials, such as liquids and gases, also have specific heats. However, the calculations concerning the heat capacities of gases are further complicated by the pressures as well as the varying temperatures imposed. Despite the tendency to ignore the variance in specific heat and to employ a constant, the values of water are given as follows in Table 3.

Table 3. Specific Heats of Water
(Value at 55° F, taken as unity)

Temp. ° F.	Spec. Heat	Temp. ° F.	Spec. Heat
20	1.0168	140	0.9986
30	1.0098	160	1.0002
40	1.0045	180	1.0019
50	1.0012	200	1.0039
60	0.9990	220	1.007
70	0.9977	240	1.012
80	0.9970	260	1.018
90	0.9967	280	1.023
100	0.9967	300	1.029

The specific heat of various foods and their containers are of interest to the refrigeration engineer in estimating the amount of heat to be extracted in cooling a refrigerator load. The heat that can be felt and detected is termed the *sensible heat,* its name denoting the heat we can sense or feel and so preventing confusion with other heats. It is the sensible heat which forms the preponderance of the heat load the refrigerating machine is called on to remove in cooling most edibles. Of course, ice manufacturing plants and other storage warehouses which freeze edibles are additionally concerned with other heat problems.

Most foods have a high water content and, therefore, a large heat capacity, averaging about 0.80. A few foods are listed in Table 4 so that the specific heats of various edibles can be ascertained and the truth of the statement regarding the water content verified. The specific heats listed are those of fresh foods before freezing.

An estimate of the heat quantities that must be removed from the food and its containers can be easily accomplished. For instance, if we have 1000 pounds of cider contained in glass bottles having a weight of 75 pounds, which in turn have been packed in pine boxes totaling 50 pounds in weight, the whole shipment being at a temperature of 80° F, the heat in

Btu that would have to be extracted to cool these materials to 50° F can be calculated as follows:

80° − 50° = 30° difference in each case

1000 pounds of cider having a specific heat of 0.90, cooled over a 30° range = 1000 × 0.90 × 30 = 27,000 Btu

75 pounds of glass with a specific heat of 0.194, cooled 30° = 75 × 0.194 × 30 = 30 = 436.5 Btu.

50 pounds of pine wood having a specific heat of 0.650 cooled 30° = 50 × 0.650 × 30 = 975 Btu, or a total of 27,000 + 436.5 + 975 = 28,411.5 Btu.

By observing a thermometer immersed in the cider and taking an average reading, we could calculate at any time during the cooling process just how much of the sensible heat had been removed. In fact, we could calculate any heat load of any material by finding the specific heat value and the cooling range, and if we desired to freeze the product (in this specific case, the cider), an additional factor would have to be considered, namely the *latent heat*.

Table 4. Specific Heats of Foods

Food	Spec. Heat	Food	Spec. Heat
Apples	0.92	Eggs	0.76
Beans	0.91	Fish	0.80
Beef	0.75	Grapes	0.92
Butter	0.60	Milk	0.90
Cabbage	0.93	Peaches	0.92
Cheese	0.64	Pork	0.50
Chicken	0.80	Potatoes	0.80
Celery	0.91	Veal	0.70
Cider	0.90	Watermelon	0.92

LATENT HEAT

One of the most mystifying laws to the layman is that of latent heat. The word itself, *latent,* expresses it aptly enough, for it means hidden or not apparent. In order to indicate clearly just what latent heat is, an example will be cited, and an illustration shown in Fig. 15.

We are aware that most substances are capable, under proper conditions, of assuming two or more physical states. For instance, lead, when cold, is a solid and when heated and molten, a liquid. Water gives us an outstanding example, for it can assume three states; that is, solid, liquid, and vapor within a relatively short temperature range. Ice, of course, represents the solid state; water, the liquid; and steam, the vaporous or gaseous state.

Fig. 15. Apparatus necessary for conversion of one pound of ice at 0° F to water at 32° F.

CHANGE OF STATE

Temperature and heat play important parts in effecting changes from one state to another. For instance, the only quantity that would convert a tumbler of water to either of the other states (that is, a solid or a vapor) would be the addition or extraction of heat.

To prove the function that heat plays in effecting a change of state, let us take a block of ice and perform an experiment with it. In order to simplify the calculations we will utilize a piece of ice that weighs exactly one pound, with a thermometer frozen in the center of it, and we will assume that the block of ice is at a temperature of zero degrees Fahrenheit. We will not go into the apparatus or computations necessary to measure the heat values applied, but we will assume the use of an imaginary Btu meter just as though there existed such a device. Thus,

35

every time a Btu of heat energy is expended it will be registered on our meter.

Prior to beginning the experiment, a graph or squared section of paper is obtained so that the results can be plotted on it. In the lower right-hand section of the paper (a part that will not be required for our log), we will put down the various findings. Setting up the apparatus is a simple matter, for we require only a vessel and a source of heat, the latter being measurable by our "meter". Place the piece of ice, weighing exactly one pound, in the vessel and indicate the start of the experiment by marking an *A* at the lower left-hand corner of the chart, as shown in Fig. 16, to denote that the ice is at zero degrees Fahrenheit, and that no heat has been added as yet. The specific heat of ice is not as great as that of water; in fact, it is just about half the value, 0.504 to be exact. For this experiment let us assume it is fifty percent, or 0.5.

Fig. 16. Diagrammatical representation of amount of heat units required to change one pound of ice at sea level and 0° F to steam at 212° F.

Let us now begin to add heat. In accordance with its specific heat capacity, we will find that every time 0.5 of a Btu is added, the temperature of the ice will increase one degree on the Fahrenheit scale. Continuing to add heat and marking off the progress on the chart, we find that up to 32° Fahrenheit, the number of Btu required to bring the ice to this degree of sensible heat would be 1 (lb. of ice) × 0.5 (spec. heat) × 32 (degrees rise) = 16 Btu. Thus, from A to B (a range of 32°) 16 Btu were required.

At 32° Fahrenheit, an interesting stage is reached. We find that the further addition of heat does not warm up the ice and the thermometer frozen in the test block of ice continues to indicate a sensible heat of 32°. Even providing the ice with a greater quantity of heat causes no observable change in temperature. The only noticeable feature is the melting of ice or conversion of a solid (ice) to a liquid (water). This important transformation is called the *change of state*, due to the fact that a solid is converted to a liquid. If we continued to add heat to the vessel containing the ice and water, we would find that when a total of 144 Btu has been added, the entire block of ice would be converted to water and that the water would have a temperature of 32° F. In other words, to change one pound of ice at a temperature of 32° F to water at the same temperature, 144 Btu are required to effect this change in state. This we will plot on our chart as line B and C.

LATENT HEAT OF FUSION

We have had to apply 144 Btu which was taken up by the ice at 32° and caused it to melt and assume its liquid state, without a single indication of any rise or increase in its sensible heat. This masked assimilation of heat is termed *latent heat* and may be said to be the amount of heat units which must be supplied to a solid in order to change it to a liquid without an increase in temperature. It is readily understood that a reversal of the process, that is, for the liquid to assume its solid state (ice), heat must be extracted from the water.

One pound of water at a temperature of 32° F requires the extraction of 144 Btu to cause it to freeze into a solid block of ice at 32° F. Every solid substance, in varying degrees, has a latent-heat value, and that amount required to convert it or bring about a change of state is termed the *latent heat of fusion*. This heat, assimilated or extracted as the case may be, is not measurable with the thermometer because the heat units are absorbed or expended in inter-molecular work. Latent heat of fusion separates the molecules from their attractive forces, so that a change of state is effected.

LATENT HEAT OF EVAPORATION

Now let us refer again to the experiment we have under way. We have converted the solid to a liquid, both at 32° F, by adding 144 Btu. Having to deal with water, and knowing that it has a specific-heat factor of 1.00, we may look for a rise of 1 degree in temperature for each Btu added. This will hold until 212° Fahrenheit is indicated on the thermometer, assuming that the experiment is made at atmospheric pressures existing at sea level. Over the 180-degree range of 32 to 212 degrees we will be obliged to add 180 Btu in order to raise the temperature of one pound of water from the former to the latter temperature. This may be plotted as line C and D on the chart, and the number of heat units recorded on the lower edge.

The further addition of heat would serve to bring out the fact that while the phenomenon of boiling would occur at 212° F, the temperature or sensible heat would not be increased. Just as ice requires a certain quantity of heat units (144 Btu per pound) to melt or convert it from a solid to a liquid without a rise in temperature, we find a similar condition existing. This time a liquid is being converted to a vapor, or steam in this case.

Careful measurements have determined that the conversion of one pound of pure water at 212° F to steam at 212° F requires exactly 970.4 Btu when carried out at the normal pressure of the atmosphere encountered at sea level. If we carefully add heat and keep count of the Btu expended, we will find that when all the water has been changed to steam, 970.4 heat units will have been used. Thus, line D and E may be plotted on the chart, and a note of the number of Btu expended jotted down on the lower right hand corner. The further addition of heat would serve only to heat the steam, such as would be possible if it had been trapped or the experiment performed in a closed vessel so that heat could be applied to it.

STEAM

Steam is the hot, invisible vapor state of water at its boiling point. The visible white vapor is really a collection of fine watery particles, formed from true steam by condensation.

Steam acts like all true vapors or gases in that it:

1. Has fluidity.
2. Has mobility.
3. Has elasticity.
4. Exerts equal pressure in all directions.

The difference in volume between water and steam at atmospheric pressure is 1646 to 1, and this wide ratio is manifested by nearly all gases and vapors. The heating of the steam generated in our experiment could be represented by the line *E* and *F*, and would be expended over a considerable range. By referring to the log of our experiment we can trace the heat quantities required to convert ice to water and steam and the action of heat on both liquid and solid. A tabulated form in the lower right-hand portion of the log represents the quantities of heat energy required to effect a change of state, rise in temperature, and other values.

From this experiment it is apparent that heat added to a substance either increases its temperature or changes its state. The heat that brings about the change of state from a solid to a liquid is known as the *latent heat of fusion*, while that required to convert a liquid to a gas is termed the *latent heat of evaporation or vaporization*. One of the interesting facts brought out previously is that a liquid, once brought to the boiling point, does not increase in temperature, but utilizes the heat energy it takes up in converting more liquid to a gas or vapor.

If we were creatures dwelling on another planet and accustomed to a normal atmospheric temperature of 250° or 300° F, we would be able to utilize water to cool substances by immersing them in a bath of water. Since heat and solid are only relative to our senses, materials immersed in a pan of water will lose heat to it, and the heat extracted will be taken up by the water and used to generate steam. Through the extraction or flow of heat from objects so immersed, the water (assumed to be under the same normal sea-level pressures) would remain at 212° F, which would be cooler than the atmospheric temperature of our hypothetical planet. Fortunately for earth dwellers, there are other liquids that boil at temperatures much below that of water.

REFRIGERANTS

Among common liquids that boil at a temperature below that of water is alcohol, with a boiling point of 173° F, while ether boils at ordinary summer heat, which is 94° F. Other substances boil at still lower temperatures. Carbon dioxide boils at −110° F, Ammonia at −20° F, Sulfur dioxide at 14° F, Methyl chloride at −11.6° F, Ethyl chloride at 54° F, *Freon-12* at 21.6° and *Freon-22* at 41.4° F, all at the atmospheric pressure encountered at sea level.

Those materials, solid or liquid, which vaporize or liquefy at comparatively low temperatures, and are suitable for use in refrigeration work, are termed refrigerants. The refrigerants are employed in specially designed apparatus so that the extraction of heat from rooms or perishables can be accomplished as inexpensively as possible.

The latent heat of vaporization of liquid refrigerants varies with the material and with the pressure at which vaporization is allowed to proceed. The latent heat of vaporization in Btu per pound of various refrigerants in common use and volatilizing at 0° F is presented as follows. There will be notes accompanying these as some of the refrigerants listed are now practically obsolete but you may run across some older equipment which uses them:

Ammonia 572.2

Ammonia is considered a hazardous material and you will encounter it only in small domestic type refrigerators, or very large commercial installations, where the hazards have been allowed for.

Methyl chloride..176.0

Methyl chloride in the right proportions with air will burn or explode. It was not used much in home refrigeration and is not being used today to any degree.

Ethyl chloride..173.4

This substance has been practically eliminated from the refrigeration field.

Sulfur dioxide..171.8

Sulfur dioxide was once used to a large degree, but it is very toxic and the only places that you will encounter it is in old domestic refrigerators.

Carbon dioxide 117.5

Not used in domestic refrigerators.

Freon-12 ....71.0

Very popular at the present time.

Freon-22 .... 100.7

Used in many domestic freezers.

Freon-502 .... 74.8

Used in large air conditioning systems.

Freon-21 .... 119.4

From this list, it is observed that ammonia is an excellent refrigerant in that it absorbs a great quantity of heat as compared to other agents. However, it is toxic and classified as a hazardous material so that its use is confined mostly to domestic absorption type refrigeration and to large industrial manufacturers of ice. Methyl chloride is combustible under the right mixtures with air, so it is seldom used any more. Sulfur dioxide, while quite popular in the past, is extremely toxic and is no longer used. It will be covered to some extent in this book, as you may come across an SO_2 system that needs repair. Carbon dioxide in solid form finds use in transporting ice cream, fish and similar products and where a great quantity of escaping gas fumes would not endanger human life. This substance is also known as dry ice.

It is very easy to understand that a vessel containing a refrigerant, which is allowed to absorb heat, will cause cooling or a refrigerating effect. Since heat (as energy) cannot be created or destroyed, it follows that heat removed from one body must show up in some form of energy in another body. Essentially, the removal of heat from one body and its transference to and dissipation by another is refrigeration.

USE OF ICE

Where ice is employed, the latent heat of fusion results in the cooling of other materials suffering such loss of heat energy that the heat is taken up by the ice and it reverts to water. The heat-laden liquid (water) is drained away, carrying its heat load out and away from the refrigerator. A graphical illustration of this cycle is shown in Fig. 17.

41

Fig. 17. An example of the ice refrigeration cycle.

TON OF REFRIGERATION

Due to the fact that refrigeration was first produced by ice, the rate of removal of heat in a cooling operation was expressed in terms of pounds or tons of ice required per unit time, usually per day. It has been found previously that one pound of ice absorbs 144 Btu when it melts. One ton of ice consequently absorbs 2000×144 or 288,000 Btu. When one ton of ice melts in 24 hours, the rate is 288,000/24 or 12,000 Btu per hour, or 12,000/60=200 Btu per minute. This rate has been officially designated as one ton of refrigeration, and is the basis for rating all refrigerating machinery.

EVAPORATION

We have previously discussed the quantities of heat required to bring about a change of state. Considerable heat energy was necessary when a change took place from the liquid to the vapor phase. The term evaporation is well-known to all. A common example of refrigeration by means of evaporation is found in the cooling of the human body by perspiration. The beads of moisture appear on the skin and, by evaporation into the air, actually maintain the body at a lower temperature than the surrounding atmosphere.

The normal internal temperature of man is 98.6° F. Everyone at some time or other has encountered atmospheric temperatures that were much higher than body heat, never realizing that it was evaporation that acted as the refrigerating element to keep the body cool. The human body is extremely sensitive and is always within a very limited range of temperature, allowing a variation of only a few degrees. For instance, if the inside body temperature were raised above 108° F, death is almost certain, yet men have lived in atmospheres where temperatures were as high as 140° F, and even higher. Their bodies were maintained at normal body heat by evaporation of moisture from their skins.

Another example of cooling by evaporation found in nature is the reduction in temperature which usually follows a summer rain. The required heat to bring about the evaporation of the rain water is drawn from the atmosphere and the earth, and thus effects the coolness we note after such precipitation.

PRESSURE

In the previous instances of boiling temperatures, the wording *"at pressures found at sea level"* was used. For instance, water boils at 212° F when it is heated in a vessel open to the atmosphere and provided the experiment takes place at an altitude at or near sea level or under equivalent pressure. Regardless of how rapidly heat may be added to the open vessel of water, no rise in boiling temperature can be secured, the only result achieved being that steam is formed with greater rapidity, commensurate with the amount of heat. If, however, pressure is allowed to enter as a factor in the conversion of a liquid to a vapor, or vice versa, it will be found that the boiling point is altered. We are all more or less familiar with the story of the man who tried to prepare his eggs on the top of a high mountain where atmospheric pressure, due to altitude, was so low that the water boiled at such a low temperature that it would not cook the eggs. From this simple example, it is apparent that the lower the pressure exerted on the liquid, the lower will be the boiling point.

By way of demonstrating the effect of pressure, let us take a sturdy flask, one capable of withstanding both heat and pressure, and pour in

43

some water. We will assume it to be a glass vessel so that we can view the process and make certain when boiling actually occurs. A stopper provided with three perforations, so that a pressure gauge, thermometer, and a valve can be inserted in the individual openings, is forced tightly into the neck of the flask. The completely assembled apparatus is then placed over a souce of heat, such as a gas flame, as depicted in Fig. 17. The exit valve is allowed to remain wide open. Soon the water will boil, and if the experiment is performed at or near sea level, the pressure gauge will register zero pounds of pressure, while the thermometer will indicate a temperature of 212° F. As long as the valve orifice is large enough to offer a free and unobstructed exit to all of the steam formed, no amount of heat will raise the temperature.

Fig. 18. Apparatus necessary for evolution of pressure-temperature relations of water.

Now let us turn the valve so that the opening is slightly restricted. The pressure will build up, and by regulating the amount of gas burned and the exit valve so that a jet of steam is released, we can obtain or impose any pressure we desire on the vessel. Let us select a pressure of about 5 pounds as the pressure desired. With proper regulation of the gas and steam valve, this value can easily be obtained and maintained. When secured, let us read the thermometer. By readjusting the steam and the gas valves, let us raise the pressure to 10.3 pounds gauge pressure. At this point we would find the water would start to boil at 240.1° F. Again we would find that no amount of heat added rapidly or otherwise to our flask would raise its temperature above that point, provided the pressure remained constant.

With heat added rapidly, more steam would be formed and we would of course be obliged to open the steam exit valve in order to maintain the pressure at the desired point. If only a little heat were added, we would have to close the valve somewhat to maintain the pressure. By referring to a table giving the properties of saturated steam, various gauge pressures and corresponding temperatures can be determined. For instance, if our boiler were of heavy steel, we could increase the pressure to a point where the gauge would indicate 100.3 pounds, at which point the boiling of the water would occur at 338.1° F. With the increase in pressure, the boiling point advances, and again, in accordance with this pressure-temperature relationship, the boiling point is lowered as the pressure is reduced.

When the pressure decreases, the boiling temperature of a liquid will decrease also. For example, water at an atmospheric pressure equal to 4800 ft. above sea level would boil at approximately 202° F. By the same token, water at 10,000 ft. altitude would boil at approximately 193.2° F.

PRESSURE-TEMPERATURE RELATIONS

Most bodies expand when heated, whether of solid, liquid, or gaseous form. If the expansion is restricted, great forces are set up by the body in the effort to expand. Every gas or vapor confined in a closed vessel exerts a certain pressure against the restraining walls of the vessel. The pressure imposed depends on the amount and temperature of the gas. The natural tendency for a confined gas is to expand when the temperature is increased, and if the gas is contained in a vessel with rigid walls, the pressure will increase. This is of such importance that a law has been formulated to express it, stating that the pressure exerted by a gas or vapor

45

in a closed vessel is *directly proportional to the absolute temperature of the gas.*

Low Pressures

Before proceeding with our experiments with lower pressures, it is best that we understand the comparative terms. As previously pointed out, objects at sea level are taken as being subjected to a certain constant pressure. This pressure is due to the weight of the atmosphere, which at sea level is found to exert a pressure of 14.7 pounds per square inch of surface.

The atmosphere is like a high sea, and if we dive into water similar conditions are encountered, for the deeper we go the greater the pressure becomes, because of the volume above. As we rise, the pressure becomes less. Modern aircraft designers take into consideration the rarefication of the atmosphere with increasing altitude and know that at a certain height the air density will not allow an airplane to rise any higher. This they term the ceiling or density at which further altitude for each particular design is not possible. Since pressure is so instrumental in a great many engineering calculations, a universal standard was necessary. Thus, the pressure existing at sea level was accepted as a standard, being termed zero pounds gauge, or 14.7 pounds absolute.

In the manufacture of gauges, these instruments are usually set to indicate zero pressure at normal atmospheric sea-level density. It must be remembered that this setting at zero is made when the actual and true pressure is in reality 14.7 pounds per square inch, the pressure that the weight of the atmosphere exerts at sea level. Nearly all steam and pressure gauges are set in this fashion and the pressures indicated by these instruments are termed gauge pressures. The usual increments are in one-, five-, and ten-pound readings so that only approximate pressures can be determined. For most commercial applications these roughly calibrated gauges are sufficiently accurate and are quite inexpensive.

For fine work; experimental and scientific determinations, another pressure scale is employed. This scale is called the absolute pressure and is based on true pressures, for zero on this scale indicates no pressure at all, or in other words, a perfect vacuum. An external view and construction details of a typical pressure gauge is shown in Fig. 19. It consists of a metal tube of elliptical cross section bent into a nearly complete ring and closed at one end. The flatter sides of the tube form the inner and outer sides of the ring. The open end of the tube is connected to the pipe through

HOLLOW SPRING TUBE

GEARS

ELLIPTICAL CROSS SECTION

PIVOT

END FREE TO MOVE

TO SOURCE WHOSE PRESSURE IT IS DESIRED TO MEASURE

FRONT VIEW

INTERNAL CONNECTION

Fig. 19. The external view and construction details of a typical pressure gauge.

which the liquid under pressure is admitted. The closed end of the tube is free to move. As the pressure increases, the tube tends to straighten out, moving a pointer through a lever-and-gear arrangement. The scale is graduated directly in pounds per square inch.

On the pressure gauge scale, the normal atmospheric pressure found at sea level (14.7 pounds per square inch) is indicated as zero pounds pressure and below that point the term of pressure is converted to vacuum or inches of mercury. A long glass tube, sealed at its upper extremity, is exhausted of air and filled with mercury as shown in Fig. 20, its lower open end dipping into a reservoir of mercury open to the atmosphere, or to the pressure imposed. Such a device is called a *barometer* and is used to measure the pressure in terms of inches of mercury in place of such a term as pounds under zero gauge pressure. Just as variance in temperature will cause a rise or fall of the mercury level in a thermometer, so will a difference of pressure exerted on the mercury in the reservoir of the barometer bring about a certain reading in terms of inches of mercury. The pressure of the atmosphere, or any other pressure exerting its force on the mercury exposed in the reservoir, drives the liquid up into the evacuated end to a height corresponding with the pressure. In a way, the barometer is much like the thermometer, for it indicates a value by the height of a mercury column, one recording in terms of temperature and the other in terms of an equivalent of pressure.

On the absolute scale, it will be observed that pressures begin from an absolute zero or perfect vacuum, and that a gauge calibrated in this manner will indicate 14.7 pounds at normal sea-level pressures. On the other hand, a gauge calibrated in the fashion employed for the commercial field of refrigeration and air conditioning will indicate this latter pressure as zero pounds, where normal atmospheric conditions prevail.

EFFECT OF LOW PRESSURES

In the previous experiments, we determined that whenever pressure was increased, the boiling point was raised. Making use of the same apparatus, let us investigate the effect of a reduced pressure on the evaporating temperature. To do so, we will again place a quantity of water in the flask, force the stopper containing a pressure gauge, thermometer, and control valve tightly in place so that no leakage can occur to mar the accuracy of our experiment, and place the apparatus over a gas flame. This time, however, instead of using the exit valve for holding the steam within the vessel to cause a pressure to build up, we will connect a vacuum line to it and use it to prevent the steam from being drawn out or evacuated, so that a certain low pressure or inches of vacuum can be maintained within the vessel.

Referring to the pressure scale, Fig. 20, we find that 30 inches of mercury is roughly equivalent to 15 pounds on the gauge-pressure scale, so that a 2-inch vacuum means a pressure of 1 pound less than the sea-level atmospheric pressure. For our first experiment, let us take a pressure of 14 pounds absolute, or 0.7 of a pound less pressure than normally exerted. From a table on the properties of saturated steam, we will find that this corresponds to 1.42 inches of mercury on the gauge-pressure scale.

By regulating the exit valve so that the proper vacuum or low pressure is obtained, we can easily arrive at a point where boiling occurs at the constant pressure we desire. At this point of 14 pounds absolute pressure, we would find that the water boils at 209.55° F. Then, we can open the exit valve a trifle more and regulate it so that a pressure of 12 pounds absolute (5.49 inches of mercury) is maintained. This pressure results in a boiling point of 201.96° F.

The fact is thus brought out that the pressure to which water is subjected has just as much to do with its boiling as does the temperature. This holds true for any liquid, for evaporation can be made to occur at any temperature above its freezing point if the pressure to which it is subjected is made low enough.

760 m. m.
29. 92 INCHES

MERCURY TUBE

ATMOSPHERIC   PRESSURE

MERCURY BOWL

Fig. 20. Balancing the pressure of air with a mercury column.

As you progress in your studies, you will find that some times you will have to put a vacuum pump on a refrigeration system to get rid of all unwanted gases and water. At sea level you should theoretically get the vacuum pump to pull down to 29.92 inches of mercury. This will not be possible as the pump you will be using will not pull a perfect vacuum. Depending on the atmospheric pressure and the efficiency of your pump, you will probably be able to pump down to between 27 and 28 inches of mercury. At a 5000 foot altitude you may reach approximately 23 inches of vacuum.

HEAT TRANSMISSION

In refrigeration, we are interested in getting the heat contained in a room or refrigerator to a medium that will effect its removal. Ice is a simple method and formerly was widely used. The more modern refrigerator is equipped with a cooling apparatus that supplants ice and

49

provides for more constant and cooler temperatures. Before we can study how heat is taken up and removed mechanically, it is imperative to learn the behavior of heat.

One of the most important laws has already been mentioned; that which refers to the flow of heat from a body of higher temperature to one having a lower sensible heat. Never of its own accord will heat or water flow up hill or in the opposite direction. Therefore, it follows that heat in a refrigerator or room will flow to the cooler object, such as ice or the cooling device.

The transmission of heat may be accomplished in three ways:

1. By conduction.
2. By convection.
3. By radiation.

Conduction

This is the transference of heat by molecular impact from one particle to another in contact. For instance, if the end of a bar of iron is heated in the fire, some of the heat will pass through the bar to the cooler portion. Heat traveling in a body, or from one body to another where the two are in intimate contact, is termed *conduction*. An illustration of this transfer of heat is shown in Fig. 21.

HEAT TRANSFER BY CONDUCTION

METAL ROD

COLD END OF ROD BECOMES HOT

BUNSEN BURNER

Fig. 21. Illustrating the transfer of heat by conduction.

Metals are usually splendid heat conductors. Each and every material has a conduction value, some good like the metals, others mediocre, and a few very poor. For instance, heat will quickly pass through a piece of copper but will have considerable difficulty in passing through a piece of cork. The materials that have very low heat conductivities are termed heat insulators. Even the very poorest conductors, or insulation materials, allow a certain amount of heat to pass through. There is no material which offers a perfect barrier or resistance to the passage of heat.

Convection

Convection is the principle used in hot-air heating. Air that is free to circulate, such as in any air body of appreciable size, will be set into motion where a difference of temperature occurs, for it will absorb heat from the warmer wall, become heated, expand and become lighter. The heated portion of the air will rise and cooler air will move into its place, which in turn will become heated. The heated portion of the air eventually moves over to the colder wall, and the heat flows from the air to the colder object. Thus, any body of air capable of motion will transmit heat by convection. Hot-air and hot-water heating systems work on the convection principle. They convey heat by bodily moving the heated substance from one place to another, as shown in Fig. 22.

The most efficient heat insulation known is a vacuum, but except for very small containers, it is structurally impractical to employ it commercially. The next best insulating medium is air subdivided into the smallest possible units so that it is still or stagnant. Air that is contained in spaces of appreciable size, such as between the double walls of refrigerators, will circulate and transmit heat by convection. Cork is an insulation material of a high order and has great resistance to the passage or transmission of heat because of its air content. The air cells in cork are of such minute size that the air trapped in them is so restricted that only a little circulation is possible and is, for all practical purposes, still air, so that little or no convection occurs.

Radiation

Heat energy transmitted through the air in the same way light is sent out by a lighted lamp, a radiant heater, or the sun, is called *radiated energy*. This is illustrated in Fig. 23. Large cold-storage warehouses, auditoriums, theaters, and homes are built with consideration of the heat evolved through radiant energy of the sun. Small household appliances rarely have to consider any radiant-heat factor, for they are used in existing structures without any change in building design and are sheltered from direct heat.

51

Fig. 22. Showing heat transfer by convection.

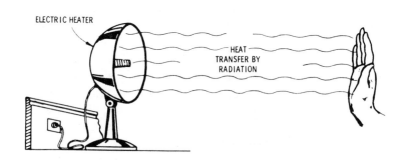

Fig. 23. Illustrating heat transfer by radiation.

CONDENSATION

In a previous experiment we illustrated that a liquid, heated to the boiling point corresponding to the pressure imposed, will assimilate heat and produce a vapor or gas. The heat taken up by the liquid is used to speed up molecular activity until a vapor is evolved. It stands to reason then, that any vapor or gas contains a considerable quantity of heat.

In accordance with the foregoing experiments, we found that heat itself always flows from the warmer to the colder body. To prove this again and illustrate just what condensation is, let us take a vessel, fill it about half full of water, and bring it to the boiling point over a gas flame, as shown in Fig. 24. As soon as steam is generated, let us take a dry, cold plate and

COLD PLATE

Fig. 24. An example of condensation.

hold it at an angle over the jet of steam as it issues from our crude boiler. It will be observed that the steam impinges on the plate and is converted again into water, the droplets forming and dripping off the edge of the plate. This was caused by the hot steam giving up its heat load (latent heat of vaporization) to the cold plate, the heat flowing from the steam to the plate.

To make sure that this is really the case, let us take the vessel we used in a previous experiment and again fill it with water, secure the stopper in place, and apply heat. This time let us set the exhaust valve so that we can maintain a few pounds of steam pressure, perhaps 5 pounds. While steam is generating and we are regulating the valve to get this constant pressure, let us build a condenser. This is easily done; all we need is about 12 feet of copper tubing. Any tubing or material will do, but copper is so easily bent and formed that it is not a difficult task to wind it in a spiral form, such as shown in Fig. 25.

Fig. 25. An example of condensation by steam vapors.

The formed coil is submerged in a pail of cold water with its open end at the bottom. The top end is supplied with a piece of rubber tubing. When the gas and steam valves have been regulated so that there is a constant steam pressure of 5 pounds within the boiler, and a fair amount of steam being exhausted from the exit, quickly connect the rubber tube on the condenser to the steam pipe. It will be observed that the pressure gauge has dropped somewhat, and after the steam has blown out the air contained in our condenser, we will see that no steam is issuing from the open end of the copper coil submerged in the water.

If we weighed the steam-generating apparatus and the condenser equipment before and after the experiment, we would find that the boiler lost weight through evaporation of water, while the condenser gained through condensation exactly what had been lost. Also, if a thermometer has been placed in the condenser water and a reading taken at the start and end of our experiment, it would have been found that the water increased in temperature. In fact, if we ignore various losses, the amount of steam condensed can be estimated from the increase in temperature of the water.

It is apparent that the steam is easily convertible to its liquid form if the heat of vaporization is extracted. Then too, our old rule would apply, not only to steam, but to all vapors and gases; that is, if we increase the pressure we will find that the gas will condense at the lower temperature found in the condenser, for the temperature of the gas will be raised far above the condenser temperature. Thus, heat will flow from the hot gas to the cold condenser, the greater the temperature difference, the faster will be the heat exchange.

In refrigeration and air-conditioning applications, the gases employed as a refrigerating medium have, for any given pressure, a corresponding temperature at which they condense or liquefy. Where both gas and liquid are present in the same vessel, and the closed container is heated to cause boiling, the temperature of both gas and liquid will be the same at the boiling point. If pressure is increased, the boiling point will be raised. Above a certain point the gas will cease to have any latent heat of vaporization, and it will remain as a gas, regardless of the intensity of the pressure imposed.

ALTITUDES AND REFRIGERATION CAPACITY

As altitude increases, there will be a decrease in the efficiency of refrigeration. This is due to the lower air pressures, and thus the cooling power of the condenser of the refrigerator and the horsepower of the motor, naturally decrease.

Do not be alarmed about this on home refrigerators, as they are oversized for sea level and work good at higher altitudes, but still with less efficiency.

Atmospheric Factors in Refrigeration

Dry-bulb temperature is the temperature that the plain thermometer will register. Wet-bulb temperature is that temperature observed if an ordinary thermometer bulb is covered with material such as linen, wet with distilled water and exposed to atmospheric evaporation.

The following table shows dry and wet bulb temperatures as related to summer air conditioning design conditions.

Table 5. Typical Outdoor Design Conditions For Air Conditioning

City	Dry Bulb Deg. F	Wet Bulb Deg. F
Birmingham, Ala.	95	78
Pheonix, Ariz.	105	76
Los Angeles, Calif.	90	70
Denver, Colo.	95	64
Washington, D.C.	95	78
Tampa, Fla.	95	78
Atlanta, Ga.	95	76
Chicago, Ill.	95	75
New Orleans, La.	95	80
Boston, Mass.	92	75
St. Louis, Mo.	95	78
New York City, N.Y.	95	75
Cincinnati, Ohio	95	78
Dallas, Tex.	100	78
Seattle, Wash.	85	65

Not only the dry bulb temperature effects the amount of refrigeration required, but the humidity, which is indicated by the wet bulb thermometer's temperature, will also affect the amount of refrigeration required, due to latent heat, required to condense the water vapor, which will collect as frost on the evaporator of the air conditioner.

Refrigerant pressure versus temperature is valuable information to have when repairing or installing refrigeration equipment. The following table is based on gauge pressure in psi at sea level.

The above table will be of some use to you in home refrigeration and air conditioning, as it may be used to determine the relative temperatures of

Table 6.   Refrigerant Pressure vs. Temperature

Temp. ° F	R-12	Refrigerant R-22	R-502	R-717 NH$_3$
−40	11.0*	0.5	4.1	8.7*
−35	8.4*	2.6	6.5	5.4*
−30	5.5*	4.9	9.2	1.6*
−25	2.3*	7.4	12.1	1.3
−20	0.6*	10.1	15.3	3.6
−15	2.4	13.2	18.8	6.2
−10	4.5	16.5	22.6	9.0
−5	6.7	20.1	26.7	12.2
0	9.2	24.0	31.1	15.7
5	11.8	28.2	35.9	19.6
10	14.6	32.8	41.0	23.8
15	17.7	37.7	46.5	28.4
20	21.0	43.0	52.5	33.5
25	24.6	48.8	58.8	39.0
30	28.5	54.9	65.6	45.0
35	32.6	61.5	72.8	51.6
40	37.0	68.5	80.5	58.6
45	41.7	76.0	88.7	66.3
50	46.7	84.0	97.4	74.5
55	52.0	92.6	106.6	83.4
60	57.7	101.6	116.4	92.9
65	63.8	111.2	126.7	103.1
70	70.2	121.4	137.6	114.1
75	77.0	132.2	149.1	125.8
80	84.2	143.6	161.2	138.3
85	91.8	155.7	174.0	151.7
90	99.8	168.4	187.4	165.9
95	108.2	181.8	201.4	181.1
100	117.2	195.9	216.2	197.2
105	126.6	210.8	231.7	214.2
110	136.4	226.4	247.9	232.3
115	146.8	242.7	264.9	251.5
120	157.6	259.9	282.7	271.7
125	169.1	277.9	301.4	293.1
130	181.0	296.8	320.8	
135	193.5	316.6	341.3	
140	206.6	337.2	362.6	

*Inches of mercury below standard atmosphere

the evaporators. In commercial refrigeration, where pressure switches instead of thermostats are used to control the temperature desired, it can be an extremely valuable tool in setting the desired temperatures. On present day home refrigeration and air conditioning, R-12 and R-22 are the most commonly used refrigerants. In this text you will notice other refrigerants have been included, as you no doubt will come across some of these in older machines that you may be called upon to service.

REFRIGERATION BY VAPORIZATION

The temperature at which a liquid boils or vaporizes is called its "boiling point." For liquids with very low boiling points, it is not necessary to supply heat by fire or other artifical means, as the heat in surrounding objects may be sufficient to cause boiling or vaporization. This is true, for example, with anhydrous ammonia, the boiling point of which is 28° below zero Fahrenheit. This is sufficiently low to cause it to boil violently when placed in an open vessel at ordinary temperatures. The absorption of heat by the vaporization of the ammonia will cause the outside of the container to become heavily frosted by moisture condensed and frozen from the air immediately surrounding the container.

Since the boiling temperature of any liquid may be changed by the pressure exerted upon it, it is easy to cause a liquid refrigerant to boil at any desired temperature by placing it in a vessel where the required pressure is maintained. The process of boiling or vaporization by which a liquid is changed to a vapor can be reversed—that is, the vapor can be reconverted into a liquid by the removal of heat. This is called "condensation". An increase in pressure will, by raising the boiling point, assist in the condensation of the vapor. Liquids used as refrigerants, on account of their initial cost, must be recovered, and the process of condensation is usually employed for this purpose.

Although there are liquids which boil at temperatures suitable for refrigeration, comparatively few possess all the requirements of a practical refrigerant. Those used commercially are *Freon* 12 (R12), *Freon* 22 (R22), *Freon* 502 (R502), and Ammonia (R717-NH_3). There are others previously used such as; sulfur dioxide, methyl chloride, and carbon dioxide. It is very questionable that you will run into any of the latter refrigerants now. For large installations, anhydrous ammonia will be used, as well as lithium bromide. The ones just mentioned, and for which a table is given for pressures and temperatures, are the refrigerants which you will come in contact with at the present day, in household and industrial refrigeration.

BASIC SYSTEMS

Refrigeration is distributed by several methods. In household and commercial refrigeration, the one you will find in most cases, is the "direct expansion system." The volatile refrigerant is allowed to expand in a pipe placed in the room or household refrigerator to be cooled, where the refrigerant absorbs its latent heat of evaporation from the material to be cooled. This method is used in small cold-storage rooms, constant temperature rooms, freezer rooms, household refrigerators, and where possible losses due to the leakage of refrigerant would be low.

In the "indirect system," a refrigerant medium (such as brine) is cooled down by the direct expansion of the refrigerant, and is then pumped through the material or space to be cooled, where it absorbs its sensible heat. Brine systems are used to advantage in large installations where the danger of the large amount of refrigerant is important, and where the rooms or series of rooms have fluctuating temperatures. It will also be found in homes that are heated with hot-water systems, except that instead of brine, ordinary water is used, with temperature controls for keeping the water from freezing. You will find these systems primarily used in air-conditioning, which is covered in another book.

For small refrigerated rooms or spaces where it is desired to operate the refrigerating machine only part of each day, the brine coils are supplemented by holdover or congealing tanks. A *holdover* tank is a steel tank containing strong brine in which are immersed direct expansion coils. During the period of operation of the refrigeration machine, the brine is cooled down and is capable of absorbing heat during the shut-down period, the amount depending upon the quantity of brine, its specific heat, and the temperature head. *Congealing* tanks serve the same general purpose but operate on a different principle. Instead of a strong brine, they contain a comparatively weak brine solution which freezes or congeals to a slushy mass of crystals during the period of operation. This mass of congealed brine, in addition to its sensible heat, is capable of absorbing heat equivalent to its latent heat of fusion.

LAWS OF GASES

Extensive investigations of the behavior of gases have shown that a given weight of gas expands or contracts uniformly 1/459 of its original volume for each degree it is raised or lowered above or below zero degrees Fahrenheit, provided the pressure on the gas remains constant. This fact is known as the *Law of Charles*. Following this same reasoning, we find that at −459° F, a gas would cease to exist. This assumption, therefore, establishes −459° F as *absolute zero*. Actually, this temperature or condition has never been attained. The *Law of Conservation of Matter* states that matter can be changed from one form into another. Temperature within a few degrees of absolute zero has been reached when liquefying oxygen, nitrogen, and hydrogen, but these (like other gases) change their physical state from gas into liquid and fail to disappear entirely at these low temperatures.

The fact that absolute zero has never been reached is also explained by another law known as the *Law of Conservation of Energy*. It has already been explained that heat is a form of energy. This law is stated as follows: Energy can be neither created nor destroyed, though it can be changed from one form into another. Table 7 gives factors for converting energy from one form into another.

Table 7. Energy Conversion Factors

1 watt hour (WH) = 3.411 British thermal unit (Btu)
1 British thermal unit (Btu) = 0.252 calorie (cal)
1 calorie (cal) = 3.968 British thermal units (Btu)
1 pound melting ice equivalent (MIE) = 144 British thermal units (Btu)
1 British thermal unit (Btu) = .00695 pounds melting ice equivalent (MIE)
1 pound melting ice equivalent (MIE) = 36.3 calories (cal)
1 calorie (cal) = 0.0276 melting ice equivalent (MIE)

Having considered the effect of temperature on a gas, the effect of pressure on gases to aid the study of refrigeration must next be considered. In 1662, Robert Boyle announced a simple relationship existing between the volume of a gas and the pressure applied to it. This has since become known to scientists as *Boyle's Law*, and may be stated as follows: At a constant temperature, the volume of a given weight of gas varies inversely as the pressure to which it is subjected. The more pressure applied to a gas, the smaller its volume becomes, if the temperature remains the same; likewise, if the pressure is released or reduced, the volume of the gas increases. Mathematically, this is expressed as:

$$PV = p \times v$$

where,

P equals the pressure on the gas at volume V.

p equals the pressure on the same weight of gas at volume v.

Boyle's Law has been found to be only approximately true, especially for the refrigerant gases which are more easily liquified. The variations from the Law are greater approaching the point of liquefaction, or condensing point, of any gas, although the material movement of air is determined by this Law.

It will be found that if the temperature is held constant and sufficient pressure is applied to a given weight of gas, it will change from the gaseous state into the liquid state. The point at which this change of state takes place is known as the point of liquefaction or condensing point.

It should now be evident that there is a definite relationship existing between the pressures, temperatures, and volumes at which a given weight of gas may exist. This relationship is used extensively in scientific work. It is known as the *Combined Law of Boyle and Charles,* and may be expressed mathematically as

$$\frac{PV}{T} = \frac{pv}{t}$$

where,

pressure P and p are expressed in the absolute pressure scale in pounds per square foot,

volumes V and v are expressed in cubic feet,

temperatures T and t are expressed in degrees on the absolute temperature scale.

When the pressure, temperature, or volume of a gas is varied, a new set of conditions is created under which a given weight of gas exists in accordance with the mathematical equation just mentioned. If a gas is raised to a certain temperature (which varies with each individual gas, no matter how much pressure is applied to it), it will be found impossible to condense it. This temperature is known as the *critical temperature*. The pressure corresponding to the critical temperature is termed the *critical pressure*. Above the critical points it is impossible to vaporize or condense a substance.

When a liquid is evaporated to a gas, the change of physical state is always accompanied by the absorption of heat. Evaporation has a cooling effect on the surroundings of the liquid, since the liquid obtains the necessary heat from its surroundings to change the molecular structure. This action takes place in the evaporator of a refrigeration system. Any liquid tends to saturate the surrounding space with its vapor. This property of liquids is an important element in all refrigeration work. On the other hand, when a gas is condensed into a liquid, the change of physical state is always accompanied by the giving up of heat. This action takes place in the condenser of the condensing unit of a refrigeration system due to the mechanical work exerted on the gas by the compressor. If gas or liquid is placed in a closed container and the temperature of the container changed, it will be found that the pressure exerted by that gas in the container is directly proportional to the absolute temperature. Thus, if the temperature is raised, the pressure increases, and if the temperature is lowered, the pressure decreases.

THE REFRIGERATION CYCLE

The refrigeration cycle is simply a means of heat extraction. In a compression system four distinct parts are required, as follows:

1. Compressor (pump)
2. Condenser-receiver
3. Evaporator (cooling coils)
4. Expansion valve (pressure reducing device)

Fig. 26 illustrates schematically the parts which make up the compression system. In this process, a refrigerant is used which can be alternately vaporized and liquified. The heat energy required to change the liquid refrigerant into a gas is obtained from the air space surrounding the evaporator. This low pressure is then drawn through the suction line and into the compressor. In passing through the compressor, the heat laden gas is raised from the low pressure in the suction line to a higher pressure, thereby raising its temperature. It is then forced from the compressor through the discharge line into the condenser where the heat is removed from the vapor by means of natural air circulation. The removal of heat from the vapor causes it to liquefy and then flow from the bottom of the condenser and into the receiver.

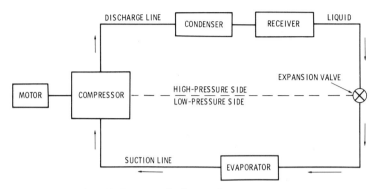

Fig. 26. Diagram of a basic refrigeration circuit.

The expansion valve is adjusted to control the flow of refrigerant into the evaporator at a rate which is sufficient to maintain a desired temperature. In this connection, it should be noted that once the refrigerant has returned to a liquid state it is again ready to be admitted through the expansion valve (or other pressure reducing device) to the evaporator. In the evaporator, the pressure is reduced, the boiling point is lowered, and vaporization takes place, resulting in extraction of heat. This action is repeated continually as long as the compressor is running.

PRESSURE

It is of the utmost importance that the refrigeration student understand the meaning of the various kinds of pressure as related to refrigeration.

Atmospheric Pressure

Atmospheric pressure is pressure which is exerted by the atmosphere in all directions, as indicated by a barometer. Standard atmospheric pressure is considered to be 14.695 pounds per square inch (usually written 14.7) which is equivalent to 29.92 inches of mercury.

Absolute Pressure

Absolute pressure is the sum, at any particular time, of gauge pressure and atmospheric pressure. Thus, for example, if the pressure gauge at one

63

particular time reads 53.7 pounds, the absolute pressure will be 53.7 + 14.7, or 68.4 pounds per square inch. The aforementioned definitions may be written as follows:

$$\text{Absolute pressure} = \text{gauge pressure} + 14.7$$
$$\text{Gauge pressure} = \text{absolute pressure} - 14.7$$

where,

14.7 is the normal atmospheric pressure.

SUMMARY

In this chapter the fundamental principles of refrigeration have been presented. Since refrigeration deals with the removal of heat from space or material substances, it is important that the student or serviceman clearly understand the method used in refrigeration as well as the nature of heat, the heat of vaporization, latent heat, and the change of state.

It should be noted that modern refrigeration is simply accomplished by a change of state from liquid to gas in various types of refrigerants, the change of its physical state being accomplished by the absorption of heat. The laws of gases have been fully treated, in addition to the energy sources utilized for pumping the refrigerant through the refrigeration system components.

The measurement of heat and a simple calculation for removal of heat from various material substances using specific heats and Btu values will further assist the reader in understanding the nature of refrigeration, its laws, and, finally, its utilization.

REVIEW QUESTIONS

1. What is meant by units of heat?

2. What is meant by conduction, convection, and radiation?

3. How does specific heat affect various substances?

4. How does latent heat affect the change of state in various fluids?

5. What is meant by heat of vaporization?

6. Name various commercial refrigerants.

7. How is refrigeration accomplished?

8. What is meant by an indirect system of refrigeration?

9. Define the Law of Conservation of Matter.

10. What is the absolute zero reading on the Fahrenheit scale?

11. How many calories are represented by 1 Btu?

12. Define Boyle's Law.

13. State the Law of Conservation of Energy.

14. State the relation between pressure, temperature, and volume in a given weight of gas.

15. Explain how the refrigeration cycle is accomplished.

16. What will be the temperature reading on the Fahrenheit scale when the Centigrade thermometer reads minus 10°?

17. Give the relation between absolute and gauge pressure.

CHAPTER 2

Refrigerants

Refrigerants are heat-carrying mediums which, during their cycle, absorb heat at a low temperature level and are compressed by a heat pump to a higher temperature where they are able to discharge the absorbed heat. The ideal refrigerant would be one that could discharge to the condenser all the heat that the refrigerant is capable of absorbing in the evaporator or cooler. All refrigerating mediums, however, carry a certain portion of the heat from the condenser back to the evaporator, and this reduces the heat-absorbing capacity of the medium on the low side of the system.

Tables showing the physical and chemical properties of refrigerants can be found in the appendix.

DESIRABLE PROPERTIES

The requirements of a good refrigerant for commercial use are:

1. Low boiling point.
2. Safe and nontoxic.
3. Ease of liquefaction action at moderate pressure and temperature.
4. High latent-heat value.
5. Ability to operate on a positive pressure.
6. No effect on moisture.
7. Mix well with oil.
8. Noncorrosive to metal.

Classifications

Refrigerants may be divided into three classes according to their manner of absorption or extraction of heat from the substances to be refrigerated.

Class 1—This class includes those refrigerants that cool materials by the absorption of the latent heat. The temperature and pressure properties of these refrigerants are shown in Table 1.

Table 1. Temperature and Pressure Properties

	Boiling Point °F	Freezing Point °F	Critical Temperature °F	Critical Pressure p.s.i.a.
Freon 14	−198.2	−312	−49.9	542
Freon 13	−114.5	−296	83.8	579
Carbon dioxide	−108.4	−69.9 triple	87.8	1,071
Freon 22	−41.4	−256	204.8	716
Ammonia	−28.0	−107.9	271.2	1,651
Freon 12	−21.6	−252.4	232.7	582
Methyl chloride	−10.76	−143.7	289.6	969.2
Sulfur dioxide	14.0	−98.9	314.8	1,141.5
Freon 114	38.4	−137	294.3	474
Freon 21	48.0	−211.0	353.3	750
Freon 11	74.7	−168.0	388.4	635
Methylene chloride	103.7	−143.0	421	640
Freon 113	117.6	−31.0	417.4	495

Class 2—The refrigerants in this class are those that cool substances by absorbing their sensible heats. They are: Air, calcium chloride brine, sodium chloride (salt) brine, alcohol, and similar nonfreezing solutions.

Class 3—This group consists of solutions that contain absorbed vapors of liquefiable agents or refrigerating media. These solutions function by the nature of their ability to carry the liquefiable vapors that produce a cooling effect by the absorption of latent heat. An example of this group is aqua ammonia, which is a solution composed of distilled water and pure ammonia.

The refrigerants in Class 1 are employed in the standard *compression* type of refrigerating systems. The refrigerants in Class 2 are employed as *immediate cooling agents* between Class 1 and the substance to be refrigerated, and do the same work for Class 3. The latter is employed in the standard *absorption* type of refrigerating systems.

THE FREON REFRIGERANTS

The *Freon* family of refrigerants is presently used almost universally in household-type refrigerators. In the past, refrigerants selected for use were chosen principally for their boiling points and pressures, and their stability within the system or unit regardless of other important necessary properties such as nonflammability and nontoxicity. Of course, there are many factors that must be taken into account when selecting a chemical compound for use as a refrigerant other than boiling point, pressure, stability, toxicity, and flammability. They must include molecular weight, density, compression ratio, heat value, temperature of compression, compressor displacement, design or type of compressor, etc., to mention only a few of the major considerations.

Chemical Properties

The *Freon* refrigerants are colorless, almost odorless substances, the boiling points of which vary over a wide range of temperatures. *Freon* refrigerants are nontoxic, noncorrosive, nonirritating, and nonflammable under all conditions of usage. They are generally prepared by replacing chlorine or hydrogen with fluorine. Chemically they are inert and thermally stable up to temperatures far beyond conditions found in actual operation.

Physical Properties

Pressures required to liquefy the refrigerant vapor affect the design of the system. The refrigerating effect and specific volume of the refrigerant vapor determines the compressor displacement; and the heat of vaporization and specific volume of liquid refrigerant affects the quantity of refrigerant to be circulated through the pressure regulating valve or other device. Table 1 covers boiling point at one atmospheric pressure, freezing point, critical temperature, and critical pressures of not only the *Freon* refrigerants, but other commonly used refrigerants as well.

Operating Pressures

Pressures of saturated vapor at various temperatures of typical refrigerants are listed in Table 1. Table 2 lists the pressure of saturated vapor under standard ton conditions. Operating pressures will vary with the temperature of the condensing medium, amount of condenser surface, whether an air- or water-cooled condenser is used, operating back pressure, presence of noncondensible or foul gas in the condenser, circulation of the condensing medium through the condenser, condition of the condenser surface, extent of superheating of the refrigerant gas, and other factors.

Table 2. Operating Pressures
(Standard Ton)

	Pressure p.s.i.g. 86° F.	Pressure p.s.i.g. 5° F.	Compression Ratio
Carbon dioxide	1024.3	319.7	3.11
Freon 22	159.8	28.33	4.045
Ammonia	154.5	19.57	4.94
Freon 12	93.2	11.81	4.075
Methyl chloride	80.0	6.46	4.48
Sulfur dioxide	51.75	5.87*	5.63
Freon 114	21.99	16.14*	5.42
Freon 21	16.53	19.25*	5.96
Freon 11	3.58	23.95*	6.24
Methylene chloride	9.44*	27.53*	8.57
Freon 113	13.93*	27.92*	8.01

*Inches mercury below one atmosphere.

Freon-12

Freon-12, CCl_2F_2, has a boiling point of $-21.6°$ F, and is extensively used as a refrigerant in both direct and indirect industrial, commercial, and household air-conditioning systems, as well as in household refrigerators, ice-cream cabinets, frozen-food cabinets, food locker plants, water coolers, etc., employing reciprocating type compressors ranging in size from fractional to 800 h.p. *Freon-12* is also used in household refrigerating systems, ice-cream and frozen-food cabinets employing rotary-vane type compressors. *Freon-12* is used in industrial process water and brine cooling to $-110°$ F, employing multistage centrifugal-type compressors in cascade of 100 tons refrigeration capacity and larger.

The health hazards resulting from exposure to *Freon* when used as a refrigerant are remote. *Freon* is in a class of special, nontoxic gases. Vapor in any proportion will not irritate the skin, eyes, nose, or throat, and being odorless and nonirritating, it will eliminate all possibilities of panic hazards should it escape from a refrigerating system. A pound of *Freon-12* liquid expands to 3.8 cu. ft. of vapor at 68° F room temperature.

69

Freon-12 is a stable compound capable of undergoing, without decomposition, the physical change to which it is commonly subjected in service, such as freezing, vaporization, and compression. The boiling point is $-21.6°$ F at the atmospheric pressure of sea level.

The low boiling point permits low temperatures to be reached without the compressor operating on a vacuum. This low but positive pressure prevents moisture-laden air from accidentally entering the system and also permits detection and location of the source of leaks, which is difficult if the refrigeration system operates at a negative back pressure.

Under ordinary conditions when no moisture is present, *Freon* does not corrode the metals commonly used in refrigerating systems. In the presence of water, it will discolor brass, steel, and copper, but there is little or no evidence of any serious corrosive action. It is only slightly soluble in water, and the solution formed will not corrode any of the common metals used in refrigeration construction. It is nonflammable and noncombustible under fire conditions or where appreciable quantities come in contact with flame or hot metal surfaces. It requires an open flame at 1382° F to decompose the vapor, and then the vapor only decomposes to form hydrogen chloride and hydrogen fluoride, both of which are irritating but are readily dissolved in water. Air mixtures are not capable of burning and contain no elements that will support combustion. Therefore, *Freon* is considered nonflammable. *Freon* is so safe and nontoxic, that the only possible way to cause death with *Freon* is to get concentrations so great as to exclude the oxygen.

Freon-21

Freon-21, $CHCl_2F$, has a boiling point of 48° F and may be used as a refrigerant in industrial and commercial air-conditioning systems; also in industrial process water and brine cooling to $-50°$ F employing single or multistage centrifugal-type compressors of 100-ton refrigeration capacity and larger. *Freon-21* has been used in fractional-horsepower household refrigerating systems and drinking water coolers employing rotary-vane type compressors, and has also been used in comfort cooling air-conditioning systems of the absorption type where dimethyl ether or tetraethylene glycol was used as the absorbent.

Freon-22

Freon-22, $CHClF_2$, has a boiling point of $-41.4°$ F, and is used as a refrigerant in industrial and commercial low-temperature refrigerating systems to $-150°$ F, and also in window-type and unit package room coolers and air-conditioning units. *Freon-22* is used in many installations where more efficient operation is desired in providing the necessary lower temperatures for low-temperature locker plants, lower temperatures that result in quicker freezing of foods, greater volume of products handled by the quick-freezing units, home or farm freezers, and countless numbers of low-temperature industrial applications.

Freon-113

Freon-113, $CCl_2F\text{-}CClF_2$, has a boiling point of $117.6°$ F, and is used as a refrigerant in most all industrial and commercial air-conditioning systems. *Freon-113* is also used in industrial process water and brine cooling to $0°$ F, employing four or more stage centrifugal-type compressors of 25-ton refrigeration capacity and larger.

Freon-11

Freon-11, CCl_3F, has a boiling point of $74.7°$ F, and has wide usage as a refrigerant in indirect industrial and commercial air-conditioning systems; also in industrial process water and brine cooling to $-40°$ F, employing single or multistage centrifugal-type compressors of 100 ton refrigeration capacity and larger.

Freon-114

Freon-114, $C_2Cl_2F_4$, has a boiling point of 38.4 F, and is being used as a refrigerant in fractional-horsepower household refrigerating systems and drinking water coolers employing rotary-vane type compressors; also in indirect industrial and commercial air-conditioning systems and in industrial process water and brine cooling to $-70°$ F employing multistage centrifugal-type compressors in cascade of 100-ton refrigeration capacity and larger.

Freon-13

Freon-13, $CClF_3$, and *Freon-14*, CF_4, are two compounds recently added to the *Freon* family, and have a boiling point of $-114.5°$ F, and $-198.2°$ F, respectively. These will undoubtedly find usage as refrigerants in extremely low-temperature industrial refrigerating systems approximating liquid-air temperatures, and will be employed in cascaded reciprocating-type compressors. The use of these two *Freon* refrigerants will require thorough investigation and research as to the behavior of

71

metals at low temperatures while under stress, lubrication of mechanical compressors, and the possible development of more efficient insulating materials.

Sulfur Dioxide

This was formerly one of the most common refrigerants employed in household refrigerators. It is also known as SO_2, and is a colorless gas or liquid. It is toxic and has a very pungent odor, and is obtained by burning sulfur in air. It is not considered a safe refrigerant, especially in quantities. It combines with water and forms sulfurous and sulfuric acids which are corrosive to metal. Sulfur dioxide has an adverse effect on almost anything with which it comes in contact. It boils at about 14° F (standard conditions), and has a latent-heat value of 166 Btu per pound.

Sulfur dioxide has the disadvantage that it must operate on a vacuum to give temperatures required in most refrigeration work. Should a leak occur, moisture-laden air will be drawn into the system, eventually corroding the metal parts and ruining the compressor. Also, in relation to *Freon* or methyl chloride, approximately ⅓ more vapor must be pumped in order to get the same amount of refrigeration. This means that the condensing unit will either have to be speeded up to give the desired capacity, or the size of the cylinders increased proportionately.

Sulfur dioxide does not mix well with oil. The suction line must be on a steady slant to the machine, otherwise the oil will trap out, making a constriction in the suction line. On many installations it is not possible to avoid traps, and on these jobs sulfur dioxide is not satisfactory. Because of its characteristic pungent odor, comparatively small leaks are readily detected. Even the smallest leaks are readily located by means of an ammonia swab. A small piece of cloth or sponge may be secured to a wire and dipped into strong aqua ammonia or household ammonia and then passed over points where leaks may be present. A dense white smoke forms where the sulfur dioxide and ammonia fumes come in contact. When no ammonia is available, leaks may be located by the usual soap bubble or oil test. The soap solution or oil is put on the tube joints and points where bubbles would be noted.

Note: Liquid sulfur dioxide on any part of the body produces freezing. In any such case a physician should be consulted immediately.

Methyl Chloride

This refrigerant was used mostly in commercial refrigerator units, and its use in household refrigerators was very limited.

It is a good refrigerant, but due to the fact that it will burn under some conditions and is slightly toxic, it does not conform to some of the strict city codes now in force. Roughly speaking, the average relative concentration by weight of different refrigerant vapors in a room of a given size that produces the same effect on a person breathing the air thus contaminated can be approximately specified as follows:

Carbon dioxide	100
Methyl chloride	70
Ammonia	2
Sulfur dioxide	1

In other words, methyl chloride is 35 times safer than ammonia and 70 times safer than sulfur dioxide. To produce any serious effects from breathing methyl chloride, a considerable quantity is required. For instance, in a room $20 \times 20 \times 10$ ft., it would be necessary to liberate about 60 lbs. of methyl chloride to produce any serious effects. Methyl chloride has a low boiling point. Under standard atmospheric pressure it boils to $-10.8°$ F. It is easy to liquefy and has a comparatively high latent value—approximately 176 Btu per pound under standard conditions. It will operate on a positive pressure as low as $-10°$ and mixes well with oil. In its dry state it has no corrosive effect on metal, but in the presence of moisture, copper plating of the compressor parts results, and in severe cases of moisture, a sticky black sludge is formed which is detrimental to the working parts of the system.

Methyl chloride is not irritating and, consequently, does not serve as its own warning agent in case of leaks as does sulfur dioxide. In some cases, a warning agent is added, such as a small percentage of acrolein (1%). Many consider the addition of 5% of sulfur dioxide a dependable warning agent, but there is some controversy as to the desirability of doing this.

Ammonia

Ammonia is a refrigerant employed in refrigerators operating on the absorption principle. It is also used in large machines for industrial and other purposes. It is a colorless gas with a pungent characteristic odor. Its boiling temperature at normal atmospheric pressure is $-28°$ F, and its freezing temperature is $-107.86°$ F. It is very soluble in water, 1 volume of water absorbing 1.148 volumes of ammonia at $32°$ F.

Ammonia is combustible or explosive when mixed with air in certain proportions (about one volume of ammonia to two volumes of air) and much more so when mixed with oxygen. Because of its high latent-heat value (555 Btu at 18° F) large refrigeration effects are possible with relatively small-sized machinery. It is very toxic, and requires heavy steel fittings. Pressures of 125 to 200 pounds per square inch are not uncommon, and water-cooled units are essential.

Carbon Dioxide

Carbon dioxide at ordinary temperatures is a colorless gas with a slightly pungent odor and acid taste. It is harmless to breathe except in extremely large concentrations, when the lack of oxygen would cause suffocation. It is nonexplosive, nonflammable and does not support combustion. The boiling point of carbon dioxide is so extremely low that at 5° F a pressure of well over 300 pounds per square inch is required to prevent its evaporation. At a condenser temperature of 80° F, a pressure of approximately 1000 pounds per square inch is required to solidify the gas. Its critical temperature is 87.8° F and −69.9° F is the triple point.

Because of its high operating pressure, the compressor of the carbon dioxide refrigerator unit is very small, even for a comparatively large refrigerating capacity. Because of its low efficiency as compared to other common refrigerants, it is seldom used in household units, but is used in some industrial applications and aboard ships. Leakage of carbon dioxide gas can be tested by making sure that there is pressure on the part to be tested and then use soap solutions at the suspected points. Leakage into condenser water can be tested with the use of "bromthymol blue." The water entering and leaving the condenser should be tested at the same time because of the sensitivity of the test. When carbon dioxide is present, the normal blue color changes to yellow.

LEAK DETECTION

The following methods of leak detection are reprinted "By permission from ASHRAE, Handbook of Fundamentals 1972".

The detection of leaks in refrigeration equipment is a major problem for both manufacturers and service engineers. Several methods of leak detection will be described.

Electronic Detector

The electronic detector is widely used in the manufacture and assembly of refrigeration equipment. The operation of the instrument depends on the variation of current flow due to ionization of decomposed refrigerant between two oppositely charged platinum electrodes. This instrument can be used to detect any of the halogenated refrigerants except Refrigerant 14. It is not recommended for use in atmospheres containing explosive or flammable vapors. Some other vapors, such as alcohol and carbon monoxide, may interfere with the test.

The electronic detector is the most sensitive of the various leak detector methods, reportedly capable of sensing a leak of $1/100$ oz. per year of Refrigerant 12, or approximately 1×10^{-6} cc per second.

A portable model is available with an automatic balancing system which corrects for refrigerant vapors that might be present in the atmosphere around the test area.

Halide Torch

The Halide torch has been used for many years as a fast and reliable method of detecting leaks of halogenated refrigerants. Air is drawn over a copper element heated by a methyl alcohol or hydrocarbon flame. If halogenated vapors are present, they will be decomposed and the color of the flame will change to bluish-green. Although not as sensitive as the electronic detector, this method is suitable for most purposes.

Authors Note: Do not breath the fumes from the halide torch.

Bubble Method

The object to be tested is pressurized with air or nitrogen. A pressure corresponding to operating conditions is generally used. The object can be immersed in a water bath and any leaks detected by observing the formation of bubbles in the liquid. Addition of a detergent to the water will decrease the surface tension, prevent escaping gas from clinging to the side of the object and will promote the formation of a regular stream of small bubbles. Kerosene or other organic liquids are sometimes used for the same reason. A solution of soap or detergent can be brushed or poured onto joints or other bubbles which can readily be detected.

Authors Note: If air or nitrogen is used for the test, the system must be thoroughly purged of these gases by means of a good vacuum pump for quite a period of time, to remove all of the air or nitrogen and any moisture, before charging the unit with refrigerant.

Detection of Ammonia and Sulfur Dioxide Leaks

Ammonia can be detected by burning a sulfur candle in the vicinity of the suspected leak or by bringing a solution of hydrochloric acid near the object. If ammonia vapor is present, a white cloud of ammonium sulfite or ammonium chloride will be formed. Ammonia can also be detected with any indicating paper which changes color in the presence of a base.

Sulfur dioxide can be detected by the appearance of a white smoke when aqueous ammonia is brought near the leak.

The presence of leaks can also be determined by pressurizing or evacuating the system and observing the change in pressure or vacuum over a period of time. This is good practice in checking on the "tightness" of a system but is of little help in locating the point of leakage.

PRESSURE-TEMPERATURE CHART

A logarithmic-scaled chart giving the relationship between pressure and corresponding temperatures in Fahrenheit degrees of common refrigerants is shown in Fig. 1. The axis of the abscissa shows the temperature and the axis of the ordinate shows the pressure in pounds per square inch gauge and absolute, respectively. To ascertain the pressure of a refrigerant at any particular temperature, follow the desired temperature until the curve of the particular refrigerant is reached. The corresponding pressure is then found on the pressure axis.

For example, the temperature corresponding to a pressure of 40 pounds gauge for sulfur dioxide is approximately 60° F, and the corresponding temperature at the same pressure for dichlorodiflouremethane (*Freon-12*) is approximately 27° F, etc.

CARE IN HANDLING REFRIGERANTS

In the foregoing discussion it has been observed that one of the requirements of an ideal refrigerant is that it must be nontoxic. In reality, however, all gases (with the exception of pure air) are more or less toxic or asphyxiating. It is therefore important that wherever gases or highly volatile liquids are used, adequate ventilation should be provided, because even nontoxic gases in air produce a suffocating effect.

Vaporized refrigerants, especially ammonia and sulfur dioxide, bring about irritation and congestion of the lungs and bronchial organs, accompanied by violent coughing, vomiting and when breathed in sufficient quantity, suffocation. It is therefore of the utmost importance that the serviceman subjected to a refrigerant gas should find access to fresh air at

Fig. 1. Showing vapor pressure of refrigerants at various temperatures.

frequent intervals to clear his lungs. When engaged in the repair of ammonia and sulfur dioxide machines, approved gas masks and goggles should be used. *Carrene, Freon-12* and carbon dioxide fumes are not irritating and can be inhaled in considerable concentrations for a short period without serious consequences. It should be remembered that liquid refrigerants will refrigerate or remove heat from anything with which

77

they come in contact when released from a container, as in the case of an accident.

HANDLING REFRIGERANT CYLINDERS

It is of the utmost importance to handle cylinders of compressed gas with care (see Fig. 2), and to observe the following precautions:

TYPE 'M'   TYPE 'S'   TYPE 'C'   TYPE 'D'
            TYPE 'MS'   TYPE 'MC'   TYPE 'MD'

Fig. 2. Various sizes of refrigerant cylinders.

1. Never drop cylinders nor permit them to strike each other violently.

2. Never use a lifting magnet nor a sling (rope or chain) when handling cylinders. A crane may be used when a safe cradle or platform is provided to hold the cylinders.

3. Where caps are provided for valve protection, such caps should be kept on the cylinders except when the cylinders are in use.

4. Never overfill cylinders. Whenever the refrigerant is discharged from or into a cylinder, immediately thereafter weigh the cylinder and record the weight of refrigerant remaining in the cylinder.

5. Never attempt to mix gases in a cylinder.

6. Never use cylinders for rollers, supports, or for any purpose other than to carry gas.

7. Never tamper with the safety devices in valves or cylinders.

8. Open cylinder valves slowly. Never use wrenches or tools except those provided or approved by the gas manufacturer.

9. Make sure that the threads on regulators or other unions are the same as those on cylinder valve outlets. Never force connections that do not fit.

10. Regulators and pressure gauges provided for use with a particular gas must not be used on cylinders containing different gases.

11. Never attempt to repair or alter cylinders or valves.

12. Never store cylinders near highly flammable substances, such as oil, gasoline, waste, etc.

13. Cylinders should not be exposed to continuous dampness, salt water, or salt spray.

14. Store full and empty cylinders apart to avoid confusion.

15. Protect cylinders from any object that will produce a cut or other abrasion in the surface of the metal.

Refrigerant Containers

You can purchase refrigerants in containers from 1 lb. up to 150 lb. If you use the 150 lb. containers, it is well for domestic use to also purchase 10 lb. containers, which are easier to handle. The 10 lb. containers may be refilled from the 150 lb. container by lowering the temperature of the 10 lb. container with ice or cold water, and connecting the two tanks with a flexible hose. Be certain that the air is purged out of the hose before connecting to the 10 lb. tank. This is accomplished by cracking the valve on the 150 lb. tank, and allowing some gas to escape, then connect the hose to the small tank. Place the small tank on a scales, turn the large tank on its side or end, and watch the scales, stopping at the point that you reach 10 lbs. plus the weight of the tank. DO NOT OVERFILL, IF ANYTHING UNDERFILL the small tank.

A SIMPLE REFRIGERATION SYSTEM

The principle of using the latent heat of vaporization of a liquid, such as sulfur dioxide, for producing refrigeration can be illustrated very easily by thinking of a refrigerator of very simple design, similar to the one shown in Fig. 3. The refrigerator is made up of a box that is completely insulated on all six sides to prevent the entrance of heat by conduction, convection, and radiation. We then place in the top of the cabinet a series of finned coils, with one end connected to the cylinder charged with sulfur dioxide. Through this end we will charge, for example, two pounds of sulfur dioxide into the coil after which the compressed cylinder will again be sealed and disconnected from the line, with the charging end of the pipe open to the atmosphere.

REFRIGERATOR

Fig. 3. A simple refrigeration cycle.

Since the liquid sulfur dioxide is exposed to the air, the only pressure to which the liquid is subjected is to atmospheric pressure, which is approximately 14.7 pounds per square inch absolute or zero pounds gauge pressure. At this pressure, as previously explained, sulfur dioxide liquid will boil or vaporize at a temperature of 14° F, or at any higher temperature. We will say, just for example, that the temperature of the room in which the refrigerator is located is 70° F. If this is the case, the temperature of the cabinet at the time of the addition of sulfur dioxide liquid will also be 70° F. The liquid sulfur dioxide in the coils will therefore immediately start boiling and vaporizing because the surrounding temperature is above the boiling point (14° F) of the liquid. As the liquid boils away, it will absorb heat from the cabinet because, for every pound of sulfur dioxide liquid vaporized, 168 Btu of heat will be extracted from the cabinet. As soon as the temperature of the cooling coil is reduced to a point lower than the cabinet temperature, the air in the cabinet will start circulating in the direction shown by the arrows in Fig. 3, because heat always flows from the warmer to the colder object.

With this method, however, the two pounds of sulfur dioxide liquid would soon be vaporized and the gas given off to the air outside the cabinet, and refrigeration would then stop until a new charge was placed in the cooling coil. Sulfur dioxide is expensive and difficult to handle and some means must therefore be used to reclaim the vapor in order to use the original charge continuously. The inconvenience of recharging the coil must also be prevented, and the refrigerator must be built so that it will automatically maintain proper food-preservation temperatures at all times with absolutely no inconvenience to the customer. This is accomplished by the compressor pulling the warm sulfur dioxide gas from the cooling unit and pumping it into the condenser where it is changed to a liquid ready to return to the cooling unit.

SUMMARY

Refrigerants, being the prime source without which mechanical refrigeration systems would not be possible, have been fully described. Thus, the characteristics of typical refrigerants have been included, giving the boiling point and heat of vaporization of each.

A considerable amount of text has been devoted to the *Freon* family of refrigerants, since these have been one of the major factors responsible for the growth in the commercial field of refrigeration. This is largely because the properties of *Freons* have permitted their use where flammable or more toxic refrigerants would be hazardous.

The storage and handling of refrigerants including cylinders, storage methods, and cylinder capacity, together with general precautions to be observed when leakage or accidents occur, are given in the last part of the chapter.

REVIEW QUESTIONS

1. Name the desirable properties of a good refrigerant for commercial use.

2. How are commercial refrigerants classified?

3. What is the purpose of calcium chloride brine as used in commercial refrigeration plants?

4. Why is ammonia employed in preference to other refrigerants in commercial plants?

5. State the boiling point and heat of vaporization for *Freon*-12.

6. In what respect does the *Freon* family of refrigerants differ from such refrigerants as methyl or ethyl chloride?

7. Name the chemical and physical properties of *Freons*.

8. Why is *Freon*-12 preferred in both commercial and household refrigeration units?

9. What is meant by the critical temperature of a refrigerant?

10. Why does a high latent heat of evaporation per unit of weight affect the desirability of a refrigerant?

11. Describe the methods of leak detection in a refrigeration system.

12. What precautions should be observed in the storage and handling of refrigerant cylinders?

CHAPTER 3

Compressors

Compressors may be divided into several types, depending on the size of the plant, the refrigerant used, and other factors, as:

1. Single cylinder
2. Multicylinder

With respect to the method of compression as:

3. Reciprocating
4. Rotary
5. Centrifugal

With respect to the drive employed as:

6. Direct drive
7. Belt drive

With respect to the location of the prime mover as:

8. Independent (belt drive)
9. Semihermetic (direct drive, motor and compressor in separate housings)

HOW THE COMPRESSOR WORKS

The passage of the refrigerant to and from the compressor is controlled by means of a discharge and suction valve located on a specially designed valve plate which forms the lower part of the cylinder head. The flapping action of the valve permits the flow of refrigerant out through the discharge valve port only, and in only through the suction valve port. Thus, when the piston moves away from the valve plate (suction stroke), a pressure reduction takes place. Because the pressure in the cylinder is now below that in the suction line, a flow of refrigerant occurs that pushes open the suction valve and permits a certain quantity of refrigerant to enter the compressor.

As the motion of the piston reverses and moves toward the valve plate (compression stroke), it increases the pressure which forces the suction valve to close. A further compression, as the piston moves close to the valve plate, opens the dischage valve and forces the refrigerant into the discharge line, thus causing what is known as the high-side or positive working pressure of the refrigerant.

RECIPROCATING COMPRESSORS

Reciprocating compressors derive their name from the reciprocating (back and forth) action of the piston. The length of the piston movement within its cylinder is known as the stroke. The capacity of the compressor, as previously noted, depends on such factors as the number of cylinders, the stroke, revolutions per minute of the crankshaft, etc. Fig. 1 shows a typical reciprocating compressor.

Fig. 2 shows a heavy duty, two cylinder industrial compressor of vertical design and low-speed operation. The motion of the piston is transferred from the crankshaft which, in turn, receives its motion from a belt wheel as shown. The prime mover for this type of compressor is usually an electric motor, although there is no reason why other types of prime movers could not be used.

The piston motion is effected by the crankshaft arm or arms to each of which is fastened one end of a connecting rod. The other end of the connecting rod is fastened to the piston by means of a piston pin. It is in this manner that the connecting rod and crank arm transfer the rotation of the crankshaft to the reciprocating or back and forth motion of the piston. The passage of the refrigerant to and from the compressor is controlled by

EXHAUST PORT

CYLINDER HEAD

PISTON

CONNECTING ROD

COUNTERWEIGHT

CRANK-SHAFT

SHAFT SEAL

Courtesy Multiplex Mfg. Co.

Fig. 1. Cutaway view of a single cylinder belt-driven reciprocating refrigeration compressor.

means of a discharge and suction valve located on a specially designed valve plate. This valve plate forms the lower part of the cylinder head.

The flapping action of the valve permits the refrigerant to flow out only through the discharge valve port, and in only through the suction valve port. Thus, when the piston moves away from the valve plate (suction stroke), a reduction of pressure takes place. Because the pressure in the cylinder is now below that in the suction line, a flow of refrigerant occurs which pushes open the suction valve and permits a certain quantity of refrigerant to enter the compressor. As the piston reverses its motion and moves toward the valve plate (compression stroke), it increases its pressure, which forces the suction valve to close. A further compression as the piston moves close to the valve plate opens the discharge valve and forces the refrigerant into the discharge line, thus causing what is known as the high-side pressure of the refrigerant. It is in this manner that the

WATER JACKET

MANIFOLD CONNECTION (DISCHARGE)

DISCHARGE VALVES

SUCTION VALVES

BELT WHEEL

OUTBOARD BEARING

MANIFOLD CONNECTION (SUCTION)

WRIST PINS

OIL-PRESSURE GAUGE

FLEXO-SEAL

CRANKSHAFT

CONNECTING ROD BEARINGS

OIL STRAINER

CYLINDER HEAD

SAFETY HEAD

COMPRESSION RINGS

CAPACITY CONTROL VALVES

PISTONS

OIL SCRAPER RINGS

CONNECTING RODS

OIL PUMP

CRANKCASE

Courtesy the Frick Company

Fig. 2. Sectional view of a heavy-duty vertical industrial reciprocating compressor.

suction stroke fills the cylinder with vaporous refrigerant, whereas the compression stroke compresses it and forces it out of the cylinder, the valves being actuated by the difference in pressure only.

Lubrication

Force-feed lubrication is usually provided in a reciprocating compressor by a reversible oil pump directly driven from the crankshaft. The pump provides lubrication for the main and connecting rod bearings, the wrist pin bushings, the shaft seal, and all other movable compressor parts. Certain large commercial compressors are equipped with a mechanical sight-feed system for supplying oil under pressure to the suction inlet and to the thrust side of the cylinder walls through special rod passages.

Shaft Seal

The function of the shaft seal in a refrigeration compressor is to prevent the gas from escaping from the compressor at the point where the shaft leaves the compressor housing. Depending on the size of the machine, a variety of leak-free shaft seals are used. A bellows-type or flexible seal usually provides protection against leaks for either pressure or vacuum. Fig. 3 shows the seal arrangement of a typical heavy-duty compressor.

Courtesy the Frick Company

Fig. 3. Sectional view of a heavy-duty compressor shaft seal.

87

Pistons and Rings

Pistons employed in the majority of compressors are made of the best grade of cast iron. They are carefully machined, polished, and fitted to the cylinders at close tolerances. There are several different designs for pistons. Some early models were not equipped with piston rings, but were lapped to the cylinder in a very close fit. At present, however, two or more piston rings are employed, their function being to insure proper lubrication and sealing of the cylinder walls. Pistons should fit the cylinder walls so that, when inserted, they do not drop of their own weight, but must be urged through with the fingers. This fit is usually obtained by allowing 0.0003'' (three ten-thousandths of an inch) of clearance per inch of cylinder diameter. Thus, a 1½'' cylinder would have a 0.00045'' clearance between the piston and the cylinder wall.

If cylinders are scored, they must be honed after reboring to assure a high degree of wall finish which maintains cylinder proportions over long periods. After rings have been fitted, the compressor should be operated overnight to permit the rings to wear in. The compressor should then be drained and flushed with petroleum spirits or dry-cleaning fluid and dehydrated before being returned to service. Typical piston rings are shown in Fig. 4.

Connecting Rods and Wrist Pins

The connecting rod forms the link between the piston and the crankshaft. One end is connected to the piston by means of a hardened, ground, and highly polished steel wrist pin. Compressor connecting rods, as shown in Fig. 5, are very similar to those employed in automobile engines except for their size.

Several different methods are used to secure the piston wrist pin and connecting rod. In some designs, the wrist pin is tightly clamped to the connecting rod, with the moving bearing surface in the piston, while others have a bushing in the connecting rod which allows the pin to turn freely in both piston and connecting rod. Again, others have the wrist pin held solidly in the piston. Connecting rods are usually made of a high grade of cast iron or drop-forged steel.

Crankshafts

The crankshafts in refrigeration compressors are commonly made of steel forgings which are machined to the proper tolerances for the main and connecting rod bearing. The double crank is employed in all two-

Fig. 4. Compressor piston rings.

cylinder compressors, while the single crank is used on the single-cylinder type. Crankshafts are equipped with counterweights and are carefully balanced to insure smooth and vibrationless compressor operation.

Some types of compressors are equipped with what is known as the eccentric shaft. This is a different application of the crankshaft principle. It employs an assembly consisting of the main shaft (over which an eccentric is fitted) and the outer eccentric strap.

Valves

The function of the valves in a compressor is to direct the flow of refrigerants through the unit. These valves are named according to the function they perform, such as *suction* and *discharge* valves. The so-called poppet type of valve has largely been replaced by the ring-plate valve. The ring-plate valve (Fig. 6) is designed for quick, positive, leak-free operation, and is usually precision machined from specially heat-treated alloy steel. Because of their location in the cylinder safety head, valves can easily be removed and replaced as an assembly. Dis-

89

Fig. 5. Compressor connecting rod construction.

Fig. 6. Sectional view of a compressor ring valve.

charge valves consist of two spring-loaded ring plates, and the suction valve consists of one ring plate.

ROTARY COMPRESSORS

The rotary compressor, although not as common as the reciprocating type, employs a slightly different method in compression of the refrigerant. The rotary compressor compresses gas due to the movement of blades in relation to a pump chamber. In one type, the blades slide on an eccentric on the driveshaft inside a stationary concentric ring known as the pump chamber. In another design, the blades revolve concentrically with the shaft inside an off-center ring (Fig. 7).

Hermetically sealed rotary compressors are used in some makes of domestic refrigerators and freezers.

Fig. 7. Interior of a rotary compressor (seal and cover removed).

A representative type of a rotary booster compressor is shown in Fig. 8. During rotation, each gas space between the sliding blades is at maximum volume as it passes the cylinder suction port. As the rotor turns, this space is gradually reduced and the trapped gas is compressed. When the space reaches the minimum volume point (the cylinder discharge port) the gas is forced into the discharge line.

HERMETIC COMPRESSORS

Hermetic or sealed-type compressors are directly connected to an electric motor; the motor and compressor operate on the same shaft and are enclosed in a common casing. Condensing units of this type are used almost exclusively in domestic refrigerators, and also in locker and home cold-storage plants, drinking fountains, ice cream and food display

91

Courtesy the Frick Company

Fig. 8. Cutaway view of a rotary booster compressor.

cabinets, soda fountains, and the like. They are made to operate on either the reciprocating or rotary principle, and may be mounted with the shaft in either the vertical or horizontal position. In a unit of this type, the revolutions per minute (rpm) obviously are the same for both compressor and motor. This factor has a very important bearing on the size and design of the unit, since it determines the type of refrigerant, the type of controls to be used, etc.

A representative hermetic horizontal compressor is illustrated in Fig. 9. In a unit of this type, the one-piece housing provides for quietness and a minimum of vibration. In addition, the seal and coupling (always a maintenance problem in open compressors) are eliminated. Still another dependable feature is the fact that the motor operates in an ideal atmosphere. Because of the fact that it is entirely enclosed, no airborne dust or dirt can reach it. Suction gas at 50° to 60° F cools the motor and shell. Together, the foregoing factors assure long, trouble-free motor operation.

Courtesy the Trane Company

Fig. 9. Sectional view of a horizontal hermetic compressor.

A different type of compressor is shown in Fig. 10. This particular unit is internally spring-mounted. The motor, located above the compressor, operates in a vertical position, whereas the compressor is horizontal. This construction permits operation of the compressor in oil, simplifying the lubricating problem. The suction intake is placed so that the suction vapor must travel through the holes in the motor rotor in order to get to the top of the shell and then to the intake tube. Certain models of this type of compressor are provided with internal thermostats. These are inserted in the motor windings and therefore measure motor temperature exactly, without allowing for the air gap between the motor and the top of the shell where overloads are normally located. This is particularly important in heat-pump applications where the ambient temperature may have considerable influence on the protection system. Because the thermostat is

93

located at the most critical point, it gives instantaneous and accurate sensing of the motor temperature and can therefore remove the compressor from the line at a safe temperature level. It is always operative when the compressor is running.

Another important feature consists of an anti-slug device, consisting basically of two assemblies. One is the centrifuge which is press fit on the crankshaft and therefore rotates at the speed of the compressor. The refrigerant is drawn in through the holes in the top. Any liquid or oil is expelled through the slots on the side by centrifugal force, and the gas (being lighter) is drawn through the slots in the hub. The second assembly collects the gas and directs it to the cylinder heads. This system always operates when the compressor is running and functions under all conditions which may affect slugging. It is not dependent upon any external component which may fail, and so is practically foolproof.

Courtesy Tecumseh Products Co.

Fig. 10. Sectional view of a vertical hermetic compressor.

CENTRIFUGAL COMPRESSORS

The centrifugal compressor is a relatively high-speed machine in which a continuous stream of gas is compressed by centrifugal force. The centrifugal compressor is, by its very nature, a large volume machine which means that a large amount of refrigeration can be handled within a relatively small floor area.

From the foregoing it follows that the centrifugal compressor is well suited when a large amount of refrigeration is required, but is not practical to use on *small* applications. To be specific in defining the term "small," it should be noted that, at the present time, the centrifugal machine has a practical lower capacity limit of approximately 700 tons when using ammonia as a refrigerant. Eventually, as smaller, higher speed compressors are designed, this lower limit will be reduced.

Centrifugal compressors are available in a single or multistage arrangement to meet specific design requirements in refrigeration applications. In cases of a staged system, however, a centrifugal compressor should be used as the low-stage unit together with a reciprocating high stage. In such a case, the centrifugal compressor could be used for capacities down to about 200 tons, depending on the temperature level. Centrifugal compressors have also been added as booster units for existing reciprocating systems to provide lower evaporating temperatures and increased capacity. A cutaway view of a centrifugal compressor is shown in Fig. 11.

Multistage centrifugal compressors are readily adapted to cycles involving intercooling, including interstage liquid flash cooling. According to *Carrier Corporation* engineers, a line of centrifugal compressors designed to be universally applicable to all gases and vapors must take into account the chemical properties of these fluids. Ammonia, and many other gases as well, is peculiar in its reaction to certain metals. For instance, the copper and copper alloys which are used in *Freon* machines have no place in ammonia compressors. Therefore, all parts of the centrifugal compressor which come into direct contact with the ammonia are of materials other than copper and copper alloys. Interstage and balancing piston labyrinths are of aluminum, and the rotating part of the shaft labyrinth is "free machining" stainless steel. The stationary part of the shaft labyrinth is cast iron with a pure lead lining. Because of its inert property, a lead coating is applied to the wheels, which are basically steel forgings. A special contact-type floating carbon ring seal from which all copper parts have been eliminated, is used on ammonia applications. This

95

same seal is used for practically all other gases, with atmospheric air being the one exception.

Courtesy Carrier Corporation

Fig. 11. Cutaway of an open centrifugal compressor with an electric motor drive.

SYSTEM PRECAUTIONS

As with all machines, centrifugal or reciprocating, there are certain precautions to be observed during installation and initial operation. All piping should be carefully cleaned of rust, scale, sand, and other foreign materials. If water has been permitted to accumulate in the piping, it should be removed and the system dried out as thoroughly as possible. Piping should be properly designed so that there are no liquid traps which might cause flooding over into the compressor. A knockout drum in the suction piping at the compressor inlet is helpful in this regard.

While a centrifugal compressor will take a slug of liquid with much less detrimental effect than will a reciprocating machine, it is well to keep in mind that the centrifugal type is *not* designed for this service. It is particularly important to note that only a little liquid entering the compressor suction will increase the weight flow tremendously. This can increase the horsepower requirement to the extent that it may cause the unit to bog down in the case of turbine drive, or to trip out on overload in the case of motor drive. Appreciable amounts of liquid flowing into the suction of the compressor can cause excessive thrust, resulting in damage to the thrust bearing, and possible damage to other internal parts if the rotor shifts axially as a result.

Piping should be designed so that strains due to weight and to thermal changes are not imposed on the compressor and turbine equipment. Close shaft alignment is necessary on rotating equipment. This is impossible if piping forces are imposed on the units.

COMPRESSOR MAINTENANCE

Normal and correct operation can be observed and determined by the temperature and sound of the compressor. The normal compressor noise consists of a slight valve ticking. Normal compressor temperature at the crankcase should not be more than 25° to 30° F above the suction inlet gas temperature. The head of the machine and the discharge pipe should not be more than 30° warmer than the temperature corresponding to the discharge pressure. This discharge temperature may vary up or down with the temperature of the suction gas. The foregoing approximation is correct for most ordinary conditions. Heating of the compressor cylinder or of the heads higher than mentioned indicates trouble in the form of broken or leaky valves and should be investigated promptly.

Oil Level

The oil level should be a part of the daily observation by the operators. Once the system is charged and the proper oil return effected, there normally should be no cause for adding additional oil. If the compressor keeps losing oil, then look for a trap in the suction piping, or a place of high lift through a vertical pipe from an evaporator from which the oil is not returning. Also check for leaks. Be sure that the oil level is at least half way up on the compressor gauge glass, but not over the top. A higher oil level is harmful as it will cause an oil bump in the compressor.

Oil Pressure

When the compressor is used on booster service at low temperatures and is handling either halocarbons or ammonia, the oil pressure gauge should read at least 10 pounds above atmospheric pressure. To set the pressure, close the suction gas stop valve until a vacuum of 27 inches is maintained while the compressor is running, then adjust the oil relief valve at the seal end, but keep this valve open at least 5 turns. If necessary, also adjust the relief valve above the pump until a pressure of 10 pounds is obtained.

Suction Strainer

The compressor will have either one or two suction traps, inside each of which is a metal strainer and cloth filter. When being installed, the filter bag must be pushed into the strainer screen until it makes contact with the bottom at the closed or spring end, and the top must then be folded back over the edge of the open end. The filter and screen are then inserted in the suction trap with the closed end next to the flanged cover which provides the spring tension for holding the screen and filter in place. The hem on the filter at the open end will prevent it from being pulled through by the suction gas when properly positioned in the bottom groove of the trap.

Valves

The plate valves inside the machine will very rarely require grinding. A visual inspection of these valves is recommended about once every six months. The stop valves are usually double seated and can be repacked under pressure by opening the valve wide until the back of the valve button sets against the bonnet. Use first grade braided packing.

Compressor Care

Compressors, especially those connected to an iron or steel pipe system, should be opened and thoroughly cleaned every two weeks of operation. To pump out the crankcase before opening a machine, the suction stop valve B (Fig. 12) is closed, the discharge valve A is kept open, and a vacuum or near vacuum is pumped on the compressor crankcase. Usually, running it several minutes will remove all of the refrigerant mixed with the oil from the crankcase. Stop the compressor and close valve A; the oil filling plug is removed next to relieve the pressure. Open slowly, making sure the pressure is first relieved. The crankcase cover plate can then be removed and the drain plug opened to drain all remaining oil. The inside of the compressor should be thoroughly cleaned, using lint-free rags.

After the crankcase has been cleaned, all of the old oil, if dirty, should be replaced with a fresh charge. It is also an excellent precaution to remove the cover plate from the scale trap and thoroughly clean it. Never use gasoline for cleaning purposes. It is recommended that the wiping rags be saturated with good clean oil such as that used in the compressor. A really thorough cleaning is recommended; halfway measures are never satisfactory and will only lead to eventual trouble.

Fig. 12. Piping diagram of a typical refrigeration system.

At the time the machine is opened to be cleaned, it is well to check the nuts on the connecting rod bolts to be sure that they are tight. Clean the oil screen in a new machine at the end of the first month of operation. Cleaning once a year will be sufficient thereafter. Before restarting the compressor, remove the oil plug in the discharge side of the cylinder block, then start the unit with the suction and discharge valves closed. The air in the compressor will then be discharged. Just as the machine is shutting down, replace the plug, after which the compressor may be put in operation in the normal manner.

COMPRESSOR MAINTENANCE AND SERVICE

Numerous refrigeration units are of the open type with a belt-driven compressor mounted separately and in which the compressor bearings are provided with shaft seals. The following information will assist in servicing this type of unit:

Shaft Seal

The function of the seal in an open-type refrigerating compressor system is to prevent the gas from escaping the compressor at the point where the shaft leaves the compressor housing. This, in the early stages of the refrigeration industry, was a difficult problem to solve. The shaft, of necessity, must revolve, and yet the refrigerant must not be allowed to escape when there is a pressure within the crankcase. Neither must air be allowed to enter when the pressure in the crankcase is below zero pounds pressure. The solution to the problem of crankshaft leaks was found in the development of a bellows-type seal, which is now an important part of most compressors.

The seal is an assembly of parts consisting of a seal ring fastened to a bellows which, in turn, is fastened to a seal flange, as shown in Fig. 13. A spring surrounds the bellows, with one end resting against the seal ring and the other against the seal flange. The face of the seal ring is lapped flat and smooth. When assembled in the compressor, the seal ring presses against a shoulder on the shaft that has also been lapped, forming a gas- and oil-tight joint at this point. The bellows provides the degree of flexibility necessary to keep the seal ring in perfect contact with the shaft. The seal flange is clamped against a gasket around an opening in the side of the compressor. The compressor shown in Fig. 2 is of the hermetic type and needs no shaft seal.

To determine if a compressor is leaking at the seal, proceed as follows: Close both shut-off valves by turning the stem to the right as far as possible. To insure adequate refrigerant pressure in the compressor crankcase and on the seal bearing face, attach a refrigerant drum

OIL CHANNEL

KEY WAY

COMPRESSOR
SHAFT SEAL
BEARING

SEAL RING

COMPRESSOR
CASE

SPRING

SEAL BELLOWS

COMPRESSOR SHAFT

METALLIC SEAL
GASKET

SEAL
FLANGE

BOLT HEAD

COMPRESSOR SHAFT
END COVER

Fig. 13. Showing a typical shaft seal assembly.

containing the correct refrigerant to the suction shut-off valve outlet port. When making this connection, there should be a gauge in the line from the drum to the compressor to accurately determine the pressure of the refrigerant in the compressor.

Test pressures for this purpose should be approximately 70 to 80 pounds. If the compressor is located in a cool location, it may be necessary to raise the pressure in the drum by adding heat. In this process, care must be taken not to exceed 100 pounds, as pressure greater than this may damage the bellows assembly of the seal. With this pressure on the crankcase of the compressor, test for leaks with a halide torch, moving the finder tube close to the seal nut, the crankshaft, the seal-plate gasket, and the point at which the seal comes in contact with the seal plate. If this process does not disclose a leak, turn the flywheel over slowly by hand, holding the finder tube close to the aforementioned parts.

After the leak has been detected, locate the exact place where it is leaking, if possible. If the leak is around the seal-plate gasket, replace the gasket. If the leak should be at the seal or nut, replace with a new seal plate, gasket, and seal assembly.

To replace the seal, observe the following instructions:

1. Remove the compressor from the condensing unit.

2. Remove the flywheel, using a puller. Leave the flywheel nut on the crankshaft so that the wheel puller will not distort the threads.
3. Remove the seal guard, seal nut, and seal assembly.
4. Remove the seal plate and gasket.
5. When assembling the seal, put a small quantity of clean compressor oil on the seal face plate and the seal.
6. To reassemble, reverse these operations, making sure the seal plate is bolted in place and the seal guard is at the top.

Compressor Efficiency Test

An efficiency test is a check on the relative amount of useful work that the compressor will accomplish. Strictly speaking, the efficiency of any machine is taken as the ratio of the power output to the power input in the same unit. This is usually written:

$$\text{efficiency} = \frac{\text{output}}{\text{input}}$$

The factors that determine the efficiency of a compressor are:

1. The degree to which the piston valve remains closed on the up stroke.
2. The degree to which the discharge valve remains tight when the piston is on the down stroke.

To test the compressor efficiency, proceed in the following manner:

1. Stop the compressor and install a compound gauge on the suction line valve and a pressure gauge on the discharge line valve.
2. Close the suction valve and operate the compressor until about 25-in. of vacuum is obtained on the compound gauge, then stop the unit and note the gauge readings. If the compressor will not pull a 25-in. vacuum or better, it is probable that air is leaking by the discharge valve and the piston valve.
3. If the head pressure drops and the vacuum-gauge reading remains practically constant, it is probable that there is an external leak at one of the points on the head of the compressor, or at the gauge and shut-off valve connection.
4. To repair either a leaky suction valve or discharge valve, remove the head of the compressor and remove the discharge valve plate carefully. In disassembling the discharge valve, caution should be used so as not to disturb the actual conditions prevailing during the

test. It is possible for some dirt, scale, or other foreign matter to get under the valve disc and on the seat, causing poor performance. If there is no evidence of dirt or foreign matter, check the seat on both the discharge valve and piston assembly for low spots or scratches. If these are found, replace the discs or the valve plate completely.

Generally the trouble can be found at either the suction valve or the discharge valve. Badly scored seats on either side of the discharge or suction valve require replacing with new assemblies. Removal of deep scores changes the valve lift, further endangering the efficiency. After the repairs are made, thoroughly clean the parts with gasoline or other solvent and reassemble them, using new gaskets. Repeat the efficiency test to insure that the trouble has been eliminated.

Stuck or Tight Compressor

The reason for a compressor becoming *stuck* is usually a result of moisture in the system or lack of lubrication. When this occurs, the compressor should be thoroughly cleaned. The compressor should be completely disassembled and the parts thoroughly cleaned and refitted. New oil and refrigerant should be put into the cleaned system.

A tight compressor will result when a cylinder head, a seal cover, or similar part has been removed and not replaced carefully, or when the screws have been tightened unevenly. This will develop a misalignment causing a bind in the moving parts, which may cause the compressor to become stuck.

Compressor Knocks

A knock in the compressor may be caused by a loose connecting rod, eccentric strap and rod, eccentric disc, piston pin, crankshaft, or too much oil in the system. A compressor knock can be determined by placing the point of a screwdriver against the crankcase and the ear against the handle. It will not be possible to determine what causes the knock until the compressor is disassembled.

Sometimes it may be possible to determine a looseness of the aforementioned parts without completely disassembling the compressor. First, remove the cylinder head and valve plate to expose the head of the piston. Now rotate the compressor by hand and press down on the top of the piston with the finger. Any looseness can be felt at each stroke of the piston. The loose part should be replaced.

It is always well to check the compressor oil level before analyzing and determining compressor repairs. Oil knocks are usually caused by adding too much oil in the servicing of the unit. It should never be necessary to add oil to a system unless there has been a leakage of oil. A low charge is sometimes diagnosed as a lack of oil. Always make sure that a low oil level is actually due to lack of oil rather than to a low charge before adding oil.

Lubrication

In conventional reciprocating compressors the lubrication is accomplished by the so-called *splash system*. Special *dippers* or *slingers* fastened to the crankshaft distribute the crankcase oil to the pistons, pin bearings, cylinder walls, crankshaft bearings, and seal. Perhaps the most important point for the serviceman to remember in connection with compressor service, is to check the amount of oil in an open-type compressor. This is accomplished as follows:

After the unit has been operated for several minutes, attach a compound gauge to the suction service valve and close the valve. The unit should be pumped down to balance the pressure (zero on the gauge) then stopped. Allow a few minutes before removing the oil filler plug and carefully measure the oil level. This level should be checked with the manufacturer's instructions concerning the oil level required for the unit in question. Only oil specifically recommended by the manufacturer for the particular unit and refrigerant should be used when adding oil.

COMPRESSOR REMOVAL

To remove a compressor from the unit, proceed as follows: Attach a compound gauge to the suction line valve. Close the suction line shut-off valve and run the compressor until 20 to 25 inches of vacuum are obtained, then close the discharge shut-off valve. Before removing any fittings, crack the suction line valve to bring the gauge reading back to zero. Before removing service valves, loosen the pressure gauge and relieve the pressure in the head of the compressor. Remove the cap screws holding the suction and discharge valves to the compressor. If the compressor is to be taken away from the premises for repairs, place service valves over both the discharge and suction-line openings. This prevents air and moisture from entering and oil from leaking out. Loosen the nuts that hold the compressor to the base and bend the tubing away from the compressor just enough to permit the assembly to be lifted out. Care should be taken not to loosen the mounting pads and washers on which the compressor rests.

COMPRESSOR REPLACEMENT

The installation of a compressor is roughly a reversal of the process of removing it. First, place the mounting pads and washers in position. Place the compressor carefully on the base in the same position it occupied before removal. Bolt the compressor in place. Replace the suction and discharge valves, using new gaskets. After the compressor and valves are bolted in place, the compressor must be evacuated to remove the air.

BELT ALIGNMENT

The importance of proper belt alignment in open units must be kept constantly in mind when service necessitates changing of a compressor, motor, or motor pulley. To determine the correct belt alignment and tension, proceed as follows:

The correct belt tension for a V-type belt is to have the tension so adjusted that it is possible to depress or raise the belt ½ in. from its original position with the fingers without undue pressure. With the compressor and motor in their respective positions and in line as nearly correct as possible, loosen the setscrew on the motor pulley so that the pulley turns freely on the motor shaft. It may be necessary on older motors to dress down the motor shaft with emery cloth. A little oil should be added to insure that the motor pulley turns freely. At this point, attach the belt with the proper tension, turn the flywheel forward several times, then backwards several times. The point at which the belt is in correct alignment will be that position where there is no in or out travel of the motor pulley on the motor shaft, whether the flywheel is turned forward or backward.

The movement of the motor pulley outward when the flywheel is turned forward, or inward when the flywheel is turned backward (or vice versa), indicates that the motor pulley is out of plane with the

105

flywheel; the motor must be readjusted so that this condition does not exist. Close observance of the aforementioned procedure will result in a greatly increased length of time the belt will remain in a serviceable condition.

RUNNING TIME CHECK

The percentage of running or operating time of the unit, assuming the unit functions normally, depends upon the amount of work being done. The amount of work needed depends upon the size of the refrigerator cabinet to be cooled, the temperature of the air that surrounds it, and the amount of heat that must be extracted from it. The amount of heat that leaks into the refrigerator depends upon the temperature of the room, the frequency and length of time of the door openings, and in the amount and temperature of the food placed in the refrigerator. Climatic conditions also affect the operations of a refrigerator. Thus, a refrigerator will operate more efficiently in a dry climate than in a climate of high humidity.

Keeping the aforementioned factors in mind, the length of the "on" and "off" period on the average well-adjusted refrigerator should be somewhat as follows:

At 75° F room temperature, the "on" period should be around 2 to 4 minutes, and the "off" period around 10 to 15 minutes. In many cases where a complaint is received that the refrigerator is operating too much, a check may very easily be made by connecting a self-starting electric clock to the motor terminals. It is evident that since the clock will operate only when the motor operates, the total running time of the compressor will be registered.

COMPRESSION RATIO

The compression ratio in a refrigerant system is defined as the absolute head pressure divided by the absolute suction pressure, or

$$\text{Compression ratio} = \frac{\text{Abs. head pressure}}{\text{Abs. suction pressure}}$$

It should be noted that the ordinary compound gauge does not register atmospheric pressure but reads zero when not connected to a pressurized system. To obtain the absolute head pressure or absolute suction pressure at zero gauge or above, fifteen pounds must be added to the gauge reading.

Example—At zero gauge or above, we have

Absolute head pressure = gauge reading + 15

Absolute suction pressure = gauge reading + 15

Example—When the low side is reading in the vacuum range, we have

Absolute head pressure = gauge reading + 15 pounds

$$\text{Absolute suction pressure} = \frac{30 - \text{gauge reading in inches}}{2}$$

The calculation of compression ratio can be illustrated by the following examples:

Head pressure = 160 pounds

Suction pressure = 10 pounds

$$\text{Compression ratio} = \frac{\text{Ab. HP}}{\text{Ab. SP}} = \frac{160 + 15}{10 + 15} = \frac{175}{25} = 7{:}1$$

Example—

Head pressure = 160 pounds

Suction pressure = 10 inches of vacuum

Absolute head pressure = 160 + 15 = 175 pounds

$$\text{Absolute suction pressure} = \frac{30 - 10}{2} = \frac{20}{2} = 10$$

$$\text{Compression ratio} = \frac{\text{Ab. HP}}{\text{Ab. SP}} = \frac{175}{10} = 17.5{:}1$$

ENERGY CONSUMPTION

The electrical energy (usually referred to as the power consumption) used by a refrigerator depends on all the factors that influence the "on" and "off" period, in addition to the size of the motor employed. Motors used generally range in size from ¼ to about ¾ horsepower, depending upon the size of the unit, type of refrigerant employed, and various other factors.

Experience has shown that in the northern half of the country, a good average range for the summer months is 25 to 30 kilowatt-hours per month and 20 kilowatt-hours for other months. In the southern half of the country, these figures (especially for the summer months) will be somewhat higher.

REFRIGERATION SYSTEM LUBRICANTS

The necessity of providing lubrication to prevent frictional contact on metal surfaces is universally recognized. The correct lubrication of refrigeration machinery demands special consideration; the employment of incorrect oil or the excessive use of even the most suitable oil will create objectionable operating conditions. An excessively heavy oil will not be distributed properly to the working parts and will therefore fail to do the required job. An excessively light oil will not cling and will not satisfactorily seal the pistons and rings from the by-pass of the gas back into the suction port and crankcase. Because of this fact, the selection of the proper oil should be made with care. It must be viscous enough to give proper lubrication. In a rotary compressor, the oil acts as a lateral seal, and efficiency depends on its proper functioning. Due to superheating at the discharge valves, the oil must be able to withstand high temperatures.

Reputable refiners who have assisted in lubrication research are now able to supply proper oils and, from experience, can advise a potential user which oil will best serve his purpose. The soundest advice on lubrication, therefore, is to use the oil that experience has shown to be the best. This information can generally be secured from the manufacturer of the machine or a refiner actively engaged in supplying oils for refrigeration work.

Lubrication of Motor and Fan Bearings

Lubrication of the motor and fan bearings usually can be accomplished by feeding the oil through a piece of wicking or felt from a reservoir directly to the bearings in continuous small quantities. It is necessary, therefore, to use an oil that is light enough in body to pass through the wick but heavy enough to form a proper film on the bearing surface at operating temperatures. It must also have a low "pour point" so that it will flow freely to the bearings when starting cold. The best oil will resist deterioration due to oxidation and polymerization, which cause stickiness and decreased lubrication. The use of properly refined oil will eliminate most of these difficulties. When one considers the small amount required, and the costly damage that may result from using an inferior oil, it can be seen that the best oil is the cheapest over a long period of time.

Compressor Lubrication

There are so many types of refrigerating compressors that it is impossible to specifically determine which lubricant fits all requirements. Even when compressors and machines are quite similar in design, they are apt to require individual consideration for proper lubrication.

Speed, clearance, temperature conditions, etc., have an important bearing and the system of operation and the refrigerant used are the determining factors. Most oils are chosen for a given job as a result of actual tests. Therefore, servicemen in the field will always profit by using the oil specified by the manufacturer of the unit being serviced or an oil known to answer the same specifications. Buy oils from reputable refiners only, preferably those who have had real experience in refrigeration problems.

Types of Lubrication

Compressors are usually lubricated by splash, although a semiforced feed is occasionally used; in certain types a separate oil pump is required. There are two systems of oil circulation—one in which the oil circulates through the low side, and the other in which it does not.

As the temperature of the gas discharged from the compressor rises, the vapor pressure of the oil carried with it will also rise, and the amount of oil that can be separated will be larger. In addition, the quantity of oil carried over can be increased by the mechanical action in the crankcase of the compressor, and foaming can be induced by the release of dissolved refrigerant during the suction stroke of the piston. This latter difficulty may be quite pronounced in a machine charged with methyl chloride. If the gas and oil from the compressor are discharged into a chamber, baffles or centrifugal force can be employed to separate the oil from the gas and return it to the intake of the machine. This system is very satisfactory when employing refrigerants that dissolve oil or in a dry system where the velocity of the gas carries the oil back to the compressor. In commercial installations, the oil usually circulates through the low-pressure side. In a dry system, the high velocity of the gas keeps the oil moving. In a flooded system, the evaporation of the refrigerant leaves the oil behind, and special arrangements are used for its return.

If sufficient care has not been exercised in removing all impurities from the oil, emulsification in the liquid refrigerant can result. The effect would be the same as excessive solubility, since it can change the liquid level of the refrigerant in the evaporator. It can also result in decreased free flow of the refrigerant and prevent proper boiling in the evaporator. This causes a temperature that disagrees considerably with the corresponding pressure.

COMPRESSOR CALCULATIONS

Mechanical Efficiency—Mechanical efficiency is defined as the ratio of the work to the gas (as may be obtained from an indicator diagram) to the work delivered to the compressor drive shaft.

Compression Ratio—Compression ratio is defined as the ratio of the absolute head pressure to the absolute suction pressure, that is:

$$\text{Compression ratio} = \frac{\text{Absolute head pressure}}{\text{Absolute suction pressure}}$$

Volumetric Efficiency—Volumetric efficiency is defined as the volume of fresh vapor entering the cylinder per stroke divided by the piston displacement.

Compressor Capacity—Capacity is the refrigeration effect that can be accomplished by a compressor and is usually measured in Btu. By definition, compressor capacity is equal to the difference in total heat content between the refrigerant liquid at a temperature corresponding to the pressure of the vapor leading to the compressor and the refrigerant vapor entering the compressor.

Performance Factor—The performance factor of a hermetic compressor is the ratio between the capacity in Btu and the power input in watts. It is written:

$$\text{Performance factor} = \frac{\text{Capacity in Btu}}{\text{Power input in watts}}$$

Brake Horsepower—The brake horsepower is a function of the power input of an ideal compressor and its compression, the mechanical efficiency, and volumetric efficiency.

As noted in the foregoing, the volumetric efficiency or displacement efficiency of a compressor is the ratio of the volume of the refrigerant actually removed, reduced to the temperature and pressure conditions of the evaporator to the volume of the piston displacement. Several factors affect the volumetric efficiency. First, the volume of the vapor is greater after it enters the cylinder, as it takes up heat from the valves and walls to become superheated. Second, the pressure in the cylinder is lower than in the evaporator, as it is necessary to have a lower pressure to cause a flow which will increase the volume. Again, the re-expansion of the gases left in the clearance spaces between the piston, cylinder head, and valves will partly fill the suction chamber, so that a full charge of fresh vapor cannot be drawn in from the evaporator. The effect of the superheating and leakage cannot be calculated and must be determined by experiment, but the clearance effect can be calculated when the clearance ratio is known. It is only practical to determine the volumetric efficiency by a calorimeter test, which is the accepted method used by practically all manufacturers today.

Theoretical Compressor Capacities

Example—A certain condensing unit is equipped with a two-cylinder compressor having a 2½'' bore by 3½'' stroke, and is operating at 500 rpm.

Calculate:

(a) Capacity in Btu/hr when using *Freon*-12 as the refrigerant, operating at 125 lbs. head pressure and 35 lbs. suction pressure.
(b) Capacity in Btu/hr when using *Freon*-12 as the refrigerant, operating at 125 lbs. head pressure and 11 lbs. suction pressure.

Solution:

The cubic inch displacement per revolution will be obtained as follows:

$$\text{Displacement} = \frac{\pi D^2}{4} \times 2 \times L =$$

$$\frac{3.1416 \times 2.5^2 \times 2 \times 3.5}{4} = 34.4 \text{ cu. in.}$$

The cubic feet displacement per hour at 500 rpm will accordingly be:

$$\frac{34.4 \times 500 \times 60}{1728} = 597 \text{ cu. ft. per hour displacement}$$

(a) From the *Freon*-12 table (Table 1), column v_g at 35 lbs. suction pressure, the specific volume of vapor equals 0.819 cu. ft./lb. Thus,

$$\frac{597}{0.819} = 729 \text{ lbs. } Freon\text{-}12 \text{ per hr.}$$

From the *Freon*-12 table, we further obtain:

Total heat h_g of vapor per pound at 35 lbs. = 82.49 Btu
Total h_f liquid per lb. at 125 lb. = 32.15 Btu
The net refrigeration effect will therefore be 82.49 − 32.15 = 50.34 Btu

From the foregoing it follows that the theoretical capacity is

$$729 \times 50.34 = 36{,}700 \text{ Btu/hr, approximately}$$

(b) From the *Freon*-12 table, column v_g, the specific volume at 11 lbs. suction pressure equals 1.514 cu. ft. per pound.

Therefore,

$$\frac{597}{1.514} = 394 \text{ lbs. } Freon\text{-}12 \text{ per hr.}$$

The total heat h_g of vapor at 11 lbs. from the *Freon*-12 table is 78.67 Btu/lb.

Table 1. Freon-12 Properties of Saturated Vapors

Temp. °F t	Pressure Abs. lb./in.² p	Pressure Gauge lb./in.² p_d	Volume Liquid ft.³/lb. v_f	Volume Vapor ft.³ v_g	Density Liquid lb./ft.³ $1/v_f$	Density Vapor lb./ft.³ $1/v_g$	Heat Content from −40° Liquid Btu/lb. h_f	Heat Content from −40° Latent Btu/lb. h	Heat Content from −40° Vapor Btu/lb. h_g	Entropy from −40° Liquid Btu/lb. °F s_f	Entropy from −40° Vapor Btu/lb. °F s_g	Temp. °F t
−40	9.32	10.92*	0.0106	3.911	94.58	0.2557	0	73.50	73.50	0	0.17517	−40
−38	9.82	9.91*	.0106	3.727	94.39	.2683	0.40	73.34	73.74	0.00094	.17490	−38
−36	10.34	8.87*	.0106	3.553	94.20	.2815	0.81	73.17	73.98	.00188	.17463	−36
−34	10.87	7.80*	.0106	3.389	93.99	.2951	1.21	73.01	74.22	.00282	.17438	−34
−32	11.43	6.66*	0107	3.234	93.79	.3092	1.62	72.84	74.46	.00376	.17412	−32
−30	12.02	5.45*	.0107	3.088	93.59	0.3238	2.03	72.67	74.70	0.00471	0.17387	−30
−28	12.62	4.23*	.0107	2.950	93.39	.3390	2.44	72.50	74.94	.00565	.17364	−28
−26	13.26	2.93*	.0107	2.820	93.18	.3546	2.85	72.33	75.18	.00659	.17340	−26
−24	13.90	1.63*	.0108	2.698	92.98	.3706	3.25	72.16	75.41	.00753	.17317	−24
−22	14.58	0.24*	.0108	2.583	92.78	.3871	3.66	71.98	75.64	.00846	.17296	−22
−20	15.28	0.58	0.0108	2.474	92.58	0.4042	4.07	71.80	75.87	0.00940	0.17275	−20
−18	16.01	1.31	.0108	2.370	92.38	.4219	4.48	71.63	76.11	.01033	.17253	−18
−16	16.77	2.07	.0108	2.271	92.18	.4403	4.89	71.45	76.34	.01126	.17232	−16
−14	17.55	2.85	.0109	2.177	91.97	.4593	5.30	71.27	76.57	.01218	.17212	−14
−12	18.37	3.67	.0109	2.088	91.77	.4789	5.72	71.09	76.81	.01310	.17194	−12
−10	19.20	4.50	0.0109	2.003	91.57	0.4993	6.14	70.91	77.05	0.01403	0.17175	−10
−8	20.08	5.38	.0109	1.922	91.35	.5203	6.57	70.72	77.29	.01496	.17158	−8
−6	20.98	6.28	.0110	1.845	91.14	.5420	6.99	70.53	77.52	.01589	.17140	−6
−4	21.91	7.21	.0110	1.772	90.93	.5644	7.41	70.34	77.75	.01682	.17123	−4
−2	22.87	8.17	.0110	1.703	90.72	.5872	7.83	70.15	77.98	.01775	.17107	−2
0	23.87	9.17	0.0110	1.637	90.52	0.6109	8.25	69.96	78.21	0.01869	0.17091	0
2	24.89	10.19	.0110	1.574	90.31	.6352	8.67	69.77	78.44	.01961	.17075	2
4	25.96	11.26	.0111	1.514	90.11	.6606	9.10	69.57	78.67	.02052	.17060	4
5†	26.51	11.81	.0111	1.485	90.00	.6735	9.32	69.47	78.79	.02097	.17052	5†

*Inches of mercury below one atmosphere.   †Standard ton temperatures.
Courtesy of Frigidaire Division, General Motors Corp.

Total h_f liquid per lb. at 125 lb. = 32.15 Btu
The net RE = 78.67 − 32.15 = 46.52 Btu/lb
That is 394 × 46.52 = 18,330 Btu/hr, approximately.

If the required refrigeration in Btu per hour is known, the size and characteristics of the compressor can be determined as follows: First

113

Table 1. Freon-12 Properties of Saturated Vapors (cont'd.)

Temp. °F t	Pressure Abs. lb./in.² p	Pressure Gauge lb./in.² P_d	Volume Liquid ft.³/lb. v_f	Volume Vapor ft.³/lb. v_g	Density Liquid lb./ft.³ $1/v_f$	Density Vapor lb./ft.³ $1/v_g$	Heat Content from −40° Liquid Btu/lb. h_f	Heat Content from −40° Latent Btu/lb. h	Heat Content from −40° Vapor Btu/lb. h_g	Entropy from −40° Liquid Btu/lb. °F s_f	Entropy from −40° Vapor Btu/lb. °F s_g	Temp. °F t
6	27.05	12.35	.0111	1.457	89.88	.6864	9.53	69.37	78.90	.02143	.17045	6
8	28.18	13.48	.0111	1.403	89.68	.7129	9.96	69.17	79.13	.02235	.17030	8
10	29.35	14.65	0.0112	1.351	89.45	0.7402	10.39	68.97	79.36	0.02328	0.17015	10
12	30.56	15.86	.0112	1.301	89.24	.7687	10.82	68.77	79.59	.02419	.17001	12
14	31.80	17.10	.0112	1.253	89.03	.7981	11.26	68.56	79.82	.02510	.16987	14
16	33.08	18.38	.0112	1.207	88.81	.8288	11.70	68.35	80.05	.02601	.16974	16
18	34.40	19.70	.0113	1.163	88.58	.8598	12.12	68.15	80.27	.02692	.16961	18
20	35.75	21.05	0.0113	1.121	88.37	0.8921	12.55	67.94	80.49	0.02783	0.16949	20
22	37.15	22.45	.0113	1.081	88.13	.9251	13.00	67.72	80.72	.02873	.16938	22
24	38.58	23.88	.0113	1.043	87.91	.9588	13.44	67.51	80.95	.02963	.16926	24
26	40.07	25.37	.0114	1.007	87.68	.9930	13.88	67.29	81.17	.03053	.16913	26
28	41.59	26.89	.0114	.973	87.47	1.028	14.32	67.07	81.39	.03143	.16900	28
30	43.16	28.46	0.0115	.939	87.24	1.065	14.76	66.85	81.61	0.03233	0.16887	30
32	44.77	30.07	.0115	.908	87.02	1.102	15.21	66.62	81.83	.03323	.16876	32
34	46.42	31.72	.0115	.877	86.78	1.140	15.65	66.40	82.05	.03413	.16865	34
36	48.13	33.43	.0116	.848	86.55	1.180	16.10	66.17	82.27	.03502	.16854	36
38	49.88	35.18	.0116	.819	86.33	1.221	16.55	65.94	82.49	.03591	.16843	38
40	51.68	36.98	0.0116	.792	86.10	1.263	17.00	65.71	82.71	0.03680	0.16833	40
42	53.51	38.81	.0116	.767	85.88	1.304	17.46	65.47	82.93	.03770	.16823	42
44	55.40	40.70	.0117	.742	85.66	1.349	17.91	65.24	83.15	.03859	.16813	44
46	57.35	42.65	.0117	.718	85.43	1.393	18.36	65.00	83.36	.03948	.16803	46
48	59.35	44.65	.0117	.695	85.19	1.438	18.82	64.74	83.57	.04037	.16794	48
50	61.39	46.69	0.0118	.673	84.94	1.485	19.27	64.51	83.78	0.04126	0.16785	50
52	63.49	48.79	.0118	.652	84.71	1.534	19.72	64.27	83.99	.04215	.16776	52

Courtesy of Frigidaire Division, General Motors Corp.

determine the refrigerating effect (RE) per pound of refrigerant under standard conditions.

Example—Determine the net refrigerating effect for *Freon*-12 refrigerant under standard ton conditions, 5° F evaporator, 86° F liquid in receiver.

Table 1. Freon-12 Properties of Saturated Vapors (cont'd.)

Temp. °F t	Pressure Abs. lb./in.³ p	Pressure Gauge lb./in.³ P_d	Volume Liquid ft.³/lb. v_f	Volume Vapor ft.³/lb. v_g	Density Liquid lb./ft.³ $1/v_f$	Density Vapor lb./ft.³ $1/v_g$	Heat Content from −40° Liquid Btu/lb. h_f	Heat Content from −40° Latent Btu/lb. h	Heat Content from −40° Vapor Btu/lb. h_g	Entropy from −40° Liquid Btu/lb. °F s_f	Entropy from −40° Vapor Btu/lb. °F s_g	Temp. °F t
54	65.63	50.93	.0118	.632	84.50	1.583	20.18	64.02	84.20	.04304	.16767	54
56	67.84	53.14	.0119	.612	84.28	1.633	20.64	63.77	84.41	.04392	.16758	56
58	70.10	55.40	.0119	.593	84.04	1.686	21.11	63.51	84.62	.04480	.16749	58
60	72.41	57.71	0.0119	0.575	83.78	1.740	21.57	63.25	84.82	0.04568	0.16741	60
62	74.77	60.07	.0120	.557	83.57	1.795	22.03	62.99	85.02	.04657	.16733	62
64	77.20	62.50	.0120	.540	83.34	1.851	22.49	62.73	85.22	.04745	.16725	64
66	79.67	64.97	.0120	.524	83.10	1.909	22.95	62.47	85.42	.04833	.16717	66
68	82.24	67.54	.0121	.508	82.86	1.968	23.42	62.20	85.62	.04921	.16709	68
70	84.52	70.12	0.0121	0.493	82.60	2.028	23.90	61.92	85.82	0.05009	0.16701	70
72	87.50	72.80	.0121	.479	82.37	2.090	24.37	61.65	86.02	.05097	.16693	72
74	90.20	75.50	.0122	.464	82.12	2.153	24.84	61.38	86.22	.05185	.16685	74
76	93.00	78.30	.0122	.451	81.87	2.218	25.32	61.10	86.42	.05272	.16677	76
78	95.85	81.15	.0123	.438	81.62	2.284	25.80	60.81	86.61	.05359	.16669	78
80	98.76	84.06	0.0123	0.425	81.39	2.353	26.28	60.52	86.80	0.05446	0.16662	80
82	101.7	87.00	.0123	.413	81.12	2.423	26.76	60.23	86.99	.05534	.16655	82
84	104.8	90.1	.0124	.401	80.87	2.495	27.24	59.94	87.18	.05621	.16648	84
86†	107.9	93.2	.0124	.389	80.63	2.569	27.72	59.65	87.37	.05708	.16640	86
88	111.1	96.4	.0124	.378	80.37	2.645	28.21	59.35	87.56	.05795	.16632	88†
90	114.3	99.6	0.0125	0.368	80.11	2.721	28.70	59.04	87.74	0.05882	0.16624	90
92	117.7	103.0	.0125	.357	79.86	2.799	29.19	58.73	87.92	.05969	.16616	92
94	121.0	106.3	.0126	.347	79.60	2.880	29.68	58.42	88.10	.06056	.16608	94
96	124.5	109.8	.0126	.338	79.32	2.963	30.18	58.10	88.28	.06143	.16600	96
98	128.0	113.3	.0126	.328	79.06	3.048	30.67	57.78	88.45	.06230	.16592	98

†Standard ton temperatures.
Courtesy of Frigidaire Division, General Motors Corp.

Solution:

With reference to the *Freon*-12 table we obtain:

Column h_f = Latent heat of evaporation at 5° F = 69.47 Btu/lb

Heat of liquid at 86° F, column h_f = 27.72 Btu/lb

Heat of liquid at 5° F, column h_f = 9.32 Btu/lb

Table 1. Freon-12 Properties of Saturated Vapors (cont'd.)

Temp. °F t	Pressure Abs. lb./in.² p	Pressure Gauge lb./in.² P_d	Volume Liquid ft.³/lb. v_f	Volume Vapor ft.³/lb. v_g	Density Liquid lb./ft.³ $1/v_f$	Density Vapor lb./ft.³ $1/v_g$	Heat Content from —40° Liquid Btu./lb. h_f	Heat Content from —40° Latent Btu./lb. h	Heat Content from —40° Vapor Btu./lb. h_g	Entropy from —40° Liquid Btu./lb.°F s_f	Entropy from —40° Vapor Btu./lb.°F s_g	Temp. °F t
100	131.6	116.9	0.0127	0.319	78.80	3.135	31.16	57.46	88.62	0.06316	0.16584	100
102	135.3	120.6	.0127	.310	78.54	3.224	31.65	57.14	88.79	.06403	.16576	102
104	139.0	124.3	.0128	.302	78.27	3.316	32.15	56.80	88.95	.06490	.16568	104
106	142.8	128.1	.0128	.293	78.00	3.411	32.65	56.46	89.11	.06577	.16560	106
108	146.8	132.1	.0129	.285	77.73	3.509	33.15	56.12	89.27	.06663	.16551	108
110	150.7	136.0	.0129	.277	77.46	3.610	33.65	55.78	89.43	.06749	.16542	110
112	154.8	140.1	.0130	.269	77.18	3.714	34.15	55.43	89.58	.06836	.16533	112
114	158.9	144.2	.0130	.262	76.89	3.823	34.65	55.08	89.73	.06922	.16524	114
116	163.1	148.4	.0131	.254	76.60	3.934	35.15	54.72	89.87	.07008	.16515	116
118	167.4	152.7	.0131	.247	76.32	4.049	35.65	54.36	90.01	.07094	.16505	118
120	171.8	157.1	.0132	.240	76.02	4.167	36.16	53.99	90.15	.07180	.16495	120
122	176.2	161.5	.0132	.233	75.72	4.288	36.66	53.62	90.28	.07266	.16484	122
124	180.8	166.1	.0133	.227	75.40	4.413	37.16	53.24	90.40	.07352	.16473	124
126	185.4	170.7	.0133	.220	75.10	4.541	37.67	52.85	90.52	.07437	.16462	126
128	190.1	175.4	.0134	.214	74.78	4.673	38.18	52.46	90.64	.07522	.16450	128
130	194.9	180.2	.0134	.208	74.46	4.808	38.69	52.07	90.76	.07607	.16438	130
132	199.8	185.1	.0135	.202	74.13	4.948	39.19	51.67	90.86	.07691	.16425	132
134	204.8	190.1	.0135	.196	73.81	5.094	39.70	51.26	90.96	.07775	.16411	134
136	209.9	195.2	.0136	.191	73.46	5.247	40.21	50.85	91.06	.07858	.16396	136
138	215.0	200.3	.0137	.185	73.10	5.405	40.72	50.43	91.15	.07941	.16380	138
140	220.2	205.5	0.0138	0.180	72.73	5.571	41.24	50.00	91.24	0.08024	0.16363	140

Courtesy of Frigidaire Division, General Motors Corp.

Therefore 27.72 − 9.32 = 18.40 Btu/lb to cool 86° F liquid to 5° F liquid

Note that 9.32 Btu of heat of the liquid at 5° F remain, as the liquid is not cooled below this point.

The net refrigerating effect is therefore, 69.47 − 18.40 = 51.07 Btu/lb, approximately.

SAFETY PRECAUTIONS

The compressor is largely air cooled. It must not be used as an air compressor because the temperatures of compression of air are extremely high and proper cooling cannot be obtained. A first rule of operation is that the compressor is designed as a gas pump. It must never be permitted to pump liquid refrigerant or oil. Oil lifting and oil pumping will occur if the compressor is operated with suction valve B (Fig. 12) closed. Liquid refrigerant getting back to the compressor causes hammering and damage to the valves. It reduces compressor efficiency, and will cause oil to pump out of the crankcase. Liquid pumping is usually caused by a defective or improperly regulated thermal expansion valve. It can come about through a leaky or improperly seating electric valve, which permits refrigerant flow during a shut-down period.

If the compressor pounds due to lifting oil, or by liquid coming back, try to correct the condition rather than permit the pounding to continue. Liquid coming back to the compressor can be checked by momentarily throttling the compressor suction stop valve B. The suction gas should be superheated and free of liquid. A heat exchanger may be needed in the suction line. Never use a torch or attempt a repair on a line containing refrigerant. Never tamper with or attempt readjustment of a factory-set safety relief valve. Never operate without the oil showing at least half full or higher in the compressor gauge glass. Never bridge an overload or any protective device because it kicks out in operation. Find the trouble and make the proper repair. Never permit the brine in an ice-making tank or the water in a water cooler to freeze. If operating at pressure below the freezing point of the brine or water in this tank, then shut down for a while until there is a heavier load. If, at any time, the plant is not in service and temperatures are below the freezing point, be sure to drain the water from the condensers, pump, compressor head jacketed control plate, seal, and connecting piping.

Halocarbon Refrigerant

Since halocarbons are practically odorless and nontoxic, it will not be necessary to wear a gas mask when servicing the equipment unless there is a high concentration of refrigerant or an open flame or spark. However, it is very essential that proper protection be afforded the eyes by the use of goggles or large-lensed spectacles to eliminate the possibility of liquid refrigerant coming in contact with the eyes and causing possible injury by freezing of the moisture in them. This protection is necessary and should be taken whenever loosening a connection in which refrigerant is confined.

SUMMARY

The compressor in mechanical refrigeration acts as the pumping medium and provides the necessary pressure difference necessary to establish circulation of the refrigerant throughout the refrigeration system. The various types of modern compressors, their component parts, construction, and operational features have been fully represented.

Such important factors as mechanical efficiency, compression ratios, volumetric efficiency, compressor capacity, brake horsepower, and performance factors, in addition to several examples on theoretical compressor capacities with accompanying tables, have been provided. Compressor troubles, their cause and remedy will serve to assist the practical operating engineer of commercial refrigerating plants.

REVIEW QUESTIONS

1. What is the function of a compressor in the mechanical refrigeration system?

2. Name various types of compressors.

3. Name the component parts of a reciprocating compressor.

4. What is meant by forced-feed lubrication?

5. What is the function of the shaft seal?

6. What is a centrifugal compressor and how does it operate?

7. What is a rotary compressor and how does it work?

8. State the difference between a rotary and hermetic compressor.

9. Define maintenance and care of compressors.

10. Define compressor mechanical efficiency.

11. What is meant by compressor ratio?

12. Define volumetric efficiency, compressor capacity, and performance factor.

13. Calculate the theoretical capacity of a compressor equipped with two cylinders, each having a 3-inch bore and a 4-inch stroke, and operating at 600 rpm.

CHAPTER 4

Domestic Refrigeration

Domestic and commercial refrigeration principles are very similar, except that in domestic refrigeration we are dealing with smaller refrigeration units. Most of them use hermetically or semi-hermetically sealed compressors. A few have service valves, but the majority of domestic refrigerators are completely sealed, without any service valves.

The process of refrigeration is most commonly accomplished by the evaporation of a liquid refrigerant, thereby extracting heat from the medium to be cooled. The refrigeration cycle is then composed chiefly of four further steps, whose purpose is to remove this heat from the evaporating refrigerant by again putting it in the liquid state in order that it may be used repeatedly in a continuous process.

The types of compressors generally used in home refrigerators may be grouped according to their construction as *reciprocating* and *rotary*. The principal components of a compression refrigeration system are:

1. Compressor
2. Condenser
3. Drier-strainer
4. Capillary tube
5. Evaporator (freezer unit)
6. Accumulator

The next paragraph refers to domestic refrigeration only and does *not* pertain to commercial or industrial refrigeration.

In modern systems, the capillary tube method is employed to control the flow of refrigerant to the evaporator, whereas older systems are generally provided with an expansion valve for refrigerant control or metering purposes.

THE REFRIGERATION CYCLE

A thorough understanding of the cycle of operation of a refrigerator is necessary before a correct diagnosis of any service problems can be made. Thus, only by a thorough study of the fundamentals will one be able to master the field of refrigeration.

A cycle, by definition, is *an interval or period of time occupied by one round or course of events in the same order or series.* The word *cycle*, as applied here, means a series of operations in which heat is first absorbed by the refrigerant, changing it from liquid to a gas, then the gas is compressed and forced into the condenser where the heat is absorbed by the circulating air, thus bringing the refrigerant back to its original or liquid state. With reference to Fig. 1, the cycle of operation consists of the following steps:

1. The compressor pumps refrigerant through the entire system. It draws cool refrigerant gas in through the suction line from the evaporator freezer coils. At the same time, it compresses the gas and pumps it into the discharge line. The compressed gas sharply rises in temperature and enters the condenser.

2. The condenser performs a function similar to that of the radiator in an automobile, that is, the condenser is the cooling coil for the hot refrigerant gas. In the condenser, the heat is expelled into the room air outside the cabinet. During this process, the refrigerant gas gives up the heat it removed from inside the cabinet and changes into a liquid state.

3. As the hot refrigerant liquid leaves the condenser to enter the capillary tube, a drier-strainer removes any moisture or impurities.

4. The capillary tube is carefully calibrated in length and inside diameter to meter the exact amount of liquid refrigerant flow required for each unit. A predetermined length of the capillary

Fig. 1. Typical refrigerant flow diagram.

tube is usually soldered along the exterior of the suction line, forming a heat exchanger which helps to cool the hot liquid refrigerant in the capillary tube. The capillary tube then connects to the larger diameter tubing of the evaporator.

5. As the refrigerant leaves the capillary tube and enters the larger tubing of the humid plate and evaporator, the sudden increase in

tubing diameter forms a low pressure area, and the temperature of the refrigerant drops rapidly as it changes to a mixture of liquid and gas. In the process of passing through the evaporator, the refrigerant absorbs heat from the storage area and is gradually changed from a liquid and gas mixture to a gas.

6.   The low-pressure refrigerant gas leaving the evaporator coil now enters the accumulator. The accumulator is a large cylinder designed to trap any refrigerant liquid which may not have changed to gas in the evaporator. Since it is impossible to compress a liquid, the accumulator prevents any liquid from returning to the compressor.

7.   As the refrigerant gas leaves the accumulator, it returns to the compressor through the suction line which is part of the heat exchanger, thus completing the cycle.

High-pressure Side

The high-pressure side of the system is that containing the high-pressure refrigerant, and consists of the condenser, capillary tube and the compressor.

Low-pressure Side

This is that part of the system where the refrigerant is in a gaseous state at low temperature and pressure. It consists of the evaporator and suction line. The low-pressure and high-pressure side of a refrigeration system are shown in Fig. 2.

COMPRESSORS

These are usually of the hermetic or sealed type, although numerous refrigerators of the older models are of the so-called open type. The open-type compressor is one in which the motor is connected to the compressor by means of a belt, whereas in the hermetic type, the motor and compressor are connected directly to the same shaft and sealed into the same compartment to furnish an airtight and dust-proof assembly. A hermetic- or sealed-type compressor is shown in Fig. 3. The hermetically sealed unit provides a compact assembly and will, as such, require less space, with the additional advantage of the elimination of a shaft seal and stuffing boxes. Also, the elimination of the belt and pulley will result in the lessening of noise and maintenance.

Fig. 2. The low-pressure and the high-pressure side of a refrigeration system.

CONDENSERS

Since the refrigerant leaves the compressor in the form of high-pressure vapor, some method must be found to bring the vapor back into liquid form. It is the function of the condensing unit to condense the vapor back to a liquid so that it can be reused in the refrigeration cycle. Condensers are of two main types—air-cooled by means of a forced fan draft, and air-cooled by natural air circulation. The most common condenser construction is shown in Fig. 4, and consists of copper tubing or coils upon which fixed fins have been inserted to assist in removing the rapid accumulation of heat. This cooling process is further provided by means of a fan (driven either by the compressor motor or by an independent motor) which forces air through the condenser.

Other condensers are made up of corrugated plates welded together, the corrugations forming tubes in which the gas is condensed. In this form of condenser, the air passes over the outside surface of the plates and removes the heat by condensation from the refrigerant gas.

Courtesy Tecumseh Products Co.

Fig. 3. Cutaway view of a typical twin-cylinder sealed compressor used in household refrigerators.

DRIER-STRAINER

As previously noted, the function of the drier-strainer is to remove moisture and impurities from the refrigeration system. A typical drier-

125

CONDENSER WITH
DIAMOND SHAPED FINS
(FAN OPERATED)

FLAT COIL TYPE CONDENSER
(FAN OPERATED)

PLATE TYPE CONDENSER
COOLED BY NATURAL AIR CIRCULATION

Fig. 4. Various types of condensers showing construction principles.

strainer, Fig. 5, consists essentially of a tubular metal container or housing, arranged for connection into the refrigerant circuit. The drying and purifing agent is usually silica gel which, in addition to a cup-shaped inlet screen, provides the filtering and drying action.

CAPILLARY TUBES

The capillary tube is essentially an expansion device used as a part of the refrigerant circuit. It consists normally of a miniature tube, the length of which depends upon the size of the condensing unit and the kind of refrigerant used. The bore or inner diameter is very small and the length varies greatly from a few inches up to several feet. Because the capillary tube offers a restricted passage, the resistance to the refrigerant flow is sufficient to build up a high enough head pressure to produce condensation of the gas. The operating balance is obtained by properly proportioning the size and length of the tube to the particular unit on which it is to be used.

HOUSING   SILICA GEL   SCREEN

Courtesy Refrigeration Research Co., Inc.

Fig. 5. Illustrating a typical drier-strainer.

The refrigerant employed has a direct influence on both the length and the diameter of the capillary tube. The size of the inner tube diameter must, in any event, be such as to keep the tube full of liquid under normal operating conditions. Because of the minute size of the tube bore, it is important that the refrigerant circuit be kept free from dirt, grease, and any kind of foreign matter, since these obstructions may close up the tube and thus make the unit inoperative. To prevent obstructions to the free flow of refrigerants, a strainer-drier is provided, as noted in the refrigerant circuit diagram (Fig. 6). If the refrigerant tube becomes plugged, the evaporator will defrost and the unit may run continuously, or the thermal relay cuts out, in which case both the suction and discharge pressures will be abnormally high.

EVAPORATORS

The function of the evaporator is to absorb heat from the refrigerator cabinet, the heat being introduced by food placed in the refrigerator, by insulation loss, and by door openings. Evaporators used in present day designs are of the direct-expansion type because of its simple construction, low cost, and compactness, and also because it provides a more uniform temperature and rapid cooling. The evaporator consists simply of metal tubing which is formed around the freezer compartment to

127

produce a cooling area for freezing of ice cubes, and to provide the desired cooling effect for the food storage compartments. Fig. 7 illustrates various types of frozen food compartments.

In operation, when the refrigerant leaves the capillary tube and enters the larger tubing of the freezer shelves, the sudden increase in tubing diameter forms a low-pressure area, and the temperature of the refrigerant drops rapidly as it changes to a mixture of liquid and gas. This cold mixture passes through the top shelf (or top freezer plate in some models), then to the bottom shelf, descending through each additional shelf until it reaches the accumulator. In the process of passing through the shelf tubing, the refrigerant will absorb heat from the food storage area and will gradually change from a liquid and gas mixture to a gas.

Courtesy Refrigeration Research Co., Inc.

Fig. 6. Illustrating capillary installation in typical refrigerating circuit.

ACCUMULATORS

The accumulator is a large cylindrical vessel designed to trap any refrigerant liquid which may not have changed to gas in the evaporator. It is in this manner that any liquid refrigerant remaining in the low side of the system is prevented from entering the suction line to the compressor.

Through the courtesy of the *"Frigidaire Division of the General Motors Corp.,"* we have diagrams of current changes in the late model refrigerators which have been designed to conserve energy over the previous models, in line with the energy conservation programs that are in order at the present time.

Fig. 7. Typical evaporator arrangement in household refrigerators.

Post Loop Condenser Design Change:

"The energy saving post loop condenser already employed in many of our current model refrigerators has been expanded in its use so that a portion of it is incorporated in the center vertical mullion. This will eliminate the need for a center vertical drier, thus conserving more

129

electrical energy. This improvement will become effective on all Side × Side model refrigerators starting with 1976 production.''

''The post loop routing for the FCI-16V model is shown in Fig. 8. From the condenser outlet the post loop tube is routed to the bottom of the food compartment and then to the center mullion. From here it goes to the top of the center mullion, makes a loop and back down again across the bottom of the freezer compartment. From here it goes around the perimeter of the cabinet shell at the front as on former models.''

Courtesy Frigidaire Corp., Div. of General Motors Corp.

Fig. 8. Model FC1-16V refrigeration system.

"For the FCI-20V and FCI-20V3 models the routing is approximately the same as it is on the FCI-16V model, see Fig. 9. The difference is the FCT-16V utilizes a static condenser system while the FCI-20V models have a forced air system."

"The basic difference between the 20V and 22V is that the center mullion post loop section enters at the top on the 22V models, Figure 10."

CONTROL METHODS

The most common method of temperature control presently employed in household refrigeration units, is the thermostatic control. The apparatus consists principally of a thermostatic bulb (Fig. 12), which is fastened to the evaporator (chilling unit) and connected to the bellows by means of a capillary tube.

The bulb and tube are charged with a highly volatile fluid. As the temperature of the bulb increases, gas pressure in the bulb-bellows assembly increases, and the bellows pushes the operating shaft upward against the two spring pressures. The shaft operates the toggle or snap mechanism. Consequently, the upward travel of the shaft finally pushes the toggle mechanism off center, and the switch snaps closed, starting the motor. As the motor runs, the control bulb is cooled, gradually reducing the pressure in the bulb-bellows system. This reduction of bellows pressure allows the spring to push the shaft slowly downward until it has finally traveled far enough to push the toggle mechanism off center in the opposite direction, snapping the switch open and stopping the motor. The control bulb then slowly warms up until the motor again starts and the cycle repeats itself.

A typical schematic wiring diagram is illustrated in Figure 12. Schematic wiring diagrams will be most helpful to you once you are able to read them. The wiring is not layed out in the manner that it is installed in the refrigerator, but you may at a glance see how each part is connected in the electrical system, and from there tracing out the wiring in the refrigerator becomes simple.

Courtesy Frigidaire Corp., Div. of General Motors Corp.

Fig. 9. Model FC1-20V, 20V3 refrigeration system.

Courtesy Frigidaire Corp., Div. of General Motors Corp.

Fig. 10. Model FC1-22V refrigeration system.

Courtesy Frigidaire Corp., Div. of General Motors Corp.

Fig. 11. Schematic wiring diagram for models D-43 and I-43.

Fig. 12. Diagram showing the temperature control.

CHAPTER 5

Absorption System For Domestic Refrigeration

Absorption-type refrigeration uses heat and a refrigerant that may be absorbed into water. Domestic systems use ammonia, and industrial systems may use ammonia or lithium bromide. More will be covered on commercial absorption refrigeration in a later chapter.

The absorption system of refrigeration differs from that of the conventional compression type mainly in that it uses heat energy instead of mechanical energy to change the condition required in the refrigeration cycle. The heat energy required may be obtained from a gas flame, an electric heater, or a kerosene flame. The principal advantages in the absorption system of refrigeration lies in the fact that, since no moving parts are involved, the repairs and maintenance cost will be minimal. The comparison between the functioning of an *absorption system* and a *compression system* of refrigeration is principally as follows: *The absorption system uses heat (usually a gas flame) to circulate the refrigerant, while the compression system employs a compressor to circulate the refrigerant.* Some absorption systems may be operated by gas or electricity. You will find many small refrigerators in travel trailers, in which the boiler or generator may be operated from either propane gas *or* electricity at the will of the owner.

There are three components of the absorption system which compare with the four components of the compression system, as follows:

1. The action of the boiler or generator compares with the stroke of the compressor.

135

2. The condenser and evaporator serves the same purpose in an absorption system as that in the compression system of refrigeration.
3. The absorber compares to the low or suction side of the pump.

PRINCIPLES OF OPERATION

After the proper heat has been applied to the generator, the ammonia will vaporize from the water. The vapor bubbles, in trying to escape, will carry water up the percolator tube. The vapor and water are allowed to separate so that the vapor is free to continue upward into the condenser. Here, with proper air circulation, the ammonia vapor will be condensed to a liquid. It then flows through a liquid trap into the evaporator. When the evaporator shelf is level, the proper slope is established in all coils to induce a gravity flow downward.

At the time the unit is charged, a small amount of hydrogen is introduced. At this point of the cycle of the unit, hydrogen flows upward into the evaporator and tends to mix with the ammonia vapor and encourages more evaporation. It is this evaporation process which produces refrigeration. Since the mixture of hydrogen and ammonia vapor is considerably heavier than hydrogen alone, the normal tendency for this mixture is to flow downward. It is encouraged to do this, and in so doing it is forced to pass upward through the absorber.

Water, which has been separated from the ammonia by heat, flows downward through the absorber. The water temperature has been reduced so that it will again absorb the ammonia quite readily, and the water and ammonia solution flows back to the generator for recirculation. Since the hydrogen has been washed free of ammonia (and lightened), it flows upward again through the evaporator. When the absorption system is working normally, all of these actions are continuous. A thermostatically controlled gas valve, with the sensor attached to the evaporator coil, will vary the heat input and consequently the amount of refrigeration that the load of the refrigerator requires.

ABSORPTION REFRIGERATION CYCLE

The freezing system of the gas refrigerator is made up of a number of steel vessels and pipes welded together to form a hermetically sealed system. All the spaces of the system are in open and unrestricted communication so that all parts are at the same total pressure. The charge includes an aqua-ammonia solution of a strength of about 30% concentration (ammonia by weight) and hydrogen.

The elements of the system include a *generator* (sometimes called boiler or still), a *condenser,* an *evaporator,* and an *absorber.* There are three distinct fluid circuits in the system: an *ammonia circuit* including the *generator, condenser, evaporator,* and *absorber;* a *hydrogen circuit* including the evaporator and absorber; and a *solution circuit* including the generator and absorber. A diagram of this circuit is shown in Fig. 1.

Courtesy Whirlpool Corp.

Fig. 1. Operating cycle of a gas-absorption refrigerating system.

Heat is applied by a gas burner or other source of heat to expel the ammonia from solution. The ammonia vapor thus generated flows to the condenser. In the path of flow of ammonia from the generator to the condenser are interposed an *analyzer* and a *rectifier.* Some water vapor

137

will be carried along with the ammonia vapor from the generator. The analyzer and rectifier serve to remove this water vapor from the ammonia vapor. In the analyzer, the ammonia passes through a strong solution which is on its way from the absorber to the generator. This reduces the temperature of the generated vapor somewhat to condense water vapor, and the resulting heating of the strong solution expels some ammonia vapor without additional heat input. The ammonia vapor then passes through the rectifier where the residual small amount of water vapor is condensed by atmospheric cooling and drains to the generator by way of the analyzer.

The ammonia vapor, which is still warm, passes on to the section of the condenser where it is liquified by air cooling. The condenser is provided with fins for this purpose. The ammonia thus liquified flows into the evaporator. A liquid trap is interposed between the condenser section and the evaporator to prevent hydrogen from entering the condenser. Hydrogen gas enters the lower evaporator section and flows upward, in counter-flow to the downward flowing liquid ammonia. The effect of placing a hydrogen atmosphere above the liquid ammonia in the evaporator is to reduce the partial pressure of the ammonia vapor in accordance with *Dalton's Law of Partial Pressures.*

Under Dalton's Law, the total pressure of a gas mixture is equal to the sum of the partial pressures of the individual gases. Consequently, in the evaporator, the partial ammonia vapor pressure is less than the total pressure by the value of partial pressure of the hydrogen. The lesser ammonia vapor pressure results in evaporation of the ammonia, with consequent absorption and cooling of the surroundings which are in a well-insulated enclosure. The cool, heavy, gas mixture of hydrogen and ammonia vapor formed in the evaporator leaves the top of the evaporator and passes downward through the gas heat exchanger to the absorber.

Since the weight of a gas is proportional to its molecular weight, and the molecular weight of ammonia is 17 and the molecular weight of hydrogen is 2, the specific weight of the strong gas is greater than that of the weak gas. This difference in specific weights is sufficient to initiate and maintain circulation between the evaporator and the absorber. The gas heat exchanger transfers heat from one gas to the other. This saves some cooling in the evaporator by precooling the entering gas. A liquid drain at the bottom of the evaporator is connected to the down-flow space of the gas heat exchanger. In the absorber, a flow of weak solution (water weak in ammonia) comes in direct contact with the strong gas. Ammonia is absorbed by water, and the hydrogen (which is practically insoluble) passes upward from the top of the absorber through another chamber of the gas heat exchanger into the evaporator. The liquid and gas flow in opposite directions.

From the absorber, the strong aqua-ammonia solution flows through the liquid heat exchanger to the analyzer, and then to the strong liquid chamber of the generator. The liquid heat exchanger precools the liquid entering the absorber and preheats the liquid entering the generator. Further precooling of the weak solution is obtained in the finned air-cooled loop between the liquid heat exchanger and the absorber. Actual unit construction includes additional refinements which increase efficiency, but the cycle described previously is typical of the cycle in any absorption refrigerator. Refrigeration is accomplished by a continuation of the cycle described herein.

THE FLUE SYSTEM

The flue system is composed of a combustion chamber where the gas-air mixture is burned. The flue extension is to provide a draft for an adequate flow of secondary air and a dilution flue to cool the products of combustion and carry them to the top of the refrigerator.

The combustion chamber is properly known as the generator flue because it is attached to the generator. A flue baffle, is normally located at the back or top end of the generator. The purpose of the flue baffle is to distribute the heat. The flue baffle should always be located in its correct position, otherwise the efficiency of the unit will be affected.

The flue extension is made of nonmetallic materials and is attached to the generator flue. On horizontal generator flue systems, a clean-out cover is located at the rear of the unit. The flue baffle may be removed through the opening when the cleanout cover is removed. When the unit is in operation, the clean-out cover must be in place for proper draft.

The dilution flue is a nonmetallic pipe which extends from the top of the generator flue extension to the top of the cabinet. Air for dilution is induced at the bottom of the loose fitting dilution flue, and mixes with products of combustion as it travels to the top of the cabinet.

GAS CONTROL DEVICES

An absorption system refrigerator requires various gas control devices for its operation, which are shown in Fig. 2.

Fig. 2. Gas components in typical absorption refrigeration system.

1. Shut-off valve.
2. Pressure regulator (for piped gas only).
3. Input control valves.
4. Gas burner.
5. Thermostat.

The gas components may easily be removed as an assembly for service when required, provided that the gas shut-off valve is closed and that the electric supply is disconnected.

Gas Shut-off Valve

The gas shut-off valve, as the name implies, is a simple *on* and *off* valve to enable the serviceman to cut off the gas supply when necessary for repairs and service.

Pressure Regulators

These are normally provided on all piped gas refrigerating units except those using straight liquid petroleum (LP) gas, since LP gas containers are equipped with individual pressure regulators. The gas pressure regulator (Fig. 3) functions to provide a constant, even gas pressure to the gas burner during all main line pressure fluctuations.

The action of the gas pressure regulator is very simple. A valve is suspended on a flexible diaphragm in a way that will cause the valve to come in contact with the valve seat when the diaphragm opens, allowing the gas to pass. The pressure of the gas will move the diaphragm up or down, controlling the amount of gas.

Pressure applied from above the diaphragm, either by spring pressure or by weights (depending upon the type of regulator), will tend to open the valve. Therefore, the outlet pressure of the regulator can be adjusted by turning a screw to change the spring tension or, on dead weight regulators, by adding or removing weight.

Fig. 3. A gas-pressure regulator.

Input Controls

The input control consists essentially of two automatically operated gas valves which are electrically controlled by means of the refrigerator thermostat. The most common type of controls are of *General Controls* or *White Rodgers* manufacture, as shown in Fig. 4. As will be noted in the illustrations, these controls function in a similar manner, although they differ in their operation. The input control consists of:

1. A felt filter to remove gum and foreign materials from the gas.
2. A diaphragm-type pressure regulator to adjust maximum flame gas pressure when the control is used on piped gases.
3. Two identical electrically operated gas valves which alternate opening and closing as the refrigerator thermostat cycles. One valve supplies gas for a maximum burner flame, while the other releases gas for an intermediate flame.

4. A bypass to supply gas for a burner pilot flame when both valves are closed due to no electrical power.
5. A plate containing two orifices to reduce the gas flow for the intermediate and pilot flames.
6. A pushbutton valve gas release for burner lighting.

When the refrigerator is to be operated on LP gas, the pressure regulator cap screw is replaced with one having a long narrow pin. The pin locks the pressure regulator in an open position. The gas is regulated at the supply tank.

Courtesy General Controls and White Rodgers

Fig. 4. Diagram showing the components of a gas input control assembly.

Gas Burners

The gas burner assembly (Fig. 5) consists essentially of a *turbulator, orifice spud,* and *thermal safety valve*. The function of the turbulator, as its name implies, is to create turbulence in the gas stream as it enters the mixing tube. This turbulence will have two effects. The one which is apparent and readily understood is that it aids in mixing primary air and gas. The second effect is that the air-gas ratio is automatically adjusted so that with proper air-shutter adjustment, the flame will be all blue on maximum flame and will have stability on pilot flame. All piped gas-burner turbulators have two grooves in the conical end, and the head of the turbulator must always fit tightly into the rear side of the orifice. All LP gas-burner turbulators have a single groove in the conical end and are spring-loaded for adjustment to the gas pressure.

Fig. 5. Diagram illustrating the typical gas burner operation.

The burner orifice required on gas refrigerators is so small that a slight change in its dimensions changes its capacity. For this reason, each orifice spud is individually flow tested and rated to capacity. All orifice spuds having the same rated capacity have a comparative flow accuracy of plus or minus 3% for orifice spuds used with piped gas. The number marked on the orifice spud is not the hole diameter or drill size. In fact, the hole diameter may vary somewhat to compensate for other factors, such as surface resistance, which would affect the capacity.

The safety valve is a protective device designed to stop the gas flow to the burner in case the flame should leave the generator flue. This valve is similar in appearance to the thermal valve on the burner. The valve

143

consists of a bi-metallic disc to which is attached a valve stem. The valve is open at room temperature. However, by overheating the valve body or the heat conductor, the disc will snap the valve closed. The safety valve will reset automatically as it cools down in temperature.

Thermostats

The thermostat is an automatic gas valve in the burner supply line which controls the flow of gas to the burner to supply either a maximum or minimum flame, depending upon control dial setting and thermostat bulb temperature. The thermostat shown in Fig. 6 identifies the different parts of the thermostat and their relative position to each other. The working parts of the thermostat are enclosed in an outer case. In the lower part of the case is a gas passage, imposed in which is a gas valve. Under the valve is a spring which pushes the valve toward a closed position.

Fig. 6. Illustrating the absorption refrigerator thermostat valve.

Above the gas passage is a power element consisting of a bulb connected to a flexible diaphragm by a capillary tube. In this system is a small quantity of gaseous hydrocarbon. Motion of the power element is transmitted to the gas valve by the valve pin. When the bulb is warm, the gas expands and exerts a pressure in the power element, which pushes the valve open. When the bulb is cold, the gas contracts, the pressure inside the power element is reduced, and the valve spring pushes the valve closed. From the foregoing description, it is evident that there is a throttling action and that the valve may be in any position between wide open and fully closed, depending upon the thermostat dial setting and the thermostat bulb temperature.

To prevent the gas supply to the burner from being completely closed off when the valve is closed, a bypass or minimum flame passage is provided. Adjustment screws on the thermostat control the volume of gas flow through the passages which supply gas for maximum and minimum flame. There is also an adjustment screw on the burner lighter valve. This valve is used only when lighting the burner. Control of temperatures inside the refrigerator is accomplished by turning the control dial. The highest number on the dial is the coldest position. A defrost position is provided for manual defrosting.

ELECTRICAL ACCESSORIES

Gas refrigerating units are capable of operating without any electrical connections, which is usually the case where the refrigerator is operated on LP gas and where electricity is not available. Electrically equipped absorption units, however, will add certain desirable features, such as:

1. Cabinet lights.
2. Automatic defrosters.
3. Air circulator fans.
4. Air circulator thermostat, etc.

Cabinet Lights

These are wired in the conventional manner, the interior light bulb being energized by the light switch when the cabinet door is opened.

Defrost Timer

These are manufactured to operate the electrical defrost heater at certain time intervals, usually once every 12 or 24 hours. Fig. 7 illustrates a typical self-starting electrical defrost timer operating upon the well-known electric clock principle. The dial on the timer should be set to the correct time of day when the refrigerator is installed, and should be reset to the correct time any time the electric service is interrupted.

The 24-hour defrost timer contains an electric switch which makes contact for a period of about 20 minutes, beginning at the 2 A.M. position on the timer. This means that if the dial is set to the correct time, defrosting will occur automatically at 2 A.M. each day. The defrost water will drain from the chill coil onto the lower surface of the chill cover and then into a drain elbow which directs the water to an evaporation pan.

Fig. 7. A typical defrost timer.

Air Circulator Fan

The air circulator fan, as the name implies, is generally designed to become operative when the ambient temperature becomes unduly high. The additional air movement provided by the air circulator will provide more cool air for cooling the condenser. At the same time, the movement of cooler air at the rear of the cabinet will reduce the amount of heat transfer to the inside of the cabinet, as shown in Fig. 8. The thermostat which controls the air circulator fan is preset to "cut in" at approximately 100° F and to "cut out" at approximately 80° F. This means that the fan will not operate until the temperature of the thermostat reaches 100°, and once it has started to operate, the fan will remain in operation until the thermostat temperature has dropped to 80° F. The electrical components are shown in Fig. 9.

INSTALLATION AND SERVICE

Before starting the refrigerator installation, turn off all other gas appliances, such as gas ranges, gas furnaces, gas water heaters, gas spaceheaters, etc. Next, shut off the gas supply at the main inlet or outlet at the gas meter.

Courtesy Whirlpool Corp.

Fig. 8. Illustrating the air circulation fan arrangement in a typical refrigerator.

Air Circulation

All refrigerators must have proper air circulation for proper operation. When the refrigerator is in operation, air enters from the bottom and travels upward through the rear section of the cabinet and out at the top, as shown in Fig. 10. This air circulation takes place naturally unless prevented by insufficient clearance or blocked air passages. Proper air circulation will usually result when the refrigerator is installed indoors directly in front of a wall allowing a minimum distance of two inches of clearance from the back of the refrigerator to the wall, and at least a twelve-inch clearance above the top of the refrigerator.

When the recommended top clearances cannot be obtained, an air duct should be installed. The air duct should have about the same width as the refrigerator and be approximately seven inches deep. This duct may

147

return the air to the room at ceiling height or may exhaust it through a suitable vent in the roof.

Courtesy Whirlpool Corp.

Fig. 9. Arrangement of electrical components used in an automatic defrost refrigerator.

Leveling

The equal distribution of the liquid within the freezing compartment requires the unit to be installed and maintained in a level position, both front, back, and side to side. The leveling of the refrigerator can conveniently be accomplished by shimming with small wooden strips or other available material.

148

Fig. 10. Illustrating the proper clearance arrangement for the proper air circulation.

Gas Line Connection

When connecting the gas line, use tubing and fittings as prescribed by local codes. Install the gas line so that the refrigerator can be disconnected at the inlet valve without damage to the controls. On liquid petroleum (LP) gas installations, a gas pressure regulator in the gas line is not needed since the regulator at the gas supply tanks should maintain constant gas pressure at the burner. Fig. 11 shows typical gas connection methods.

The installation shown will not permit the refrigerator to be disconnected between the gas cock and the thermostat. Therefore, to disconnect the refrigerator, it is necessary to shut off the gas supply from the meter. By using a tubing connection between the gas cock and gas filter, or between the gas filter and thermostat, the controls can be installed in a way which will permit the refrigerator to be easily disconnected for service.

149

Fig. 11. Gas connections for vertical and horizontal installations.

Piping Material—Standard weight wrought-iron pipe coupling with the American Standard wrought-steel and wrought-iron pipe shall be used in installation of appliances supplied with utility gases. Threaded copper or brass pipe in iron-pipe sizes may be used with gases not corrosive to such material. Gas pipe shall not be bent. Fittings shall be used when making turns in gas pipe.

The connection of steel or wrought-iron pipe by welding is permissible. Threaded pipe fittings (except stopcocks or valves) shall be of malleable iron or steel when used with steel or wrought-iron pipe and shall be copper or brass when used with copper or brass pipe. When approved by the authority having jurisdiction, special fittings may be used to connect either steel pipe or wrought-iron pipe.

Piping Material for LP Gases—Gas piping for use with undiluted LP gases shall be of steel or wrought-iron pipe complying with the American Standard for wrought-steel and wrought-iron pipe, and brass or copper pipe, or seamless copper, brass, steel, or aluminum tubing. All pipe or tubing shall be suitable for working pressure of not less than 125 pounds per square inch. Copper tubing may be of the standard K or L grade, or equivalent, having a minimum wall thickness of 0.032 inch. Aluminum tubing shall not be used in exterior locations or where it is in contact with masonry or plaster walls, or insulation.

Defective Material—Gas pipe or tubing and fittings shall be clear and free from burrs and defects in structure or threading, and shall be thoroughly brushed, chipped, and scale blown. Defects in pipe, tubing, or fittings shall not be repaired. When defective pipe, tubing, or fittings are located in a system, the defective material shall be replaced.

Gas pipe, tubing, fittings, and valves removed from any existing installation shall not be used again until they have been thoroughly cleaned, inspected, and ascertained to be equivalent to new material. Joint compounds shall be applied sparingly and only to the male threads of pipe joints. Such compounds shall be resistant to the action of LP gases.

BURNER ORIFICE CLEANING

The burner orifice must be cleaned when the refrigerator is installed and each time a new orifice is installed. This is necessary because the orifice is so small that a particle of dirt can materially change its operating characteristics. Dirt in the orifice can also cause an unstable flame and burner outage. Clean the orifice after it has been installed on the burner body. The air-shutter barrel, mixing tube, seal screw, and turbulator must be removed.

Use a round toothpick, as shown in Fig. 12, sharpened on an emery board, to clean the orifice. Insert in the orifice and rotate with light pressure. Blow away any dust but do not rub the orifice with the fingers. If the burner is installed on a refrigerator, use a mirror for observation.

BURNER AIR ADJUSTMENT

The entire flame must be free of all yellow after the adjustments have been completed. Do not make this adjustment if there is an abnormal flue draft. On models having a filter above the condenser, check to be sure the filter is clean and in place. Tighten the air-shutter barrel securely on the burner body. This position should result in a yellow flame. Rotate the air-shutter barrel in the opposite direction, increasing the primary air until the yellow just disappears from the entire flame. From this point, continue to turn the air-shutter barrel in the same direction until the flame becomes noisy, unstable, or until the stop is reached. The desired setting for proper combustion is approximately midway between these two points.

Fig. 12. Showing typical method for orifice cleaning.

CLEANING THERMOSTAT VALVE

To clean the thermostat valve, turn off the gas supply to the refrigerator. If the cleanout cap is inaccessible, disconnect the flare nuts at the inlet and outlet of the thermostat. Remove the screws attaching the thermostat to the cabinet, and turn the thermostat until the cleanout cap is accessible. Be careful not to pinch or break the capillary. Remove the cleanout cap and gasket, spacer, valve spring, and valve. Clean the valve seat in the thermostat, using the eraser end of a pencil covered with a clean cloth. Blow into the valve seat to make certain all dirt has been removed. Clean the flat face of the valve with a clean cloth.

GAS PRESSURE ADJUSTMENT

The maximum flame gas pressure must be determined and adjusted for each refrigerator when installed or reinstated in a different location. The pressure was determined at the time the burner orifice spud was selected. The maximum flame provides the correct heat input for the greatest capacity of the refrigerator unit. The unit cooling capacity will be decreased if the heat input is more or less than the specified rating.

Adjustment to the intermediate and pilot flame gas pressure is not necessary, and provision for such has not been made. The orifice plate on the input control meters the gas flow for the lower flame settings. To adjust the maximum flame gas pressure, proceed as follows:

1. Make sure the appliance cord is connected to a power source and the thermostat is turned to a numbered position. (Maximum flame valve must be open in the input control.)
2. Remove the seal screw from the side of the burner and connect a water-filled manometer, as shown in Fig. 13.
3. Light the burner and allow the generator flue to warm.
4. *On piped gas:* Remove the seal screw and turn the adjusting screw in the pressure regulator on the input control counterclockwise to decrease the pressure; or clockwise to increase the pressure.

FLAME STABILITY

After completing the control adjustments, always check the flame for stability. Disconnect the appliance cord from the power source, thus reducing the flame to a pilot. If the pilot flame appears jumpy or wobbly, rap the bottom cross rail of the cabinet several times lightly, using a wrench. If the flame is unstable, the rapping will cause the flame to go out. If the flame is stable, the rapping will not affect it. An unstable flame is usually caused by incorrect primary air adjustment, loose turbulator, and/or dirt in the orifice. When the test for flame stability has been completed, reconnect the appliance cord.

Fig. 13. Showing manometer method of gas pressure for piped and L.P. gas.

REMOVING AND REPLACING REFRIGERATOR UNIT

To remove and replace the refrigerator unit, proceed as follows:

1. Remove all trays, baffles, and other parts from the freezing compartment to permit its removal through the freezing compartment opening at the rear of the cabinet. If the unit has a separate food compartment cooling coil, remove the grille and thermostat bulb.
2. On models with a permanently insulated section between the freezer compartment and food compartment, remove the trim around the freezer compartment, remove the rubber plug from the drain hole, and remove the screws on the sides of the freezer.

3. Move the refrigerator away from the wall and remove the rear panel, dilution flue, and top louver.

4. On models having pushbutton defrost, disconnect the electric cord from the wall receptacle. Carefully pry out the complete pushbutton assembly from the trim. Remove the side and top-door opening trim. Remove the defrost thermostat feeler from the clamp on the freezer by loosening the screws from the inside of the freezer. Remove the thermostat support screws from the liner. Pull the assembly downward out of the cabinet. Now remove all the wires from the thermostat terminals. The defrost heaters, when used, may be removed from the freezer coils after the unit is out of the cabinet by prying off the retaining clamps. Install the heater on the replacement unit while the unit is out of the cabinet by reversing the above proceedings.

5. Remove the unit retaining bolts and, on models with a gas heat exchanger (or tubes) embedded in the cabinet insulation, remove the screws from the cover plate. Carefully lift the unit out of the cabinet.

6. Remove and install on the replacement unit the styrofoam insulation and freezer drain pan, if present. When replacing the unit, it is necessary to disconnect the front chill coil supports in order to remove the plastic drain chute from beneath the freezer section. After the drain chute has been removed from the inoperative unit, reconnect and securely tighten the supports on both units.

7. Install the replacement unit by reversing the removal procedure. Make sure that the gasket and/or sealing compound around the freezing compartment opening is properly placed in position to assure a satisfactory seal. On models with embedded gas heat exchangers, place sealing compound around the edge of the cover plate and attach the plate securely to the cabinet. The tubes protruding through this plate should also be carefully sealed.

8. On models equipped for automatic defrosting, the drain tube should be properly sealed around the tube and cover plate on the cabinet side.

REFRIGERATOR MAINTENANCE

The following pointers should be observed for proper adjustments and maintenance of gas refrigerators.

Lighting the Burner

Lighting the burner is a rather simple operation provided the manufacturer's instructions accompanying the unit are followed. The lighting procedure is generally as follows:

1. Be sure the gas valve is turned on.
2. Push the lighter button.
3. Light the gas at the end of the burner tube.
4. Continue to push the lighter button until the burner valve clicks open and the burner flame ignites. (Should the burner flame go out, wait five minutes before relighting.)

Do not leave the burner lighted unless all final adjustments have been made. The burner flame should burn a blue color and must enter the flue opening.

Cold Control

The position of the thermostat dial should depend upon the refrigeration load. When the food load is heavy, turn the dial toward a colder position, and when the food load is light turn the dial to a less cold position. A colder setting will be required in summer than in winter. Experience and observing the effect of the various thermostat settings will usually provide the operator with the required knowledge after a short period of time.

Stopping Refrigeration

When necessary to discontinue refrigeration for any length of time, turn off the gas supply. Remove the ice cube trays and empty and dry them. Dry the interior of the refrigerator. Leave the door partly open to ventilate the cabinet interior and keep it fresh.

Refrigerator Too Cold

Low refrigerator temperature may be caused by any or a combination of the following:

1. If the minimum flame is set too large, the thermostat will not be able to control the temperature in the food storage compartment due to the fact that some refrigeration will be taking place even when the thermostat valve is completely closed.

2. Dirt on the valve seat will keep the valve open slightly. Gas passing through the valve will cause the minimum flame to be too large, giving the same effect as just described.
3. The refrigerator is designed for operation at normal room temperatures. Location in unheated rooms will cause foods to freeze.
4. The highest number on the control dial is the coldest setting. If the food storage compartment temperature is too cold, turn the dial to a warmer setting.

Refrigerator Not Cold Enough

If any of the basic adjustments are incorrect, the efficiency of the unit may be impaired. Therefore, the things to check on a complaint of a refrigerator not cold enough include the basic adjustments for proper air circulation, level, and right burner flame. In addition, the following items may cause a complaint of poor refrigeration:

1. Check the door gasket seal and repair if necessary.
2. If a "built-in" refrigerator is improperly vented, there may not be enough air circulation over the condenser to cool the unit efficiently. Check for proper air circulation.
3. Check for leaky cabinet seals when the service history shows that the unit has been changed, or if there is evidence pointing to wet insulation.
4. On models which depend on circulation of air around the evaporator for cabinet cooling, the efficiency of the unit will decrease as ice builds up on the evaporator. Some models located in extremely humid climates may need defrosting several times a week for peak efficiency.
5. Dirty condenser fins will prevent the air from circulating through the fins for proper cooling, and as a result, the efficiency of the unit will decrease in proportion to the amount of dust and lint on the fins.
6. The flue must be clean and have proper draft. A visual inspection of the flue will show whether the flame is entering the flue.
7. The highest number on the control dial is the coldest position. If the complaint is that the refrigerator is not cold enough, the control dial may not be set to a cold enough position.
8. If the maximum flame pressure setting is in excess of the recommended pressures, the unit may be overloaded and will not perform as efficiently as it should.

9. Dirt in the orifice spud may partially close the opening and limit the amount of gas which can be burned.

No Refrigeration

A complaint of no refrigeration can be caused by the following:

1. Check for burner flame outage.
2. If the evaporator is warm and the control dial is set on the coldest position, it should not be difficult to adjust the maximum flame screw to obtain the correct pressure at the burner. If this is impossible, change the thermostat.

Unit failure will seldom be experienced. Before condemning a unit, be sure that the following requirements have been fulfilled:

1. Clean the condenser and be sure that there is enough clearance above the refrigerator for proper air circulation.
2. Level the unit front to back and side to side, checking from the bottom shelf of the freezer section.
3. Adjust burner.

Burner Flame Goes Out

The burner will react to adverse conditions more noticeably when it is operating on minimum flame. Any difficulty with flame outage can usually be discovered by looking at the minimum flame and by checking minimum flame pressures. Burner outage is usually caused by the following:

1. If the burner is spaced too close to the generator, the minimum flame will be pulled away from the heat conductor, causing the heat conductor to cool off and close the automatic shut-off valve. The burner space illustrations are shown in Fig. 14.
2. Check to be sure that the correct orifice is in the burner and use a manometer to set the recommended pressure.
3. Dirt in the orifice spud may partially close the opening. A visual inspection of the orifice and cleaning are necessary to correct this condition.
4. Always check for flame stability after working on a burner, or when there is a complaint of burner outage. To test flame stability,

Fig. 14 Correct and incorrect burner-to-generator spacing arrangement for proper flame.

turn the control dial within its operating range until the flame is maximum size. Then gradually turn the dial in the opposite direction until the flame begins to reduce in size. Observe the flame as it reduces from maximum flame to minimum flame. If the flame is jumpy or distorted as it passes from maximum to minimum size, it is probably unstable. A further check for stability can be made by snapping the fingers or tapping two wrenches together. An unstable flame will react to the sound waves.

Flame Burns Outside of Generator Flue

The flame must burn inside the generator flue at all times for proper combustion and proper refrigeration. If the flame is allowed to burn outside the generator flue for even five minutes, the cabinet seal may be damaged and the insulated section which was heated by the flame may produce odors due to the insulation being scorched. To correct this complaint, replace the insulation and reseal the cabinet at the points affected by the flame. The following items may keep the flame from entering the flue:

1. The flue may be restricted by objects dropped into it, or by deposits which are the result of combustion. (sulfur deposits, carbon, etc.)

159

2. Clean out covers, flue extensions, flue elbows corroded, loose, or missing.

3. A natural draft will not be established in the flue until it has become warm. Turning the thermostat to the coldest position and setting the air adjustment for a hard flame will help the flame to enter the flue. Blowing through a short length of tube into the generator flue opening will also cause the flame to enter, unless the flue is stopped up. Final air adjustment should be made after the flame enters the flue.

4. A burner spaced too far away from the generator flue will cause the flame to back out when it is controlling at minimum flame.

5. If the minimum flame is too small, the generator flue may cool off, stopping the natural draft. This can also occur if the refrigerator is located in an extremely cold room even with the correct minimum flame.

6. The refrigerator should not be vented to another room, to the attic, or to the outside. Downdrafts caused by improper venting may force the flame out of the flue.

Manual Defrosting

Most complaints of "incomplete defrosting" or "defrosting takes too long" are usually caused by too large a minimum flame. To correct this condition, check to be sure that the burner has the correct orifice spud and that the minimum flame pressure is correct. Dirt on the thermostat valve seat may cause the minimum flame pressure to be incorrect. If it is impossible to obtain the correct minimum flame pressure by turning the minimum flame adjustment screw, clean the thermostat valve and valve seat.

Automatic Defrosting

A list of the most common causes of defrost problems on 24-hour automatic, or on push-button defrost models is combined in this section because of the related nature of the two types of automatic defrost.

1. Connect an ohmmeter or test cord to the heater and check for continuity or for a short. If the heater is found to be defective, replace it.

2. The defrost heater must have good contact with the section of the evaporator which is to be defrosted. Check to be sure that the heater clamps are holding the heater in contact with the section of the evaporator which is to be defrosted.

3. Failure of the electrical power supply to the heater could cause complaints of no defrosting. This could be caused by an electrical short, the service cord not plugged in, or by a blown fuse.

4. Place a long screwdriver against the timer case and press an ear to the handle. If the power cord is plugged in and the clock is running, the gear train sound should be audible. To test the timer electrically, disconnect the power cord and remove the timer. Check for continuity between terminals 1 and 4. This is the clock circuit and it should have continuity at all times. If no continuity exists, replace the timer. Check for continuity between terminals 3 and 4. Continuity should exist when the dial is turned to the 2 A.M. position. If no continuity is found, replace the timer.

5. The push-button defrost control is a "fail-safe" mechanism and will not start a defrost cycle if the power element has lost its charge. If the refrigerator is located at an altitude greater than 5000 ft. above sea level, the defrost mechanism must be recalibrated to accomplish a complete defrosting cycle.

Odors Inside the Refrigerator

Odors inside the refrigerator may be the result of storing odorous unwrapped foods, or by infrequent cleaning. Odors will also develop if a refrigerator has been in use and is stored for any length of time with the door closed. If, by any chance, the flame has burned outside of the generator flue, it is possible that a phenolic odor due to burned insulation may develop in the cabinet. To remove this odor, the section of the insulation which has been burned must be replaced.

To remove any of these odors, first remove the source, then wash with a solution of baking soda and water (approximately one tablespoon to a quart of water). In extreme cases, use activated charcoal to remove the odor.

Odors Outside the Refrigerator

Odors outside the refrigerator are usually the result of poor combustion caused by the flame burning outside the flue, touching the generator flue, or wrong primary air adjustment. Gas leaks can also cause odors outside the refrigerator. Use a soap-bubble solution to find gas leaks.

Frost Forms Rapidly

Rapid formation of frost can be caused by a bad door seal, uncovered foods or liquids, high humidity, excessive usage, or a bad cabinet seal.

161

Keep in mind that the moisture which forms the frost must enter the cabinet in one of the ways mentioned previously. To reduce the amount of frost, the user can be cautioned against excessive door openings on hot humid days. All foods and liquids should be covered before placing in the refrigerator.

Loose Gas Connection

All gas connections should be leak tested with a soap-bubble solution before putting the refrigerator into operation. Gas leaks, no matter how small, should be corrected. Be sure that all seal washers and seal screws are tightened on burners, thermostats, etc.

Inoperative Light Switches

A visible check of the light switch may be made when the power cord is plugged in by manually pushing the button in and out to see if the switch is sticking, and to see if the light is going off and on.

TROUBLE CHARTS

The chart listed on the following pages gives the more common troubles encountered in the repairs of absorption-type household refrigerators. The trouble and possible cause are given with the method used for remedying the defect.

ABSORPTION REFRIGERATION TROUBLE CHART

Symptom and Possible Cause	Possible Remedy
No Refrigeration	
(a) Burner	(a) Check to see that the flame provides proper heat input.
(b) Electrical power (when available)	(b) Check for 115 volt a.c. at front terminal board.
Partial Refrigeration	
(a) Unlevel unit	(a) Check unit for levelness. Adjust levelers at bottom corners if necessary.
(b) Air filter	(b) Check air filter above condenser. If dirty, wash in detergent and instruct user. If missing, replace and check for dirty condenser.
(c) Air circulation	(c) Check to see that cabinet fans are operative.

ABSORPTION REFRIGERATION TROUBLE CHART (Con't)

Symptom and Possible Cause	Possible Remedy
(d) Thermostat	(d) Check to see that thermostat is on numbered position calling for refrigeration.
(e) Usage	(e) Instruct user regarding excessive door openings, excessive ice usage, improper loading and storage, and commercial usage.
(f) Improper Defrosting	(f) Check to see that evaporator air chambers are defrosted (air flow not restricted). Check to see that defrost heater circuit is not energized.
(g) Flue baffle	(g) Check for omission.
(h) Vapor seal	(h) Check door gasket for good seal. If interior liner sweats or frosts, check and reseal all cabinet openings for good vapor seal.
(i) Cabinet lights	(i) Check to see that lights do not burn when door is closed.
(j) Gas pressure	(j) Check to see that maximum flame gas pressure is correct. Be sure maximum flame valve is open on input control.
(k) Burner orifice	(k) Disassemble burner and clean orifice. (Dirty orifice will restrict gas flow and reduce heat input).
(l) Inefficient unit	(l) Performance test unit before replacing.
(m) Burner thermal valve	(m) Check to see that valve opens during burner lighting.
(n) Hi-Temp safety valve	(n) Check to see that valve has not failed in "closed" position.

Burner Outage (Customer complaint may be "no refrigeration.")

(a) Burner valve heat conductor	(a) Check to see that heat conductor impinges on intermediate flame. (If necessary, turn thermostat dial to "Off" position to obtain intermediate flame.)
(b) Burner orifice	(b) Disconnect appliance cord from power supply. Check pilot flame for stability.
(c) Air circulation	(c) Check to see that absorber fan is operative. Check to see that air filter above condenser is not dirty or restricted. (Note: A restricted filter will increase generator draft, pulling flame from burner. This will give flame characteristic of being noisy.)

ABSORPTION REFRIGERATION TROUBLE CHART (Con't)

Symptom and Possible Cause	Possible Remedy
(d) Input control	(d) Check to see that both valves are operative and that there are no internal leaks. Check to see that orifice plate is for type of gas used and orifices are not restricted or oversize.

Improper Defrosting (Improper defrosting will probably not be noticed by user. The complaint would likely be "not enough or no refrigeration").

(a) Defrost timer	(a) Check to see that motor runs and defrost heater circuit is energized at 2 A.M. position.
(b) Defrost heater	(b) Check for resistance and proper thermal contact with unit defrost liquid trap. (Note: heater will glow if contact is inadequate.)
(c) Defrost switch	(c) Check for thermal contact with freezer evaporator. Replace switch if inoperative. If stuck "closed" heater will be energized too long during defrosting and will cycle on the safety switch. If stuck "open", freezer fan will not operate at any time.
(d) Safety switch	(d) Check for thermal contact with heater head conductor. Check for continuity.

Operates Too Cold

(a) Thermostat	(a) Check numbered position. Set on lower number if possible. Instruct user. Check that capillary tube is inserted full length of well evaporator. Check callibration.

Exterior Condensation

(a) Location	(a) Instruct user regarding very damp location (basement, for example).
(b) Insulation	(b) Check adjacent insulation area for void. Fill void with fiberglass or equivalent.

Interior Condensation

(a) Heaters	(a) Check adjacent antisweat heaters for resistance. Replace if inoperative.

ABSORPTION REFRIGERATION TROUBLE CHART (Con't)

Symptom and Possible Cause / **Possible Remedy**

(b) Usage

(b) Instruct user regarding uncovered foods and liquids.

(c) Vapor seal

(c) Check and adjust door gasket for good seal. Check and reseal any cabinet openings (unit, evaporator, cabinet, fans, etc.).

Odor

(a) New appliance

(a) Inside storage compartment will have "new" odor for a few minutes when door is first opened. Instruct user this is normal. When burner is lighted and heated for first time, odor will occur; this is normal. Advise user.

Inoperative Freezer Fan.

(a) Defrost switch

(a) Check to see that thermal switch in series with fan motor has cooled to "closing" temperature. Advise user it is normal for fan to be temporarily inoperative because of switch following installation, defrosting, and restarting when warm. Check for thermal contact and continuity when cold.

(b) Fan motors

(b) Check to see that motor is operative.

Noisy Burner

(a) Dirty air filter

(a) Wash filter. Do not adjust burner primary air until filter has been checked.

(b) Burner primary air

(b) Rotate burner air-shutter barrel to reduce primary air. Flame must not be yellow when adjustment is completed.

(c) Excessive generator draft

(c) Check for obstruction in area of condenser air intake.

CHAPTER 6

Thermoelectric Cooling

This a good point at which to make you aware that there is such a type of refrigeration as thermoelectric cooling, which may also be used as a heat pump.

The following information is "Reprinted by permission of ASHRAE Handbook of Fundamentals, 1972":

"PERFORMANCE OF THERMOELECTRIC COUPLES—A complete circuit consisting of two dissimilar thermoelectric materials is referred to as a couple. A typical couple is shown in Figure 15. The two thermoelectric materials are represented by n and p. The n-type has a negative Seeback coefficient and an excess of electrons. The p-type has a positive coefficient and a deficiency of electrons. The current is shown in the conventional direction; electrons actually flow in the opposite direction. While there are a total of four connections between the thermoelectric materials and the copper straps, there are only two thermoelectric junctions, the upper or cold one, at which heat is absorbed, and the lower or hot one, from which heat is evolved. If the direction of the current reverses, the upper junction would evolve heat and the lower would absorb it."

The following is "Reprinted by permission from ASHRAE 1975 Equipment Handbook:"

For many years the practical application of the thermoelectric effect was almost exclusively restricted to thermocouples for the measurement of temperature, because the metals exhibited a comparatively small Seeback effect. However, the Seeback effect in semiconductors can be considerably greater. The advent of the transistor and other semiconductors devices has stimulated research pertaining to the properties of semiconductors in general, and from this has come materials in which the thermoelectric effects are of sufficient magnitude so that the fabrication of useful devices has become a reality."

ADVANTAGES AND APPLICATIONS OF THERMOELECTRIC COOLING

"The heating and cooling functions of the condenser and evaporator in a mechanical system can be interchanged by reversing the direction of refrigerant flow. This cannot be acheived without considerable difficulty and expense. Since the motor and compressor involve rotary and reciprocating motion, wearing parts and noise may also present a problem. The containing of the refrigerant requires a hermetic system which must be leakproof. A further inherent limitation of this system is that it cannot be readily made in miniature to economically provide a small amount of refrigeration."

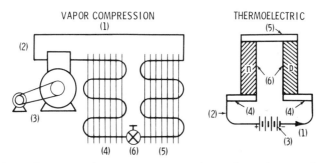

Fig. 1. Comparison of vapor compression cooling to thermoelectric cooling.

"The heating and cooling functions of a thermoelectric system can be easily interchanged by reversing the polarity of the direct current applied to it. Since there are no moving parts, there is nothing to wear out and nothing to generate noise. There is no refrigerant to contain, and the pressure-tight tubing has been replaced by electrical wiring. Since the cooling capacity of a single thermoelectric couple is very small, it is practical to make a very low refrigeration capacity. The thermoelectric system can also be made in large capacities by using many couples."

167

"Capacity control in a thermoelectric system can be achieved by varying the voltage applied to the couples. The current flow is thus changed and the capacity changes."

"Another advantage is that a thermoelectric system will operate under zero gravity, or at many times the force of earth gravity, and will operate in any position. This is important in considering cooling applications for use in space travel. A thermoelectric system also operates at temperature up to 220° F."

Fig. 2. Energy level analogies.

"Along with the advantages there are some disadvantages. One is that, at this time, a thermoelectric system requires more power than a compressor system to produce a given cooling capacity."

"Since a thermoelectric system has no moving parts and nothing to wear out, it might be expected to have an infinite life. Individual couples and some devices have operated without a noticeable change for some years. However, there are some things that, if not controlled, may cause degradation of performance. It is known that copper will diffuse into a thermoelectric material and adversely affect its thermoelectric properties. The solders for joining thermoelectric material to copper straps must be carefully selected so that they do not diffuse into the thermoelectric material and eventually result in a joint failure. These effects can be minimized, and probably eliminated, by nickel plating the

ends of the couple legs prior to soldering. The solders must also be selected with regard to the temperature range of operation of the system.''

COMPRESSOR   THERMOELECTRIC COUPLE

Fig. 3. Thermoelectric materials parameters and compressor characteristics analogies.

''*Thermal expansion and contraction of the materials on the opposite faces of a system will induce stress. If the function of a particular system is to alternately cool and heat, there will be a reversal of these stresses. Materials must be selected to withstand this.''*

''*For many applications, the advantages of thermoelectric cooling will outweigh its chief disadvantage of a relatively low coefficient of performance. For some cooling requirements, a thermoelectric system is the only practical means.''*

''*Some of the applications of thermoelectric cooling are: air-conditioning and refrigeration systems for use on submarines, temperature control of electronic components, cooling of scientific instrument components, and small specialty appliances.''*

''*The U.S. Government has sponsored the development of many thermoelectric systems. Among these are: an 8500-Btu unit for possible use as a refrigerating method for frozen food storage rooms on nuclear submarines, a 9-ton air conditioner for a small submarine, temperature control of crystal oscillators in frequency standards to improve their life and sensitivity, a 27,300 Btu direct transfer air/water system, and a 24,000 Btu direct transfer water/water system that has undergone trials on a U.S Navy destroyer.*

Thermoelectric applications in the scientific instrument field include: immersion cooler for cooling liquids in a beaker, thermocouple reference junction at 32° F, instrument for measuring dew point, and a microscope stage cooler."

"There are also applications in commercial and appliance refrigeration. Some of these are: a drinking water cooler for use on Diesel locomotives, a small bottle-type cooler for offices, and a small ice maker for use in hotel rooms."

"The latter unit was capable of freezing one tray of water (approximately one pound) in 6 hours. Small portable refrigerators with less than one cubic foot internal volume have been marketed by two manufacturers. Both of these could be operated from 120-volt alternating current source, or from a 12-volt direct-current automobile electrical system. A 2 cu ft unit can be used either as a refrigerator or a warming oven has been produced. The cabinet air temperature is 35° F on the cooling cycle and 160° F on warming."

CHAPTER 7

Refrigeration Service Equipment and Tools

This chapter covers much of the equipment and tools needed to service both domestic and commercial refrigeration systems, thus they will be applicable in much of your servicing of refrigeration work.

Domestic hermetically sealed units have been provided with several types of methods of adding refrigerants or drying out refrigerants that have accumulated moisture.

Some require a special adapter for connection to the hermetically sealed compressor, some have a service valve, and some require a special adapter, connecting the suction line to the compressor with a valve incorporated with it. This will be explained in Chapter 8.

It is the intent of this chapter to make you aware of the facilities that are available to you for refrigeration servicing.

Domestic refrigerators have a certain number of ounces of refrigerants that the unit is to be charged with. Practically 100% of commercial refrigerators have service valves on both the discharge and suction sides of the compressors. An exception to this may be self-contained water coolers, which may or may not have service valves.

SERVICE VALVES

Service valves referred to in service operations are named according to their function, as:

1. Compressor suction. $\left.\right\}$ Service shut-off valves
2. Compressor discharge

3. Receiver service.
4. Evaporator liquid-line service.
5. Evaporator suction-line service, etc.

The compressor suction and discharge valves (Fig. 1) are located at the compressor inlet and outlet connection, respectively. These are usually

Fig. 1. Exterior and interior view of a typical compressor suction or discharge valve.

dual valves, commonly called two-way valves, which open or close with the valve stem screwed "all the way out" or "all the way in". In this manner, turning the valve stem will control two outlets from one valve. The side port opening may be used for charging, dehydrating, testing, etc., after the proper connections have been made. The side port is closed by turning the valve stem, which is covered by a brass cap. The port opening is sealed with a small plug which should never be removed until the valve is checked. The main opening of the valve is closed by turning the stem all the way in (to the right).

After making gauge, charging, or dehydrating connections, the valve should be opened to the side port by turning the valve stem one and a half turns. Do not turn the stem all the way because this would close the main opening and stop the flow of refrigerant. Fig. 2 illustrates the various

Fig. 2. Showing service shut-off valve in open and closed position.

positions of the service shut-off valves. The valve stem is provided with a packing gland which may be tightened in case of a leak around the stem. The port plug and valve cap should always be replaced after adjustments are made to prevent any gas from escaping. The receiver service valve is usually located at the receiver on the liquid line to the evaporator, and is generally a single-seated type valve, having one inlet, one outlet, and no opening for a gauge.

The evaporator liquid- and suction-line service valves are used only on evaporators having a float-valve header. Frequently, the valves are equipped with a built-in strainer to remove foreign particles before entering the evaporator, and is located at the evaporator header in the liquid-line connection between the receiver and the evaporator. The evaporator suction-line service valve is also located at the evaporator header in the suction line, and is connected between the compressor and the evaporator.

TO CRACK A VALVE

The term *crack*, as used in servicing refrigeration units, means to open slightly and close again. Thus, to *crack* the liquid-line shut-off valve, for example, open it slightly to allow a tiny spurt of liquid to flow through, and then close it. The same procedure applies to cracking any valve or valve port in the system. This is important when purging the gauge on the charging line of air, and to avoid releasing an excessive amount of refrigerant.

THE COMBINATION GAUGE SET

A test gauge set, or testing manifold (Fig. 3), consists of two tee valves built into one valve body, with the valve stems extending out on each end for wrench or hand-wheel operations. The tee valves are constructed so that the valve works only on the leg which is attached to the tee. The opening or closing of the stem only affects the one opening; the other two openings on the valve will remain open.

Importance of Gauges and Combination Gauge Set

Since most household refrigeration units are not equipped with gauges, a serviceman must insert his own for the various testing and service operations that must be dealt with in solving day-to-day problems. The number of important operations and tests that may be performed by the use of a test gauge set are numerous, the most important being:

1. Observation of operating pressure.
2. Charge refrigerant through compressor, etc.
3. Purge receiver.
4. Charge liquid into high side.
5. Build up pressure in low side for control setting or to test for leaks.
6. Charge oil through the compressor, etc.

Courtesy Mueller Brass Co.

Fig. 3. Typical combination gauge set.

Types of Gauges Used

The gauges used in refrigeration servicing are:

1. The low-pressure (compound) gauge.
2. The high-pressure gauge.

The standard type of low-pressure gauge is often called a compound gauge because its construction permits a reading of both pressure in pounds per square inch and vacuum in inches of mercury as shown in Fig. 4. A standard compound gauge dial is graduated to record a pressure range of from 30 inches of vacuum to 60 pounds per square inch. The high-pressure gauge is equipped with a dial for measuring pressures of from 0 to 300 pounds per square inch. Gauges used in refrigeration service should read approximately double the actual working pressure.

SEALED SYSTEMS

The first step when charging a sealed system with refrigerants is to have the right charging and evacuating equipment. This will enable the serviceman to charge the unit with the exact amount of refrigerant required. Although there are several charging methods used the sight-glass charging cylinder is recommended. This method of refrigerant charge will be accurate regardless of ambient temperature.

Fig. 4. Typical compound pressure and plain pressure gauge.

Evacuating and Recharging

Generally, every refrigerator unit is properly charged with refrigerant upon delivery, the main service due to refrigerant troubles usually consists in evacuation and recharging. A step-by-step procedure after the charging equipment has been installed, as shown in Fig. 5, will be as follows:

1. Discharge the system and adapt the proper replacement components.
2. Evacuate the system thoroughly.
3. Charge accurately through the low side with the proper refrigerant.
4. Leak test carefully.

Figs. 6 to 10 are schematic illustrations of typical external vacuum-pump and charging-cylinder connections through a gauge manifold and the compressor process tube.

Fig. 5. Showing typical charging and evacuation arrangement when installed.

Evacuation of Sealed System

It is a good rule, whenever a sealed system is opened and the refrigerant charge removed, to evacuate the system and install a new service drier before recharging the system. To evacuate the refrigerator, proceed as follows:

1. Use a good external vacuum pump and change the oil often for efficient operation.
2. Install a service valve on the low-side process tube of the compressor.
3. Connect the vacuum pump hose to the service valve. Leave the valve in a closed position.
4. Start the vacuum pump, open its discharge valve, and slowly open the service valve on the system compressor.
5. If an extremely efficient vacuum pump is used, crack the suction valve on the pump for the first minute, then slowly open it. This procedure will prevent the oil in the system from foaming and being sucked into the vacuum pump in large quantities.
6. Next, pull a vacuum for about 20 minutes, which should give a reading of 500 microns or 29.6 inches of mercury on the compound gauge. At the end of the 20-minute vacuum time, the vacuum pump should be valved off and the micron gauge left in

177

Fig. 6. Showing arrangement of charging equipment when installed ready for use with all valves in closed position.

the system. *It is important to observe the gauge.* If the gauge reading rises, there could be a leak in the system.

7. Close the service valve on the system compressor, then stop the vacuum pump.

8. Connect a sight-glass charging cylinder hose to the service valve (Fig. 15) and purge the charging hose. Induce a charge of refrigerant into the system until the low-pressure side reads 30 to 40 psig, then leak test the low side. After the low-side leak tests are completed, run the system compressor for a few minutes.

9. Leak test the high side of the refrigerant system.

10. After completion of the leak tests, purge this temporary refrigerant

Fig. 7. To evacuate lines, open gauge set and gauge manifold. Close vacuum pump valve and check to see that vacuum holds.

charge out of the low side. This refrigerant charge will assist in removing moisture within the system.

11. Repeat the vacuum pump procedure for another 30 minutes, which should give a reading of 500 microns or 29.6 inches of mercury. If the discharge tube of the vacuum pump is in a container of refrigeration oil, there should be no bubbles in the oil after completion of the foregoing procedure, thus indicating that the system has been evacuated properly.

You have noticed that the text material states 29.6 inches of mercury in many instances. Remember that this is at sea level. In the early portion of this book, the inches of mercury to be expected

Fig. 8. Open charging tool, vacuum pump valves and evacuate assembly for about 20 minutes. Close vacuum pump valve and check to see that vacuum holds for about 5 minutes.

at other altitudes was give. This value is approximately 1 inch of mercury less, per 1000 feet of altitude. Thus at 5000 feet, you would observe 24.6 inches of mercury.

Charging of Sealed System

The equipment for charging refrigerant is the same as that used for evacuation. The charging procedure is normally as follows:

1. The first step in charging a system is to have the proper equipment so that the refrigerant can be accurately determined within ± ¼ ounce. Although there are several methods of charging, the use of

Fig. 9. To charge unit close vacuum pump line and open charging cylinder valve; allow refrigerant to be drawn into compressor.

a sight-glass charging cylinder is preferred, since this method is accurate regardless of ambient temperature.

2. Always charge through the low side of the system, either through the process tube or the suction line. Charge the system with the compressor off, since most of the refrigerant will enter in liquid state. Allow about 5 minutes after the charge has entered the system before starting the compressor. The efficiency of charging can be increased by elevating the charging cylinder; however, refrigerant should always be introduced slowly.

3. Connect the charging hose of the charger to the service valve on the compressor.

181

Fig. 10. After the compressor has been fully charged, close charging valve on compressor and remove charging tool. Check charging screw for tightness.

4. Open the charger valve and purge the hose. As soon as the refrigerant has stabilized in the charging cylinder, check the pressure reading of the cylinder and rotate the sleeve to correct the setting for pressure and the type of refrigerant being used. Control the flow of refrigerant with the service valve, not the charger valve.

5. When charging the refrigeration system, some bubbles may appear in the charging cylinder. These can be eliminated by closing the service valve and tipping the charging cylinder upside-down momentarily. Then continue with charging until the correct charge has entered the system.

6. If, at any time, it is required to raise the internal pressure of a charging cylinder, use a bucket of warm water (not over 125° F) and place the cylinder in the bucket. *Caution: Never place a flame on any refrigerant cylinder.* This can cause hydrostatic pressure to build up to dangerous levels and, in extreme cases, cause the cylinder to rupture like a bomb. Extreme care should always be taken when handling refrigerant cylinders.

7. When it has been ascertained that the system has the correct refrigerant charge, use the pinch-off tool and make the final joint. If possible, leave enough round tubing between the compressor and the pinch-off crimp in order that the process tube may be used at a later date.

EXPANSION VALVE UNITS

Because numerous serviceable refrigerators are equipped with expansion valves for refrigerant control, the more common service operations on these units are described in the following paragraphs:

Use of Combination Gauge Set to Add Oil

With reference to Fig. 11, oil may be added to the compressor as follows:

Connect a piece of ¼'' copper tubing between the oil supply and valve *E*. Pour the refrigeration oil into a clean dry bottle. Put the compound gauge on the suction line valve and close the liquid line at the receiver. When a sufficient amount of oil has been added, close the valve at point *A* and open valve *D*. The compressor may now be put into normal operation. *Caution: Make sure that the end of the oil-supply tubing is kept submerged below the oil level in the container, so that air will not be drawn into the system during the operation.*

To Charge Refrigerant Through Low Side

With reference to Fig. 12, refrigerant may be added as follows:

Connect the suction-line valve port and the refrigerant tank with a piece of ¼'' copper tubing with gauges connected as shown. The cylinder should be placed on a suitable scale in an upright position, with the scale reading noted. The ¼'' tube between the manifold and the charging drum must be long enough so as not to affect the scale reading. Purge the air out of the charging line by turning on the gas at the tank valve and cracking the valve unit at valve B. When the line has been purged, open the suction-line valve and start the compressor. A back pressure of about 5 pounds is maintained in charging with sulfur dioxide (SO₂) and about 10 pounds for *Freon* (*F*-12).

If the cylinder is cold, apply heat to the gas cylinder by means of a hot cloth or a pail of hot water; this will assist in the transfer of refrigerant to the unit. When the required amount of refrigerant has been added, close the valve on the gas drum and allow the pressure to fall to zero on the gauge, then close the valve at the suction line. Disconnect the lines and replace the plugs. Put the unit into normal operation.

Fig. 11. Schematic diagram illustrating the connections for adding oil to compressor.

Fig. 12. Schematic diagram illustrating connections for adding refrigerant through low side.

To Charge Refrigerant Through High Side

In some cases where a large amount of refrigerant is to be added, it is advantageous and time saving to charge the refrigerant as a liquid into the high side of the system instead of pumping it into the low side. In such cases, it is recommended that only a known required quantity of refrigerant be contained in the charging cylinder. To accomplish a charge, a reference to Fig. 21 will be of assistance. Proceed as follows:

Connect the refrigerant cylinder by means of ¼" copper tubing to the manifold, as shown. The compressor should be stopped and the condenser allowed to cool to room temperature. The charging line should be purged in the manner previously described. The refrigerant cylinder should be heated, inverted, and securely fastened, after which the cylinder valve is opened. While the cylinder is being

185

discharged, it may be necessary to supply additional heat to maintain the flow of refrigerant to the unit. When the required amount of gas has been added, close both the valve on the gas container and valve *A*. Valve *B* should be opened to by-pass the pressure from the charging line to the low side. Finally, after the charging line records zero, valve *B* should be closed, after which the manifold connection to the service drum is capped and the unit put into operation.

Fig. 13. Schematic diagram illustrating connections for adding refrigerant through high side.

To Evacuate Air From Entire System

During normal operations, both the back and head pressures are above atmospheric pressure, and the only way in which air can enter the system is during service operations or due to a leak in the system. When the high side of the system has a leak, it will expel gas until atmospheric pressure is reached and the back pressure will drop below atmospheric pressure.

When the unit is operated below atmospheric pressure, air will be drawn into the system. Air in the system will cause excessive high-side pressure and will nearly always stop the unit by overloading the motor, which will trip the overload cutout.

When evacuating air from a system, it is important to first ascertain that all leaks have been stopped, otherwise air will reenter the line. To purge air from the system, refer to Fig. 14 (which shows a diagram of the connections) and proceed as follows:

Connect the suction side of the compressor to the compound gauge. The compressor discharge service valve should be connected to a glass jar filled with water, by means of ¼'' copper tubing. To prevent any loss of oil which may escape during the pumping operation, the oil should be carefully measured and returned to the compressor after the purging operation. Check the compressor discharge-valve port to make sure it is closed, after which the compressor suction-valve ports should be opened to both the gauge set and the evaporator. The receiver liquid line should be checked to make sure it is opened, and the compressor should be put under intermittent operation until a maximum amount of vacuum is being obtained, as registered by the compound gauge.

If, after the unit has been stopped for a period of four or five minutes without any change in the gauge reading, the unit is free from air, and the glass jar end of the copper tubing should be connected to the middle leg of the manifold. At this time, a charged refrigerant gas drum should be connected to the manifold at valve *B*, and the line connection from the gas drum purged. When the gas-drum valve and manifold valve are opened, the gas charge enters the system and the pressure is built up to a positive value of about 10 pounds. When the required amount of refrigerant has been added, close the valve on the gas drum and add any oil which may have escaped during the purging operation. After this operation, a final check for leaks should be made and the unit put back into normal operation.

Purging Air From Condenser

Air in the condenser will cause excessive high-side pressure resulting in long operating periods, with an accompanying waste of power. The unit should be shut down during this test and during the purging operation. Allow the condenser to cool down to room temperature. With reference to the diagram, proceed as follows:

Fig. 14. Schematic diagram showing connections for purging air from entire system.

Attach a purging line (¼″ copper tubing) to the service connection valve (*E*) on the manifold, as shown in Fig. 15. If the refrigerant is sulfur dioxide (SO_2), place the end of the tubing into the bottom of a container in which 1½ pounds of concentrated lye (NaOH) has been dissolved in a gallon of water. Then crack valve *B* (loosen and immediately tighten) and purge slowly for several minutes. Observe the pressure drop registered on the pressure gauge, and when it has returned to normal, the purging should be discontinued, valve *B* should be tightened, the purging line removed, and the unit returned to normal operation.

Evacuating the Entire Refrigerant Charge

When it becomes necessary to remove the entire refrigerant charge from the unit (as in the case when an exchange of the compressor or any

Fig. 15. Schematic diagram showing the connections for purging air from the condenser.

other component of the condensing unit is necessary), it is often desirable to salvage the refrigerant by evacuating (pumping) it into an empty service drum. With reference to Fig. 16, the procedure is as follows:

To pump the gas into an empty cylinder or drum, put the cylinder in a bucket of cold water (preferably ice water), as shown. Connect the empty cylinder to the middle leg of the manifold by means of ¼'' copper tubing. The compressor discharge service valve should be closed, and valve B should be opened; the valve on the cylinder should also be open. After the compressor suction-line service valve and the receiver liquid-line valve are opened, the compressor is put into operation. Keep the cylinder as cold as possible until the operation is

189

complete. The compound gauge attached to the suction shut-off valve will show when the charge has been completely removed. If, during this pumping operation, the pressure gauge indicates an abnormally high pressure, the unit should be stopped, allowing the compressor to cool down. An indication of a complete evacuation of the system will be obtained by closing the cylinder valve. If an observation of the gauge indicates a constant high vacuum for any length of time, the evacuation is complete and the service attachments, including the gas cylinder, may be removed from the unit.

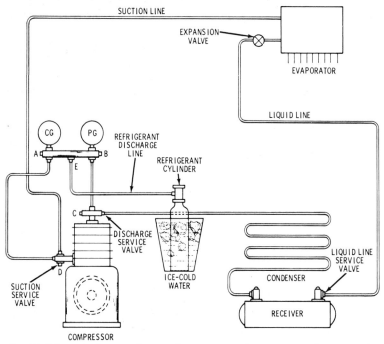

Fig. 16. Schematic diagram showing the connections for evacuating the entire refrigerant charge.

LEAK TESTING

The old soap-bubble and oil method of detecting leaks in refrigerators using modern refrigerants is, at best, only a makeshift one to be used only in the absence of proper detectors. There are several types of leak detectors available, such as the *halide torches,* or an *electronic* type which is presently manufactured by the *General Electric Company* (Fig. 17). The halide type, however, is probably the most widely employed and is sufficiently sensitive for household refrigeration systems.

Leaks in a refrigeration system will usually result in an undercharge of refrigerant. To add refrigerant would only be a temporary solution without first locating and repairing the leak, since adding refrigerant will not permanently correct the difficulty. The leak must be located and repaired, if possible, after which the entire system must be evacuated and recharged with the proper amount of refrigerant. Whenever a new charge of refrigerant is added, it is necessary to install a new drier. Any leak, regardless of its size, must be located before a determination can be made of the operative status of the system components. Do not replace a component because the system is short of gas unless a nonrepairable leak is found.

Fig. 17. General Electric type H-5 electronic portable type
leak detector unit with tiny hand-held proportioning probe.

If the analysis indicates a leak, find it before opening the system. It is better to locate the leak before discharging than if the surrounding air is contaminated with refrigerant from a newly opened system. The presence of oil around a tubing joint usually indicates a leak, but do not let this be the determining factor. Always check the area with a leak detector to make sure.

191

The Soap-Bubble Method

Soap bubbles can be used to detect small leaks in the following manner: Brush liquid detergent over the suspected area and watch for the formation of bubbles as the gas escapes. Sometimes, if the leak is of a slight intensity, several minutes must elapse for a bubble to appear. *Caution: Use the bubble method only when it is certain that the system has positive pressure.* Using it where a vacuum is present could pull moisture or soap bubbles into the system. A joint which is suspected of leaking can be enclosed in an envelope of cellophane film. Tightly tape both ends and any openings to make it gas-tight. After about an hour, pierce one end of the film for the probe and pierce the other end for air to enter. If a response is obtained, the joint should be repaired.

The Halide Leak Detector

When testing with the halide torch, make sure the room is free from refrigerant vapors. Watch the flame for the slightest change in color. A very faint *green* color indicates a small leak. The flame will be unmistakably *green* to *purple* when large leaks are encountered. To simplify the leak detection, keep the system pressurized to a minimum of 75 psi. This is easily accomplished for the high side by merely running the compressor. To pressurize the low side, allow the entire system to warm up to room temperature.

The halide leak detector method has been used for many years by air-conditioning and refrigeration servicemen, and consists of a combustible gas supply with a small burner on top. A tube from the bottom of this burner forms the probe. When this probe is moved near a large leak, the refrigerant going into the tube combines with the flame to cause perceptible change in the flame color. Disadvantages of using the halide torch include the following:

1. It is slow and relatively insensitive.
2. In a brightly lighted area, it is almost impossible to locate leaks smaller than about six ounces per year.
3. If the area is darkened, a smaller leak can be located; however, the change in the color of the flame is difficult to interpret because other gases and dust cause the flame to change color.

The Electronic Type Leak Detector

Although leak testing with a halide torch is considered satisfactory in most instances, a more reliable test is made by means of the electronic type tester illustrated in Fig. 17. This easy to use instrument reduces the guesswork from leak testing because it is more sensitive, faster responding, and is capable of detecting a leak even though the surrounding air may be contaminated. The electronic detector contains an internal pump which draws air and leaking halogen tracer gas (halogen gases are chlorine, bromine, fluorine, and iodine, or any of their compounds) through a probe and hose as the probe is passed over leaking joints, gaskets, welds, etc. The air-gas mixture is drawn between two electrodes in the control unit—the ion-emitter and collector.

The presence of halogen gas in the interelectrode space causes an increase in interelectrode conductance that is proportional to the amount of halogen gas present, which results in greater positive-ion current. In the electronic detector, this emission current is electronically amplified and the amplifier output causes the leak indicating lamp to flash. A knob on the control unit permits the user to adjust the sensitivity in order to detect leaks in the range from 0.5 ounce per year to approximately 10 ounces per year.

The leak detector is specifically designed for use in atmospheres containing a relatively high concentration of halogen gas (up to 1000 ppm). Operation in such an atmosphere is made possible by the proportioning valve on the tip of the sending probe. This proportioning valve permits the user to regulate the ratio of contaminated atmospheric air and pure air (drawn in through an activated charcoal filter located within the control unit) that passes over the sensitive element. In this fashion, the concentration of halogen in the atmosphere is reduced to a tolerable level without affecting the sensitivity of the leak detector.

The leak detector contains an automatic balance circuit so that a constant or slowly changing contamination in the atmosphere will not cause the lamp to flash. Refrigerant gas will cause the lamp to flash only when there is an abrupt increase in the amount of gas going into the tip of the probe, such as when the probe tip passes near a leak.

To make a leak test by means of the electronic detector, proceed as follows:

1. Plug the test detector into a 120-volt ac outlet.
2. Turn the sensitivity knob to the right and allow a 1-minute warm up.
3. Check the operation by turning the sensitivity knob quickly from one position to another. This should light the probe lamp.
4. Probe for leaks, starting with maximum sensitivity.

193

5. If the probe lamp lights twice for each leak, reduce the sensitivity.
6. Recheck the suspected leaks for confirmation.

After replacing a component in the refrigerating system, always leak test all joints before recharging. The extra time it takes is negligible compared to the loss of a charge due to a faulty connection. Be sure to clean the excess soldering flux (if used) from the new joints before leak testing, since it could seal off pinhole leaks that would show up later.

Service instruction charts have been worked out which will be of help in locating and repairing faults likely to be encountered in day-to-day service work.

SERVICE TOOLS

Some of the tools illustrated will be used mostly with commercial refrigeration work, while some application will also be found in home refrigeration servicing.

A careless repair job is dangerous and has no place in refrigeration service work. Properly chosen tools will not only save time but will, in addition, insure a neat, dependable, and satisfactory job which will result in mutual benefits to the serviceman and customer alike.

There are many types of tools on the market. Some of them are so poorly constructed that they are expensive at any price. A good tool warrants a fair price, because it is properly designed and will stand up under service. A list of special tools is given in this chapter.

TEST GAUGE SET

One of the handiest aids in servicing small machines is the test gauge. Most small units are not equipped with gauges; the serviceman must apply his own so that the pressures existing in the high and low sides can be determined. A test gauge set consists essentially of two *tee* valves and a special *tee* as shown in Fig. 18. The tee valves are constructed so that the valve stem works only on the leg attached to the tee. Therefore, opening or closing the stem affects only this one opening; the other two openings of the valve remain open at all times. This fact is plainly shown in the sectional view of the valve in Fig. 19. This unit can be used to obtain pressure readings on both high and low sides when adding oil, adding or recharging the refrigerant, purging off an excess of refrigerant or air, or when pumping the refrigerant from one side to the other.

COMBINATION TESTING OUTFIT

Fig. 18. Combination gauge set. High-pressure side gauges are usually scaled for pressures up to 300 pounds per square inch. The low-pressure gauge is of the compound type and is used to measure inches of vacuum and pressure in pounds per square inch.

Fig. 19. A sectional view of a refrigerator tee valve.

INSPECTION MIRRORS

Magnifying mirrors are used by many servicemen for observing leaks in refrigerating machines where soap and water or oil is used as a leak detector. The long handles permit these mirrors to be used in difficult or cramped quarters. Some compressor assemblies are designed so that these mirrors offer the only means of seeing certain bolts, inserting other pins, or aligning parts.

PIPE CUTTERS

Pipe can be cut in two ways—by hack saw or by pipe cutter. The former method is often slow and laborious. Pipe cut by a hack saw is not beaded or provided with a thickened lip, as is the case when pipe cutters are used. If a three-wheel cutter is used, as shown in Fig. 20, the outer edge of the pipe must be filed down before a die can successfully be applied to the pipe.

When a single cutter wheel is used in conjunction with two rollers, the outside of the pipe stays close to its original diameter, but the inside of the pipe is given a lip or burr that must be removed with a pipe reamer. If this burr is allowed to remain, it will decrease the free internal diameter and will offer a resistance to the passage of the fluid that is to flow through the pipe. Such burrs collect dirt and often result in subsequent binding or clogging at such obstructions.

Fig. 20. An adjustable three-wheel pipe cutter for use on ⅛'' to ¼'' pipe.

Pipes severed with a hack saw do not require reaming, and the die can be applied without the necessity of filing the outer edge. In effecting a cut with a hack saw, it is important that a square cut be made. If an angular cut is made, the die will not cut straight threads. When such a threaded end is screwed into a fitting, the pipe will point out at an angle, and it will be difficult to assemble a run of pipe with such threads, especially close or short lengths. Pipe cutters offer the quickest severance and, if properly applied, result in a square cut so that the die will follow the pipe without trouble.

TUBE CUTTERS

The larger installations make use of steel or iron pipe for connections, but the smaller machines utilize copper tubing. Copper tubing can be easily installed, because it can be bent around obstructions and eliminates many elbows. Since tubing is made in fairly long lengths, most small installations can be completed by the use of a single length of tubing. For instance, the evaporator can be connected to the receiver by means of a

¼ " tube in one single piece, even when the two units are separated by a distance of 100 feet. Therefore, sharp bends are eliminated, and all chances of clogging at constrictions or bends are eliminated, while the friction factor is greatly reduced.

Copper tubing can be cut by means of a saw and flaring block, as shown in Fig. 21. The tube is held in the flaring block for the purpose of securing a square end. It is important to secure a square end, otherwise the operator will be obliged to make careful use of a file. A much easier and quicker method is to use a tubing cutter, as shown in Fig. 22. This tool is a small version of the pipe cutter shown previously.

Fig. 21. Method of cutting tubing with hacksaw in a flaring block.

BENDING TOOLS

Where accurate bends of a specified radius must be made, a bending tool should be used. This bending tool, shown in Fig. 23, consists of a stationary arm on which is located the radius guide. The radius guide is usually provided with a pin, so that the bending arm can be attached and used to make the bend. The bending shoe is also grooved and is furnished with a shoe block or a grooved wheel. The tool makes a perfect bend without crimping or flattening the tube.

197

Fig. 22. Illustrating the use of a tube cutter to cut tubing.

PINCH-OFF TOOLS

Most of the larger apparatus is equipped with sufficient valves to enable any portion or section to be closed off and worked on without loss of the refrigerant charge. Some of the smaller and simpler systems have only a few valves. This is an economy measure so that the cost of the apparatus can be kept to a minimum. While the elimination of valves makes a simpler, lower priced piece of equipment, it makes servicing somewhat harder for the serviceman. Where there are only a few or no valves in a system, the pinch-off tool offers a means of closing the copper line so that the defective or worn part can be removed and replaced.

A simple pinch-off tool is shown in Fig. 24. Notice the jaws for the purpose of pinching or squeezing a tube shut. These jaws do not come together when the two blocks are flush; if they did, the tubing would be severed. The pinching jaws are so designed that the tubing is pinched shut but not mashed. The round opening is used to round up the pinched section when the repair or replacement has been completed.

VALVE KEYS

It will be found that some refrigeration apparatus is equipped with valves that have no handles. The stems are provided with a square end, so that a handle can be applied or a key fitted to the valve stem, as shown in Fig. 25. When opening or closing such a valve, a *valve key* or *tee wrench* is sometimes used. Common sizes have 3/16″, 7/32″, ¼″, and 5/16″

Fig. 23. Bending tubing by means of a specially designed bending tool.

Fig. 24. Typical pinch-off tool. Block is cut away to show working principles.

199

openings. These sizes can be obtained in the *tee* form or with a square end so the tool can be used with a ratchet wrench.

Fig. 25. Various valve keys.

PACKING KEYS

Some valves are packed with a resilient material that serves as a seal to prevent the refrigerant from escaping around the stem. A packing nut can be tightened by means of a wrench. When tightening external packing nuts, use a 12-point wrench if one is available, or make use of the proper open-end wrench.

The packing nut is generally the internal type and is provided with a slot across its face for tightening. Unfortunately, the stem of the valve projects through this nut and makes it necessary to use a special gland or packing-nut key, such as shown in Fig. 26. This key is hollow in the center, permitting the valve stem to extend through the tool, and is provided with two teeth that engage the slot in the packing nut.

Fig. 26. Typical packing keys.

FLARING TOOLS

A typical flaring tool is shown in Fig. 27. It consists of two bars held together by a wing nut and bolt; the bars are provided with holes for the various sizes of tubing. A yoke containing the forming die is slipped on the bars and the handle is rotated to produce a flare. This tool has been very widely used due to its simplicity, ease of performing the flaring operation, and because the flares produced by it are uniformly excellent in producing a tight seal.

Before applying pressure with the forming tool, place a drop of oil on the end of the tube. Be sure to use only refrigeration oil. Any other type of oil may enter the tube and give trouble by reacting with the refrigerant at a later date. The proper use of the flaring tool is illustrated in Fig. 28.

Fig. 27. Flaring tool for use on various tube diameters.

GASKET TACKING MACHINE

The serviceman should notice the condition of the packing around ducts and cabinet doors. If it is worn or collapsed so that air is permitted to pass, the packing strips should be renewed or replaced. One of the best labor saving devices is the use of a tacking or stapling machine when a wooden door or strip is provided for the attachment of the gasket material.

MAKING A HAND BEND

Occasionally a serviceman will find it necessary to remove a battered, leaking, plugged, or defective piece of tubing and replace it with a new

Fig. 28. Proper use of flaring tool: (A) Shows tubing in place; (B) Teeth in the die, holding tubing securely; (C) Indicates compressor making a perfect flare; (D) Black section shows flare formed at top of tubing.

piece or section. If no mechanical bending tool or spring is available, the bend must be made entirely by hand. There are five different procedures that can be used in bending copper tubing. They are as follows:

Procedure No. 1

1. Make sure the piece of tubing is straight.
2. Mark the tubing where the bend is to be made.
3. If possible, use the old piece of tubing as a guide, or use a piece of wire that has been bent to the desired form and cut to the proper length.
4. Grip the tubing with both hands, one on each side of the mark; place the thumbs directly underneath the point at which the tubing is to be bent.
5. Draw the tubing with both hands, bending it over the thumbs.
6. Bend slowly and watch for indications of buckling or kinking.
7. Keep moving thumbs and bend slowly so that the curve takes shape evenly.
8. Never attempt to make a very close bend with the hands.

Fig. 29 shows three types of bends used in refrigeration.

Fig. 29. Three types of bends as required in refrigeration service.

Precautions

If the bend to be replaced is a close one, use iron or any stiff wire to measure off a wide swing for the replacement; a close bend cannot be made by hand and should not be attempted.

Inside and Outside Springs

An inside spring can be used for making a bend in the tubing. The inside spring can only be used where the bend is to be or near the end of the tubing. If a bend is to be made in the middle or at any distance from the end, an outside spring or bending tool must be employed.

A No. 4 screen-door spring with the paint removed can be used for an inside bending tool. The spring is prone to gather dust, dirt, and grime between its coils, and this can result in dirt being introduced into a system. Make sure the spring is clean before using it. One end of the spring should be enlarged or a ring of heavy wire attached to one end, so it can be removed from the tubing. An inside spring will give a closer bend than an outside spring. An inside spring can only be used successfully on ½″ or larger sizes of tubing because an inside spring is much harder to remove when the diameter is less than ½″ in size. A tool kit can be purchased that will include not only the proper tube cutter and reamer, but will also include the tube bending springs.

Procedure No. 2

1. Straighten the piece of tubing. Ream the end.
2. Mark the tubing at the point where the bend is to be made.
3. Insert a clean, oiled spring, making sure that the spring extends well beyond the bend point.
4. Grasp the tubing with both hands, and bend over the thumbs. Move the thumbs as the bend progresses, so that a bend of the desired degree is obtained.
5. Bend a little beyond the degree desired, and then bring it back to the desired radius. This will loosen the spring so it can be removed.

Precautions

When a very close bend is required at the end of a piece of tubing, and no mechanical bending tool is available, use both inside and outside springs at the same time. Be sure that the spring is clean. Do not drop it on the floor, or place the spring where it can gather dirt between its convolutions. This dirt will drop out in the copper tubing and be introduced into the refrigerant system, where it will cause operating defects. Do not attempt too close a bend. Move the thumbs after each slight bend and examine the bend.

When a bend collapses or flattens, and the spring cannot be removed easily, tap the tube round with a hammer. This will release the spring. If done carefully, the tube bend will not be badly battered and can be placed in use if it is not subjected to machine vibration. Use a flat piece of iron or wood between the hammer and tube so it will not be dented or battered. Oil the spring with refrigeration oil. Do not use any other type of oil. If no refrigeration oil is on hand, use the spring dry. Oil makes the spring easier to remove and the bend somewhat less strenuous to form.

Outside bending springs are made in all sizes for use on standard tubing. Servicemen have more or less adopted the outside tube spring. A minature pipe bender is used on production jobs where a great number of the same sized bends are required, but this tool is not used to any great extent in the field. Outside springs should be used to acquire practice in the forming of bends, as shown in Fig. 30.

Procedure No. 3

1. Mark the tubing where the bend is desired.
2. Make another mark on the tubing one half the length of the spring.

3.   Slide the outside spring up to the second mark. The first mark (where the bend is desired) will be covered, but it is known to be at the center of the bending spring.

Fig. 30. Typical outside spring and method of application to tubing.

Precautions

Determine the size of the bend required, and make the bend in one smooth motion. If bent too much or too little at a time, the repeated working will harden the tubing and make bending difficult. Such a bend can crystallize and break the tubing if subjected to vibration in use. Some springs have plain ends, while others are provided with a flare or funnel-shaped end so that the spring can be easily slipped on and off the tube. When pulling the spring off the tube, grasp the spring and pull on the flared end. Springs will slip on and off much easier if they are slightly oiled. Wipe each tube after bending to remove any oil squeezed from between the convolutions of the spring. Use the springs with care or they will become sprung and kinked. A spring should last for years with proper care.

MAKING A RELIEVED BEND

This procedure is for making a bend that will hug the corners of a wall when rounding a corner, as shown in Fig. 31. Notice that any ordinary 90°

205

Fig. 31. Use of relief bend when close application to a wall is desirable.

bend will not touch the walls and is likely to be dislodged by any moving object coming into contact with it. The relieved bend will hug the wall tightly and permit proper attachment. Use this type of bend for a neat, satisfactory jog.

Procedure No. 4

1. Make a 90° bend as outlined in the previous procedure.
2. Move the spring to one leg and make a relieved bend.
3. Do the same with the other leg.
4. With both legs relieved, try the bend on a corner and determine whether it hugs both walls properly.

USE OF RESEATING TOOLS

The reseating tool is used for resurfacing the ends of flared fittings which, due to excessive use or abuse, would otherwise have to be replaced with new fittings. Some fittings are specially designed. Through continued use, or more often through abuse, the flared end or *nose* becomes rounded, scored, or grooved to the extend that it cannot be made gas tight. Through the use of a reseating tool, these special (as well as ordinary) fittings can be reconditioned and used again with the assurance that they are as good as new, so far as the flare nose is concerned.

There are several types of reseating tools on the market. One type is equipped with a cutter guide, removable bushings, cutter feed, and automatic tension, and is an excellent tool for use in the shop where considerable resurfacing is done, such as in the case of general overhauling work.

Procedure No. 5

1. Insert the damaged fitting in an adapter of proper size. Screw the fitting into the adapter from the rear and allow it to run in until the fitting seats on the shoulder. Use a 12-point wrench where a hex shoulder is available. Do not damage the threads on other parts of the fittings.

2. Place a drop or two of refrigeration oil on the reamer body and insert the reamer in the proper adapter.

3. With a firm, light pressure, rotate the reamer clockwise so that the end of the fitting is smooth.

4. If the fitting nose is badly scored or ridged, it may take several minutes to cut a clean, smooth face.

5. Remove the reamer and inspect the face at regular intervals. When a perfect seat is obtained, do not cut away any metal.

6. When the face is smooth, remove the reamer and take the fitting out of the adapter.

7. If the end is clean and satisfactory, proceed with the operation; if not, make further use of the reseating tool.

8. With the flare nose clean and smooth, run a strip of cloth through the fitting to remove metal shavings. Do not blow through the fitting to clear it of shavings. Moisture in the breath can condense in the fitting.

Precautions

When using a reamer, a light, steady pressure must be applied. If too heavy a cut is attempted, or if the tool is applied too lightly, the reamer may skip or chatter and produce a wavy surface almost impossible to remove by subsequent treatment with hand tools. A lathe can be employed to obtain a true, smooth surface, but this course is only open to the man in the repair shop and not to the serviceman. Bear in mind that an *SAE* flared fitting must have a projecting nose. If the operation continues to cut away an excessive amount of metal, the connection will not be air tight and a new flare must be made.

COPPER TUBING

It is necessary, especially in smaller systems, to use dehydrated tubing that is factory sealed. This prevents moisture entering the system at the time of installation. Never permit unused tubing to remain unsealed in stock. A safe procedure is to seal the ends immediately after cutting, or flatten the tube end while the stock is not in use.

A table showing the sizes and dimensions of various copper tubing used in the refrigeration industry is shown below:

Table 1. Copper Tubing Sizes and Dimensions

Sizes (O.D.)	Stubs (Gauge)	Wall Thickness (in.)	Weight (lbs. per 1,000 ft.)	Approx. No. ft. per lb.
1/8	20	0.035	38	25.4
3/16	20	0.035	65	15.4
1/4	20	0.035	92	11.0
5/16	20	0.035	119	8.5
3/8	20	0.035	145	7.0
7/16	20	0.035	171	6.0
1/2	20	0.035	198	5.0
5/8	20	0.035	251	4.0
3/4	19	0.042	362	2.8

The most popular type of fittings used to make the connections on copper tubing installations is the flared type. This is due to the ease of assembly, maintenance of tight joints, and the fact that such a fitting can be taken apart and used several times without requiring replacement. Valves are made in a variety of forms. For the small unit, packed and bellows-sealed valves are employed; for larger units the packed variety is used almost exclusively because of the larger sizes employed.

UNION SEAT

PACKING NUT

STEM

BODY

PLUG   PACKING GLAND   PACKING

Fig. 32. A forged brass valve showing assembly and parts.

A valve formed of forged brass is shown in Fig. 32, and serves to illustrate the manner of construction employed in small valve design. A comparison of the two types of valves is illustrated by examining the two angle valves shown in Fig. 33. One employs the usual packing and the other makes use of the corrugated sheet metal seal or syphon bellows. Both valves are back-seating, so that when fully opened, the valve itself is sealed against any possible leakage. New safety-code specifications, which are being adopted in almost every city, require that fittings, fastenings, valves, and safety devices conform to certain standards. One of the requirements in flare-tube fittings is that intermediate bushings are used.

209

BELLOWS   PACKING

Fig. 33. Sectional views of two types of valves.

Two types of safety devices are illustrated in Fig. 34; one is rupturable, and the other depends on heat to effect a pressure release. The rupture devices usually utilize a thin silver disc of a certain diameter and thickness which, in the event of high pressure being generated due to any cause in the system, will rupture and release the refrigerant, leaving the unit without any chance of its being destroyed or strained in any way. The fusible type of safety device is usually made in the form of a plug intended to be screwed into a boss or fitting, and is operative only by heat. Regulations require the use of such plugs on refrigerant storage or charging drums, receivers, and isolated portions of apparatus that are filled with refrigerant. In the event of fire, the gas will be released without danger of explosion.

One type of connector developed and adopted by the trade is the capillary-soldered type. It is lighter, simpler, stronger, and more economical than the threaded types. These special fittings are made in the shapes and sizes familiar to the trade, and differ only in the method of joining the tubing. The tubing is inserted into the fitting until it rests against a shoulder. Then wire solder is applied through a small feed hole in the fitting under heat from a blowtorch. The solder is drawn into the fitting and evenly distributed by capillary attraction and results in a firm joint. The method of effecting a joint is illustrated in Fig. 35, which shows the wire solder being fed into the hole in the side of the fitting (a *tee* in this instance) while a blowtorch heats the fitting. A sectional view of the completed joint indicates the capillary channel holes and shoulder; it also shows the larger surface between the fitting and tube joined by the solder.

Fig. 34. Rupturable safety devices to relieve excessive pressure. The plugs melt at excessive temperatures.

Fig. 35. Soldering a fitting in place.

SILVER BRAZING

The use of high melting point alloys has increased considerably in recent years, particularly in supermarkets, locker plants, air conditioning, or wherever a large number of joints is used. The inherent mechanical strength of the joint and the reduction of leak possibilities have made this a desirable type. In addition, code requirements in some areas have forced the use of these alloys. In order for brazing alloys to melt and flow properly, red heat (1150-1450° F) is required. Copper will react with the oxygen in the air at these temperatures to form a scale of copper oxide on the inner walls of the tubing and fittings. The scale is broken off into flakes by the turbulence of flowing liquid refrigerant. The flakes quickly break up into a fine powder which plugs driers, strainers, and capillary tubes.

Copper oxide is so finely divided that a filter bed of other materials is usually required in driers before the filtration system in the drier can remove it, thus accounting for the approximately forty-eight hours needed to remove this material. Driers are sometimes returned as defective because they have shown an undesirable pressure drop on a new system. In conjunction with other dirt, a substantial accumulation of copper oxide is often the explanation. Chemical analysis readily confirms the condition. The drier is not defective. It has done exactly what it was designed to do—take out the dirt.

If the air in the line being brazed is replaced with an inert gas, such as dry nitrogen, the formation of copper oxide can be eliminated. The line should be purged thoroughly and a slow steady flow of nitrogen maintained by means of a pressure reducing valve. High-pressure gases (nitrogen or carbon dioxide) should never be connected directly to a system, but must be metered through a pressure reducing valve for the protection of the user. Too high a rate of flow is undesirable because of the cooling effect of the gas. The inert gas method will keep the inside of the copper lines and fittings clean and bright. When it is not completely successful, the trouble is usually attributed to applying heat before all of the air has been displaced.

On large copper pipe, the end of the tubing should be covered with some type of elastic material, such as a rubber balloon, and a small slit made in the material to provide an escape valve. In this manner, the line can be kept full of nitrogen without use of excessive quantities. The inert gas method, wherever practical, is recommended for repair work. The nitrogen clears the line of refrigerant vapor preventing acid formation due to refrigerant breakdown as well as preventing copper oxide formation.

Many servicemen find it impractical to transport heavy nitrogen cylinders in and out of buildings and have solved the problem by permanently mounting the cylinder on their service truck. The cylinder is equipped with a pressure reduction valve and flexible lines long enough to reach most of their jobs. Purge hose or a similar type of material is satisfactory for the flexible line. The line pressure is held to an absolute minimum, with the gas flow just high enough to maintain the required slow, steady rate.

CHAPTER 8

Domestic Refrigeration Operation & Service

In order to enable the serviceman to render intelligent and efficient service, it is necessary that he understand the operation of the different parts that make up a complete refrigeration system. While some service problems can be detected and eliminated in a few minutes, others may require considerable work before being properly diagnosed and corrected. The importance of a thorough diagnosis of the faulty unit cannot be too strongly emphasized. Do not simply guess at the remedy to be applied until the actual trouble has been determined.

SEALED SYSTEM OPERATION

The correct operation of the sealed system is dependent upon the proper functioning of each part that makes up the system. If the system does not operate correctly due to abnormally long running periods, or warmer than normal interior temperatures, the trouble may be caused by one of the following conditions:

1. Restricted capillary tube.
2. Incorrect refrigerant charge.
3. Partial restriction in evaporator.
4. Defective compressor.

Restricted Capillary Tube

A restriction in the capillary tube may be caused by moisture in the system, kinked tubing, or foreign particles lodged in the lines. Each of these conditions will cause similar symptoms to occur. The evaporator will have little or no frost formation, and the compressor will run for an extended period of time and eventually cycle on the overload protector. Moisture in the system will usually freeze at the outlet of the capillary tube where it joins the evaporator tubing. The tubing in the immediate vicinity of the freeze-up may be heavily frosted, but the remaining evaporator tubing will be free from frost.

If, when the refrigerator is inspected, the compressor is running but the evaporator is not frosted, stop the compressor and remove the frame. Apply heat to the end of the capillary tubing; usually a match held under the tubing at this point will be sufficient. If an ice block is present, the application of heat will melt the ice and a gurgling sound can be heard as the refrigerant surges through the tubing. If, after applying heat to the capillary tube, the restriction has not been eliminated, check for kinks in the tubing or an undercharged system. In the event the application of heat melts the ice block, the drier should absorb and retain enough moisture to eliminate future freeze-ups.

If the freezing reoccurs, and the unit is under warranty, the entire unit should be replaced. If it is not under warranty, there are several procedures that you may take to remedy the problem.

Usually small domestic units have no service valves. There is a procedure, which works in most cases, which involves attaching a valve to the suction line. This valve clamps over the suction line, and has neoprene washers which press against the suction line forming a non-leaking gasket. You then screw down the valve, which has a needle point that punctures the suction line. The refrigerant gas in the unit is then released through this valve. Connect a vaccum pump to the valve, and a line from the discharge side of the pump should be inserted in a container of refrigeration oil. When the vaccum pump is running, bubbles will pass through the oil as long as any gas or water vapor is being pumped out. A vacuum gauge should be installed, so that you may check the actual vacuum that is being developed.

The door of the refrigerator should be left open, and heat should be applied to the tubing and the evaporator. When the bubbles stop showing up in the oil, the vacuum pump should be left operating for an additional ½ hour. The purpose of the application of heat and the long pumping time, is to be certain that any moisture in the system will be removed.

Then close the valve that you have installed on the suction line, or if the unit happens to have service valves, close the suction line completely. Disconnect the vacuum pump, and connect a tank of refrigerant to the suction valve. Insert a compound gauge assembly between the tank of refrigerant and the suction valve. Crack the valve on the refrigerant tank slightly, and loosen the flexible connection hose that you should be using from the compound gauge assembly to the suction valve on the compressor. This will allow some refrigerant to escape into the room, thus purging air and moisture from the compound gauge and the refrigerant lines to the compressor. Open the suction valve at the compressor and allow enough gas to enter so that the gauge shows 1 to 2 lbs. pressure, then close all valves.

Install a new drier-filter, discarding the old one. This can be accomplished by cutting the lines close to the old filter, and using flare nuts and nut fittings, or by means of connecting the lines by slipping one over the other and using a *Phasco*, fluxless brazing rod. The lines only need be heated cherry red to apply the *Phasco* rod. Silver soldering may be done instead of using a *Phasco* rod, but is more complicated to use. When cutting the old drier out, remember that you have a pound or so of pressure within the unit.

After the drier-filter has been changed, reconnect the vacuum pump as outlined before, and run it for ½ hour after bubbles stop appearing in the oil bath. Close the valve that you have put on the suction line, or the service valve on the compressor, (if it has one) before stopping the vacuum pump.

You are now ready to recharge the system with refrigerant. Use a recharging cylinder with a glass front, marked in ounces for the most commonly used refrigerants. The nameplate on the refrigeration unit will tell you how many ounces of what type refrigerant to use. With the same compound gauge arrangement that you used before, connect to the recharging cylinder instead of the compressor. Close the valve at the bottom of the cylinder, crack the valve on the refrigerant tank, turn the tank upside down and crack the connection at the recharging cylinder to purge air and moisture out of the lines. Then tighten the connection at the recharging cylinder, open the valve at the bottom of the cylinder slightly and you can see the liquid refrigerant entering the cylinder by looking through the glass front. There is a purging valve at the top of the recharging cylinder, to remove air and moisture from the cylinder. Add a few ounces more than is required to charge the system. Close the refrigerant tank valve and recharging cylinder valve. Disconnect the charging hose from the compound gauge, being careful as liquid refrigerant will come out and you may freeze your hands or get it in your eyes.

Connect the end of the charging hose to the suction valve on the compressor, leaving it slightly loose. Open the valve at the bottom of the recharging cylinder slightly, to purge the recharging line, then tighten the connection of the charging line at the compressor. You are now ready to recharge the system.

Note how many ounces of refrigerant you have in the cylinder and subtract the number of ounces required to properly charge the system. This will give you a stopping point on the recharging cylinder.

Start the refrigerator, and open very slightly the valve on the recharging cylinder. Remember that you are adding liquid refrigerant to the compressor, and this must be done slowly to prevent oil from leaving the crankcase with the refrigerant and going into the system. In other words, **ADD THE NEW REFRIGERANT SLOWLY.** When you have added the correct amount as shown on the glass front, shut the suction valve at the compressor and then the valve at the recharging cylinder. Leave all equipment in place and observe the operation of the refrigerator.

If all is well, you should see frost start to form at the inlet to the evaporator and slowly expand, eventually covering the entire evaporator. You will also notice a hissing sound as the liquid enters the evaporator.

If the refrigerator frosts back to the compressor, you have overcharged it, and should slowly purge a little of the gas out to correct this situation.

The chances of having repaired the cause of the restricted capillary tube are quite good. If the capillary tube is still plugged, the following information will be useful. The vacuum pumping and recharging of the system will not be repeated. An illustration of the refrigeration tubing, etc, is shown in Fig. 1.

A kinked capillary tube will stop the flow of refrigerant and the evaporator will be free from frost. The compressor will run continuously and eventually cycle on the overload protector. Check the capillary tube along its entire length and, if possible, straighten the tubing enough to alleviate the difficulty. Foreign particles lodged in the capillary tube will also stop the flow of refrigerant, and the system will exhibit the same symptoms as with a kinked tube. If checks have been made to eliminate the possibility of a moisture freeze-up or a kinked capillary tube, it may be assumed that a foreign particle is lodged at the entrance to the capillary tube.

Fig. 1. Schematic diagram showing arrangement of refrigerant tubing in a typical household refrigerator.

A restriction usually occurs in the capillary tube because of its small diameter. The symptoms of a restricted capillary are:

1. Lack of frost on the evaporator. The compressor may operate for a short period of time and then cycle on overload.
2. Moisture freeze-up may cause a restriction in the line, and it usually occurs at the outlet end of the capillary tube. Normally, a frost buildup can be detected in this area, but insulation wrapped

around the tubing may conceal or limit the amount of frost accumulation. Expose the discharge end of the capillary tube, and apply heat at this point. If there is enough head pressure, and if the restriction is caused by moisture freeze-up, a gurgling noise will be heard as the heat releases the refrigerant through the tubing. It is possible that this moisture will be absorbed by the drier and remedy the trouble. However, if the freeze-up recurs, the drier must be replaced. If this does not remedy the problem, replace the heat exchanger.

3. A kink in the capillary tube will reveal about the same symptom as a moisture freeze-up except for the accumulation of frost. Check the entire length of the capillary tube and, if possible, straighten the kink to relieve the restriction. Check the unit operation, and, if the trouble persists, replace the defective part.

4. If there is no freeze-up or kink in the capillary tube, it can be assumed that a foreign particle is causing the restriction.

If you find that the capillary tube is plugged, and the unit is under guarantee, replace the entire unit. If it is out of guarantee, there is a high-pressure hand pump available for blowing out capillary tubes. Cut the inlet line, allowing the refrigerant to escape and then apply pressure with the hand pump, blowing out the restriction. This operation is mentioned assuming that the lines are copper and not aluminum lines. If this does not work, replace the heat exchanger, making sure that the capillary tube attached is of the proper dimensions for the refrigerator.

It is a good practice to replace the drier-filter when repairing the capillary tube. When repairs are completed, attach a valve as was previously discussed. Use the vacuum pump as before and recharge the unit.

Incorrect Refrigerant Charge

A refrigerator system that is *undercharged* with refrigerants will produce various conditions depending upon the degree of undercharge. During normal operation, a system that is fully charged will have frost covering the entire evaporator and accumulator. Any degree of undercharge or a gradual leak of refrigerant will be noticed first by the absence of frost on the accumulator. The compressor will run for long periods of time, and the food compartment temperature may be colder than normal. As the leakage of refrigerant progresses, the last few passages in the evaporator will be free of frost. The compressor will run continually,

219

since the temperature of the evaporator at the control contact location fails to descend to the control cut-out point.

A system that is *overcharged* will have a frostback condition on the suction line under the cabinet during the *on* period. The frost will then melt and drip on the floor during the *off* period. This situation in a slightly overcharged system may be remedied by wrapping the suction line with insulation tape or its equivalent. Should moisture continue to drip on the floor, replace the sealed system.

Overcharge—The symptoms of an overcharged refrigerator are usually as follows:

1. A frostback condition on the suction line to the extent that it can be noticed on the line back to the compressor unit.
2. If the overcharge is great enough, the control contact may not get cold enough to cut-out the compressor, resulting in long periods of operation.
3. Suction and head pressure will be high at both cut-out and when the system has stabilized.
4. An overcharged system must be evacuated and recharged with the proper amount of refrigerant.

Undercharge—When too little refrigerant is present in the refrigerator, the system is undercharged. This defect can be diagnosed as follows:

1. In an undercharged system the refrigerant system will be colder than normal. The freezer compartment may or may not be cold, depending upon the amount of undercharge.
2. The compressor may run continuously because the cold control contact fails to descend to the control cut-out point.
3. Suction charge will be low (possibly a vacuum) depending upon the degree of undercharge.
4. Head pressure will be lower than normal at cut-out or when the system has stabilized.
5. An undercharged system must be evacuated and recharged with the proper amount of refrigerant.

Partial Restriction in Evaporator

An accumulation of moisture or foreign particles may freeze or lodge in the evaporator tubing and cause a partial restriction at that point. This condition is usually indicated by the evaporator being heavily frosted for a few passes ahead of the restriction, and bare of frost on the first few passes behind the restriction. A restriction of this nature tends to act as a second capillary tube. Increased pressure in the evaporator line toward the high side will cause warmer temperatures; decreasing the pressure as the refrigerant passes toward the low side will cause colder temperatures. The evaporator tubing on the high side of the restriction will be free of frost, and the tubing on the low side will be heavily frosted.

If the restriction occurs on the low side of the control feeler-tube contact point, the compressor will run continuously since the control cutout point is never reached. Should the restriction occur on the high side of the control feeler-tube contact point, the compressor will cycle frequently, but running time will be short and the food compartment temperatures will be warmer than usual. A partial restriction in the evaporator tubing will require replacement of the sealed system.

Defective Compressor

A compressor which is not pumping adequately will produce very little cooling effect. The evaporator may be covered with a thin film of frost, but the evaporator temperature will not descend to the cut-out temperature of the control, even with continuous running of the compressor. Place your hand on the evaporator surface for 2 or 3 seconds, then examine the surface. If all of the frost has melted where the evaporator was touched, suspicions of a defective compressor will be confirmed and the sealed system must be replaced.

An inefficient or defective compressor will affect the operation as follows:

1. A compressor that is not pumping correctly will produce very little cooling effect. All cooling surfaces may be covered with a thin film of frost.
2. A popular checking method is to place a hand on the accumulator for two or three seconds and then examine the surface. If all of the frost is melted at the contact surface, the operating pressures should be checked.
3. If high-side pressures are lower than normal, and low-side pressures higher than normal, suspicions of a defective compressor are confirmed and the compressor unit must be replaced.

221

ELECTRICAL TESTING

In normal operation the compressor is actuated by the thermostat. When the thermostat contacts close, the motor running winding and the relay coil are energized. The heavy surge of current that will be drawn by the motor attempting to start, will create an electromagnetic force strong enough to pick up the relay armature and close the starting contacts of the relay, which will energize the running windings. A fraction of a second later, as the motor comes up to full speed, the running current decreases to normal value and allows the relay armature to drop and open the starting contacts.

The starting winding which was needed to provide the extra torque to overcome the inertia of the compressor is now out of the circuit. The motor continues to operate on only the running winding. The overload protector has a set of bimetal contacts which are normally in the closed position, and are in series with a resistance heater. If there is difficulty in starting the motor, or if the compressor becomes too hot for any reason, the excess heat will cause the contacts to open and interrupt all electrical power to the compressor motor.

Testing With Volt-Ohmmeter

Test lamps, as shown in Fig. 2 were formerly used to a great degree for electrical troubleshooting. They can be dangerous because of bulbs breaking, and the receiving of electrical shocks. Now practically all service men use volt-ohmmeters. The continuity tests are done on the ohmmeter scale. These instruments have self contained batteries of low voltage.

Fig. 2. Typical test lamp arrangement.

Starting Winding

To test for continuity in the motor starting winding, set the ohmmeter on the low ohms scale, and place the test leads on the terminals S (starting) and C (common) as shown in Fig. 3. If the ohmmeter stays on zero ohms, the starting winding is open.

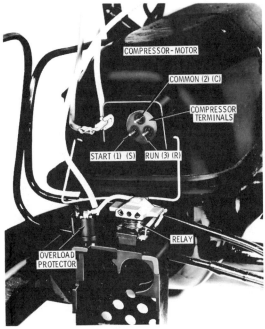

Courtesy Hotpoint Co.

Fig. 3. Compressor motor terminal arrangement in a typical household refrigerator.

Running Winding

To test the running winding, place the ohmmeter test probes on compressor terminals R (running) and C (common). If the ohmmeter reads zero, the running winding is open. If either the starting or running windings are open, the entire unit must be replaced.

Overload Protector

Before an accurate continuity check can be made on the overload protector, it must be determined that the unit has not been on overload during the past hour. If the unit is on overload, the protector contacts will be open and there will be no circuit continuity. If, however, the unit has not cycled on the overload, the continuity check will determine whether or not the overload protector is inoperative. To test the protector, place

223

the ohmmeter probes on protector terminals C (common) and terminal 3 or Line, depending upon the wiring arrangement. If the ohmmeter registers zero, either the overload protector or the resistance heater is open.

The overload protector is actuated with every start, and therefore is a very likely source of trouble.

Units With Capacitors

Capacitors are used on the larger refrigerators, and are in series with the starting winding to give more starting torque. They occasionally go bad. The best test for capacitance is to use a good capacitor checker, or temporarily replace it with a known good capacitor. Sometimes it is well to replace the overload protector when replacing a capacitor, for preventive maintenance.

COMPRESSOR TESTING WITH DIRECT POWER

Before proceeding with this test, be sure the compressor is level. To test the compressor without any cabinet wiring in the circuit, disconnect the compressor lead cord from the junction block and connect it directly to the electrical wall outlet. If necessary, use a short extension cord to reach the outlet. If the compressor starts, the trouble is not in the compressor, relay, or overload protector. Plug the compressor lead cord back into the junction block, and check the temperature control and all cabinet wiring for an open circuit or poor connections. If the compressor does not start, check the overload protector.

The overload protector is mounted on the compressor under the terminal cover. The protector trips open when the compressor is overheated and/or when excessive current is drawn. Cycling on the overload may be the result of poor air circulation around the compressor and condenser. The compressor should not attempt to start until the refrigeration system has become equalized, which takes about 3 to 5 minutes after previous run. At least 100 volts are required at the compressor terminals during the starting interval. A typical compressor motor circuit is shown in Fig. 4. Quite often the overload device is not a separate piece as shown in Fig. 4, but is included in the starting relay.

If the compressor repeatedly starts and runs for 5 or 6 seconds and then cycles on the overload protector, the starting relay contacts may be stuck closed and the excess current is tripping the overload. Disconnect the wire from compressor terminal S, and make a momentary short between

224

wire terminals R and S (see Figs. 3 and 4). If the compressor starts and runs without cycling on the overload protector, replace with a new complete starting relay.

To check for an open overload protector, short across the terminals (see Figs. 3 and 4). If the compressor starts after shorting the protector terminals, replace the overload with one having the same part number. If the compressor does not start, look for other trouble such as a low line voltage, defective starting relay, and defective operating protective relay or compressor. Other possibilities are a blown fuse, broken service cord, inoperative thermostat, faulty capacitor, etc.

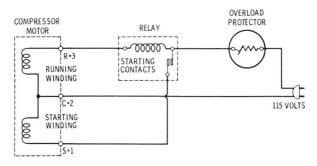

Fig. 4. Showing a schematic wiring diagram of a compressor motor.

Starting Relay

The starting relay is usually mounted under the compressor terminal cover and consists essentially of a magnetically operated switch with starting contacts. The magnet coil of the relay is in series with the running winding of the compressor motor. When current is applied to the motor, the magnet coil raises the relay plunger and closes the starting contacts, thus connecting the starting winding in parallel with the running winding. As the motor approaches running speed, the current in the running winding and in the relay coil diminishes and the plunger *drops out,* which opens the starting winding circuit. The compressor then continues to operate on the running winding only.

To check for open relay starting contacts, short between the wires at the compressor terminals. If the compressor starts, the relay is defective. If the compressor does not start, check the relay magnet. To check the relay magnet coil, disconnect the power lead from the relay screw terminal.

225

Short between the wires at the compressor terminals and, at the same time, touch the wire terminal with the disconnected power lead. This by-passes the relay and connects the line directly to the compressor, but through the overload protector. If the compressor does not start, a check for open motor windings or electrical grounds can be made with an ohmmeter. Test between compressor terminals *common* and *run*, and between *common* and *start*.

CABINET LIGHT AND SWITCH

The cabinet light (or lights) is energized by the door-operated switch. If the cabinet light does not go on when the door is opened, it may be due to a burned out light bulb, a defective light switch or socket, or faulty wiring.

CONDENSER COOLING FAN

The condenser cooling fan (when used) is connected in parallel with the compressor. Therefore, if the compressor operates but the fan does not, the fan is either defective or disconnected. An excessive fan noise complaint may arise if one of the fan blades has been bent out of alignment. If any irregularity is noted, replace the fan blade. Another cause of excessive fan noise may be loose fan bracket mounting screws. A typical fan assembly is shown in Fig. 5.

Fig. 5. Fan motor and assembly components.

Replacing Controls

To replace controls on most models, proceed as follows:

1. Disconnect the electrical power cord.
2. Pull the control knob off of the shaft.
3. Remove the control mounting screws and work the control out of the food liner opening.
4. Disconnect the control leads from its terminal. When installing a new control, make sure that the capillary tube is firmly clamped against the stainless steel plate. If the control capillary tube is not located correctly, the control will not be able to sense the proper temperature, and erratic refrigeration operation will result.

Calibrating Controls

If the cut-in and cut-out temperature readings obtained by the use of a remote reading thermometer are not within 1 or 2 degrees of those specified in the operating temperature chart supplied by the manufacturer, recalibrate the control.

The temperature control may be calibrated to vary the *cut-in* and *cut-out* temperatures by turning the range adjustment screw. This screw, shown in Fig. 6, raises and lowers these temperatures by an equal amount. The procedure is as follows:

1. Turn the range adjustment screw right to raise and left to lower the *cut-in* and *cut-out* temperatures (Fig. 6). Each ¼ turn of the screw will raise or lower the temperatures approximately 2 degrees. *Do not turn this screw more than one full turn as erratic operation may result beyond that point.*
2. Connect the line cord and recheck for proper *cut-in* and *cut-out* temperatures.

REPLACING SYSTEM COMPONENTS

When component replacements are necessary, careful checks must be made in order that identical parts are obtained from the manufacturers. Before actually opening a sealed refrigeration system, be sure that all the necessary tools are close at hand. The system should be opened only long enough to make the repair or component replacement. If the system is allowed to be open for an extended period of time, the motor winding in the compressor will pick up moisture, which will be very difficult to remove during compressor evacuation.

Fig. 6. Temperature adjustment on Ranco and Cutler Hammer temperature controls respectively.

Compressor Unit Replacements

When cutting into a system for compressor replacement, proceed as follows:

1. Pinch off the suction and discharge lines as close as possible to the compressor. This will prevent excess oil spilling when cutting the lines.
2. Clean the paint from the suction line and cut the suction line close to the pinched area with a tube cutter. Place a fitting on the cabinet suction line and plug it to prevent entrance of air and dirt.
3. Clean the paint from the discharge line at the pinched area. Then cut the discharge line close to the pinched area with a tube cutter. Install a fitting on the discharge line and attach a compound gauge with a proper fitting.

4. Remove the inoperative compressor by taking out the four spring clips holding it to the cabinet cross rails. Install a replacement compressor, making sure the rubber grommets from the inoperative compressor are remounted. The replacement compressor suction, discharge, and charging lines are equipped with flare nuts and plugs.

5. With a tube cutter, cut the compressor suction and discharge tubes as close to the flare nuts as possible, and install the proper fittings. Connect the compressor suction line to the cabinet suction line. Plug the compressor discharge line. *Note: Be sure to clean the paint from the tubing before cutting any lines.*

6. Remove the original dryer and install a replacement drier. Compression fittings are provided for the inlet to the dryer. The output end of the drier connects directly to the capillary line and will have to be silver soldered.

7. Remove the plug from the compressor charging line and connect a test can of refrigerant to the charging line via hoses and a manifold. Be sure to purge the lines.

8. Charge the system to 10 psig (pounds per square inch gauge) as read on the condenser line gauge. Close all valves and leak test all connections.

9. Replace the test can with an exact measured charge of refrigerant.

10. Remove the plug from the compressor discharge line fitting. Apply power to the compressor and evacuate the system to at least a 25-in. vacuum (as read on the condenser gauge) for a minimum of 15 minutes. Then replace and tighten the plug on the compressor discharge line fitting and disconnect the power from the compressor.

11. Charge the unit to 2 psig as read on the condenser line gauge. This is to break the vacuum in the unit.

12. Remove the gauge from the condenser tubing and the plug from the compressor discharge tube. Then connect the fitting between the compressor discharge line and the condenser line. Make the connection immediately while the unit still contains an atmosphere of refrigerant.

13. Begin to charge the unit and apply power to the compressor. While the compressor is running, leak test the high side.

14. After the system has been charged, pinch off the charging line and insert the plug previously removed back on the charging line.

Note: Leave the pinch-off tool on the line until the plug is reinstalled.

15. Disconnect the power, permitting the low-side pressure to equalize. Leak test once more, and recheck all joints and fittings. Repaint all tubing surfaces previously sanded.

Evaporator Replacement

To replace an evaporator in a sealed system, proceed as follows:

1. Disconnect the electrical line cord.
2. Cut off the suction line from the compressor close to the heat exchanger after the refrigerant charge is released. Place the fitting on the line and plug the line to prevent entrance of air and dirt.
3. Clean the paint from the compressor discharge line about two inches beyond the joint, and cut the tubing with a tube cutter. Install the fittings on the lines. Plug the compressor discharge line and attach a compound gauge with the proper fitting to the discharge line going to the condenser.
4. Clean the paint from the compressor charging line and, with a tube cutter, cut the line as close to the end as possible. Then install the fitting and plug on the open end of the line.
5. Remove the rear tube cover from the cabinet back to expose the heat exchanger tubing that is imbedded against the foam insulation.
6. Cut the heat exchanger line near the top of the cabinet back where it enters the liner. This will permit easier removal of the heat exchanger through the cabinet when removing the evaporator.
7. Remove the screws holding the evaporator to the top of the cabinet. Slowly pull out the evaporator and heat exchanger from the cabinet, as an assembly.
8. Take off the control cover and thermostat from the inoperative evaporator and install the new evaporator.
9. Insert the heat exchanger and evaporator through the liner opening on the cabinet until the evaporator is properly positioned. Then remount the evaporator assembly and top cover.
10. At the rear of the cabinet, carefully reposition the heat exchanger.
11. Connect the cabinet suction line to the compressor suction line.
12. Evacuate and recharge the unit as outlined in other parts of this chapter.

Drier Replacement

A new drier must be installed when any system component is replaced or whenever the system is opened. To replace the drier, proceed as follows:

1. Remove all paint and scale from the refrigerant lines for a distance of about three inches at both ends of the old drier (use steel wool or fine emery cloth).
2. Cut the lines approximately one inch from the old joints. To cut the capillary tube, score the walls with a knife or file. Make a uniform cut around the entire tube, then break it off.
3. Make an offset about one inch from the end of the capillary tube to prevent its penetrating too far into the new drier.
4. Immediately solder the new drier into place. Use silver solder and the proper flux.

Replacing Heat Exchanger

To replace the heat exchanger, proceed as follows:

1. Purge the charge by cutting the process tube on the compressor and install the service valve, leaving it open.
2. Remove the defective heat exchanger.
3. The capillary tube will have to be unsoldered from the evaporator inlet. Be careful not to restrict the evaporator inlet. Remove heat as soon as the solder liquefies.
4. Unsolder the suction line from the evaporator outlet; clean, then cut the suction line about four inches from the compressor.
5. Unsolder the drier from the condenser outlet. Hold the outlet tube down and wipe the solder off of the tube immediately after the drier is removed.
6. Swedge the compressor suction line so the replacement suction line will fit inside.
7. Install the replacement heat exchanger, making all solder joints except on the condenser outlet.
8. Install the new drier on the condenser outlet.
9. Make an offset on the end of the capillary tube and solder it in the outlet of the new drier.
10. Pull a vacuum for about twenty minutes, then add enough refrigerant to produce a pressure of 35 to 40 lbs.

11. Leak test the low side, then start the compressor and leak test the high side.
12. Purge the charge and continue with the vacuum as outlined.
13. Charge the system and reassemble the refrigerator, making sure to locate the system lines properly and seal where necessary.
14. Test run the refrigerator to be sure it is operating properly.

Replacing Condensers

Most refrigerant condensers are made of copper-coated steel tubing. If a leak occurs, a piece of copper tube can be spliced in place, in which case the condenser does not need to be replaced. When a condenser is in need of replacement, proceed as follows:

1. Open the sealed system at the compressor process tube first. After the refrigerant charge is released, install a service valve on the process tube, leaving the valve open.
2. Before cutting the tubing lines going to the condenser, check the ends of the replacement condenser tubing. If there is no excess tubing on the connector lines, unsolder the connector tubes from the old condenser.
3. Score and break the capillary tube about one inch from the old drier. Remove the old condenser and install the replacement. Depending on the type of condenser to be replaced (fan-cooled or static), it might be preferable to make the joints on the replacement before installing it in its place (clean all tubing before soldering the joints).
4. After all other joints are made, make an offset in the capillary tube and install the new drier. *Note: When soldering these joints, make sure to leave the service valve on the compressor process tube open.*
5. Start evacuation procedures.
6. Refer to the evacuating, purging, and charging procedures as outlined in this chapter.
7. After reassembling the refrigerator, make a test run to be sure that it is operating properly.

EVAPORATOR COOLING FAN

The evaporator cooling fan (when used) is designed to blow cold air over the evaporator compartment area to maintain an even temperature throughout. The air circulation also prevents frost accumulation on food packages and freezer racks. If the fan fails to operate or runs erratically, the reduced air circulation will cause unsatisfactory temperatures in the evaporator compartment.

Most troubles of this sort are usually due to defective wiring, a faulty fan-motor switch or a bent motor shaft. The switch and fan motor may easily be checked out by means of an ohmmeter or the conventional test lamp probes. If, after making the foregoing checks, the evaporator fan fails to operate, the trouble is either in the wiring or the fan motor itself.

Refrigerator fan motors as a rule do not have a way to oil themselves. The bronze bearing has felt outside of it, which presumably oils the motor for life. However, this is often not the case. An overall statement is impossible to make, but if the steel portion around the bearing is enlarged, you can usually figure that there is a felt enclosure around the bearing. In this case, carefully drill a 3/32'' hole into the steel enlarged portion, and with an oil can, put in 3 or 4 drops of 20 or 30 weight automobile oil, being careful not to over-oil. Place a little oil around the shaft while turning, and it will usually free up. These motors are of the high-impedance type and will stand a stall without burning up. Insert a small self tapping screw into the hole that you have drilled to retain the oil.

TEMPERATURE CONTROL SERVICE

There are generally two types of temperature control devices, and their functions depend upon:

1. Temperature standard controls serving solely as *on* and *off* switches for the compressor.
2. Temperature combination controls with automatic defrost features which govern the defrost operation as well as the compressor operation.

Most problems resulting from temperature control defects will show up in the refrigerator compartment, or in both compartments, but never in the evaporator alone. Attention given to the refrigerator compartment and control adjustments is important because the evaporator compartment will react favorably to nearly all corrective measures. Unsatisfactory temperatures in the evaporator compartment only are usually the result of a poor door gasket seal, poor air circulation in the evaporator, or an inoperative fan, defrost heater wire, or defrost timer. All dual temps use a constant *cut-in* type temperature control. Changing the control knob setting will change the *cut-out* temperature, but will not affect the *cut-in* temperature.

A constant *cut-in* simply means that the control setting only determines the temperature at which the electrical contacts open, or cut-out, interrupting power to the compressor. The temperature which causes the contacts to close always remains constant, regardless of control setting. The constant cut-in temperature is approximately 37 degrees. Therefore, whenever the temperature in the fresh-food section causes the control contacts to open, the frost that has accumulated on the cooling coil will begin to melt and drain off. All of the frost will be dissipated since the control contacts will not close until the control capillary senses a temperature of 37 degrees, which is above the frost point. If defrosting of the cooling coil is not achieved, it is usually due to a high setting of the control dial which resulted in constant running of the compressor.

Checking Operating Temperatures

Cabinet and evaporator temperatures can be varied considerably by changing the control setting. If no change occurs, check the control *cut-in* and *cut-out* temperatures. To check control operating temperatures, proceed as follows:

1. Scrape away the frost on the inside of the evaporator adjacent to the feeler-tube thermal connection.
2. Using a few drops of water, freeze the bulb of an accurately calibrated remote reading thermometer to the evaporator.
3. Set the control at *normal*. Close the cabinet door and allow the compressor to run through 2 or 3 complete cycles.

REFRIGERATION TROUBLE CHART

Symptom and Possible Cause	Possible Remedy
Unit Dead, Will Not Run	
(a) Blown fuse.	(a) Check and replace if defective.
Rating of power source too low.	Check source with voltmeter. Should check within 10% of 115 volts.
(b) Broken service cord.	(b) Check voltage at relay. If no voltage at relay, but voltage indicated at out-

REFRIGERATION TROUBLE CHART (Con't)

Symptom and Possible Cause	Possible Remedy
	let, replace service cord or plug on cord.
(c) Inoperative thermostat.	(c) Insert jumper across terminals of thermostat. If unit runs and connections are all tight, replace thermostat.
(d) Faulty capacitor	(d) Check capacitor using test cord and 150 watt light bulb. If bad or shows signs of leakage, replace capacitor.
(e) Inoperative relay.	(e) Use starter cord and check unit. If unit runs with normal wattage and above conditions are correct, replace relay.
(f) Stalled unit.	(f) Use starter cord and check unit. If unit runs with normal wattage and above conditions are correct, replace relay.
(g) Broken lead to compressor, timer or thermostat.	(g) Replace or repair broken leads.

Compressor Cycles on Overload Protector

(a) Voltage too high or too low.	(a) Check line voltage. Must be within 10% of nameplate rating. Connect refrigerator to proper power source.
(b) Relay or overload protector inoperative.	(b) Check continuity and replace if inoperative.
(c) Motor winding open.	(c) Check continuity of motor windings. If open, unit must be replaced.
(d) Lead broken or connections loose.	(d) Check for broken leads and loose terminal connections. Make necessary corrections.

Unit Hums and Shuts Off

(a) Low voltage.	(a) Check electrical source with volt-wattmeter. Under a load, voltage should be 115 volts $\pm$ 10%. Check for possible use of extremely long extension cord, or several appliances on same circuit.
(b) Faulty capacitor.	(b) Replace capacitor.
(c) Inoperative relay.	(c) Replace relay.
(d) Stalled compressor.	(d) Check with test cord, if compressor will not start—replace.

235

REFRIGERATION TROUBLE CHART (Con't)

Symptom and Possible Cause — Possible Remedy

Unit Runs Excessively

(a) Frequent door openings. High ambient temperature and humidity.

(a) Advise user on proper location and use of refrigerator.

(b) Poor door seal.

(b) Check to see if cabinet is level. Make necessary front and rear adjustments.

(c) Gasket not sealing

(c) Check door gasket. Realign door and when indicated, replace door gasket.

(d) Interior light burns constantly.

(d) Check light switch operation. If light does not go out when door is closed, replace the light switch.

(e) Insufficient air circulation.

(e) Check position of refrigerator. The rear and sides must be several inches away from walls. Locate in correct position.

(f) Gas leaking from unit.

(f) Check unit for gas leak. If a leak is found, replace the unit.

(g) Loose connection of thermostat bulb to cooling coil.

(g) Check connection and make necessary adjustment.

(h) Faulty thermostat.

(h) Turn thermostat to "off" position. If unit continues to run, replace the thermostat.

(i) Refrigerant charge.

(i) Too much or too little gas. Discharge, evacuate and recharge with proper charge.

(j) Restriction or moisture.

(j) Replace component where restriction is located. If moisture is suspected replace drier-filter.

(k) Placing sudden load on unit.

(k) Explain to user that heavy loading will cause long running time until temperatures are maintained. This running period may be several hours after heavy loading of the cabinet.

High Noise Level

(a) Loose compressor mounting bolts.

(a) Replace mounts. This type noise usually occurs during starting or stopping of the unit.

(b) Blower motor.

(b) Check for motor wheel rubbing against housing. Readjust. Replace motor if noise is excessive.

(c) Vibrating unit tubing.

(c) Run the hands over various lines. This can often help determine the location

REFRIGERATION TROUBLE CHART (Con't)

Symptom and Possible Cause	Possible Remedy
	of the vibration. Gently reform tubing to eliminate vibration problem. Check discharge line from compressor to top of condenser to eliminate possible rubbing of line against condenser.
(d) Cabinet not level.	(d) Check level, and if necessary, make the appropriate side-to-side and front-to-rear adjustments to level the cabinet.
(e) Location of cabinet.	(e) An out-of-level floor may be a factor in causing noise. Also certain types of flooring transmit vibration more readily than others. Investigate type of floor construction where cabinet is located.
(f) Compressor	(f) Only after all external sources have been checked should the compressor be changed for noise.

Food Compartment Temperatures Too Warm

(a) Inoperative thermostat.	(a) Determine product temperatures. If slightly high, adjust control for longer running time. If control is discharged, broken or if recording indicates switch settings are off, replace switch.
(b) Poor door seal.	(b) Level cabinet. Adjust door seal. Adjust tie rods in door.
(c) Repeated door openings or overloading of shelves in food compartment.	(c) Instruct user regarding blocking normal air circulation in the cabinet.
(d) Inoperative fan switch.	(d) Check leads and control. Replace if discharged, broken or if other conditions would warrant replacing the control.
(e) Inoperative fan motors.	(e) Replace motor.
(f) Lights stay on.	(f) Adjust door to engage plunger. If switch is faulty, replace switch.
(g) Food compartment air flow control.	(g) Check for obstruction in the passage from the freezer to the food compartment. Check for proper sealing around the control and the bellows.
(h) Conventional refrigerator needs defrosting.	(h) Excessive frost accumulation on freezer slows down heat transfer to refrigerant in the freezer. Excessive

237

REFRIGERATION TROUBLE CHART (Con't)

Symptom and Possible Cause	Possible Remedy
	frost also resists normal air flow around the freezer. Suggest that user defrost more often.
(i) Improper bulb spacer.	(i) Check for proper bulb spacer, use thicker spacer to lower switch cut-off point and lower cabinet temperature by increasing the running time.
(j) Seasonal control when used in closed position, or drop tray all the way back.	(j) Change seasonal control or drip tray to the open position to allow more air circulation around the freezer.

Food Compartment Temperature Too Cold

(a) Inoperative or erratic thermostat.	(a) Turn control dial to highest setting and by-pass thermostat. If unit starts and temperature goes down to normal level, replace the thermostat. Do not condemn the control until other factors are considered. If control does cycle the product regularly but foods are too cold, adjust the control warmer. If the unit does not cycle, replace the control.
(b) Very cold ambient temperature.	(b) If in an extremely cold room, the freezer temperature and cabinet temperature tend to equalize causing cabinet temperature to be below 32°, refrigerator should be moved to a warmer location.
(c) Improper bulb space.	(c) Use thinner bulb spacer to raise the effective switch cut-off point and raise cabinet temperatures by reducing running time.

Freezing Compartment Too Warm (Automatic Defrosting)

(a) Inoperative thermostat.	(a) Note running time and freezer package temperatures, adjust control for longer running time. If control is broken or discharged, replace control.
(b) Poor door seal.	(b) Level cabinet. Adjust door hinge wings. Check for interference from packages or foot pedal.

REFRIGERATION TROUBLE CHART (Con't)

Symptom and Possible Cause	Possible Remedy
(c) Light stays on.	(c) Adjust door to engage plunger. If switch is inoperative, replace switch.
(d) Blower motor inoperative.	(d) Check to see that the blower motor is free. Check voltage to motor. Replace if inoperative.
(e) Inoperative timer.	(e) Timer may not allow unit to go into defrost causing ice to build up on coil. This cuts down on the air flow. Replace timer if inoperative.
(f) Coil iced up.	(f) Check drain heater, defrost heater, limiter switch, and timer. These items could cause the coil to ice up.

Water On Floor, Suction Lines Frosted

(a) Overcharge of refrigerant.	(a) This frost back should stop after first couple of cycles. Purge slightly from high side charging port. If too much has been purged, evacuate and recharge system.
(b) Freezer blower motor (when used) inoperative.	(b) Check for voltage supply to motor. If none, check wiring. Remember this motor only runs when the compressor runs.

Oil On the Floor

(a) Terminal stud leak.	(a) Check with leak detector. Tighten stud. Evacuate and recharge with proper charge.
(b) Charging screw loose.	(b) Check with leak detector. Tighten screw. Evacuate and recharge.
(c) Broken line.	(c) Replace component or repair with line connector, if possible. Evacuate and recharge.

Frost in Freezer (Automatic Defrosting)

(a) Water spilled when ice cube trays are placed in the freezer.	(a) Caution user that water spilled will not readily be removed. Tap bottom of cold storage liner and ice will break loose.
(b) Poor door seal.	(b) Level cabinet. Adjust door hinge.

239

REFRIGERATION TROUBLE CHART (Con't)

Symptom and Possible Cause	Possible Remedy
(c) Improper loading.	(c) Check for interference from packages or door opening pedal. Packages should not block the air flow at the top in back of the cold storage liner.
(d) Inoperative freezer blower motor (when used).	(d) Check to see that the blower wheel is free. Check voltage to motor. Replace if defective.

Excessive Moisture Inside Food Compartment (Temperature Normal)

(a) Poor door seal.	(a) Level cabinet. Adjust door hinges. Adjust tie rods in door.
(b) Normal operation.	(b) During humid weather, moisture will accumulate on cold surfaces within refrigerator, just as it accumulates on a glass containing iced drink on a hot summer day. This is perfectly normal and is particularly noticeable when the refrigerator door is open frequently.

CHAPTER 9

Household Cabinet Defrosting Systems

Since the accumulation of frost on the evaporator has a direct relationship to the ambient temperature and humidity, the successful performance of any refrigerator depends upon maintaining cabinet and freezer temperatures within a range that will promote the most favorable conditions for keeping perishables and frozen foods. It should be noted that each time the refrigerator door is opened, some of the heavy cold air within the cabinet escapes and mixes with the warm room air. This process creates a low-pressure area within the cabinet, which draws in warm room air.

The rapid increase in cabinet air temperature results in an accumulation of excess moisture on the evaporator and cold surfaces within the refrigerator. An additional result of frequent door openings is that the temperature control will almost immediately energize the unit, which will run until the cabinet air temperature has been lowered to a degree corresponding to the setting of the temperature control switch. Naturally, with a greater number of door openings, the unit will operate with greater frequency and for longer periods of time, with a corresponding higher power consumption and an increase in frost accumulation on the evaporator. One method of minimizing refrigerator door openings is to remove all the required food and dishes at one time, instead of opening the refrigerator door for each dish required during preparation of a meal. When all the necessary food is removed at one time and placed on an adjacent work counter surface, door openings may be decreased as many as seven to ten times that usually required.

MANUAL DEFROSTING

Periodic defrosting is essential to the efficient and economical operation of any refrigeration system. The defrosting operation should be performed when the frost accumulation on the cooling unit becomes approximately ¼'' thick, and regardless of the thickness of the frost, the unit should be defrosted and the moisture removed once every two weeks, or more often if conditions require. If large quantities of moist food are stored, or if the weather is exceptionally humid, it may be necessary to defrost more often.

NEED FOR DEFROSTING

The need for defrosting is readily apparent if the following factors are considered:

Cabinet air temperatures are controlled by means of a temperature control, the control bulb of which is attached directly to the cooling unit and frequently insulated from the circulating air by a rubber insulator. The bulb is therefore directly affected by the temperature of the cooling unit, but is affected only slightly by the temperature of the air in the cabinet. Since the cooling unit is the coldest spot inside the cabinet, moisture in the form of frost deposits on the outside surface of the cooling unit is increased considerably.

Position of Control or Switch

On refrigerators equipped with a defrosting switch, all that is necessary to defrost the unit is to turn the control to the *defrost* position (it is not necessary to turn it off or allow the door to remain open). Allow the switch to remain in this position until all the frost has melted. To defrost refrigerators not equipped with a defrosting switch, all that is necessary is to set the *on* and *off* switch to the ''off'' position.

Hot Water Defrosting

In order to hasten the defrosting period, particularly in warm weather when it is desired to preserve the ice cubes and allow only a small rise in cabinet air temperatures, pans or containers of hot water can be placed in the cooling unit after turning the control switch to the ''off'' position. The unit will then defrost in about 20 to 30 minutes.

Defrosting Tray

It is obvious that before a defrosting operation, the defrosting tray should be empty and correctly located to receive the water from the melting ice of the cooling unit. After the defrosting is completed, the defrosting tray should be emptied, the interior of the refrigerator cleaned, and fresh water placed in the ice-cube trays.

AUTOMATIC DEFROSTING

Automatic defrosting is accomplished by providing an electric heater (or heaters) consisting of a conventional ohmic resistor encased in tubing which, by various control methods, supplies the heat necessary to prevent excessive moisture accumulation on the evaporator. The defrost heater operates in a cyclical manner. The heater may be energized on cut-out periods of the unit only, or it may be operated by a control timer mechanism starting a defrost cycle once every 12 or 24 hours, as shown in Fig. 1.

The freezer and fresh-food compartments usually have a common drainage system. During a defrost cycle in the freezer compartment, the water runs down special drain tubes to a collector or pan under the cabinet where the water is evaporated into the room air.

Operation

There is no mystery or magic connected with the principle of automatic defrosting. In this connection it should be noted that the various manufacturers of refrigerators employ different terms for automatic defrosting, such as NO-FROST, FROST-PROOF, FROST-FREE, etc. A 12-hour system means a refrigerator defrosting system (Fig. 2) is used to automatically defrost the evaporator once every 12 hours. In a defrost system of this type, operation of the compressor is controlled by the temperature control switch during both the defrost and refrigeration cycle. Should the defrost control switch be in a defrost position during a thermostat cycle (compressor off), the defrost cycle will not start until the temperature control switch closes and starts the compressor.

The defrost controls usually employed are essentially single-pole, double-throw switching devices in which the switch arm is moved to the

243

Fig. 1. Wiring diagram of an electrical circuit used in a typical automatic defrost refrigerator.

defrost position by an electric clock, as illustrated schematically in Fig. 3. The switch arm is returned to the normal position by a power element which is responsive to changes in temperature. Although there are several methods to accomplish automatic defrosting depending upon the manufacturer of the refrigerating unit, a typical system operates as follows:

A small (usually 10-watt) liner heater operates during the off cycle of the compressor. This heater supplies a small amount of heat to the thermostat bulb, insuring a reasonably short *off* cycle if the refrigerator

INTERNAL CONTACTS

Courtesy Hotpoint Co.

Fig. 2. A defrost timer showing contact and terminal arrangement.

is installed in a cold room or on an unheated back porch. When the thermostat in the fresh-food compartment has energized the compressor, the refrigerant circulates through the cooling coil in the food compartment and passes through a restrictor into the cold coil in the freezer compartment. The refrigerant expands and evaporates, absorbing heat from the fresh-food and freezer compartments.

A fan in the freezer compartment circulates the air from the bottom, up through the coil, and out through the fan. This air-flow pattern is unchanging and always goes from the bottom, up under the coil cover plate, through the coil, and out through the orifice in the coil cover

245

Fig. 3. Pictorial view of an electric clock motor and contact arrangement in a typical automatic defrost refrigerator.

plate. Moisture is removed from the air as it passes through the coil and is deposited on the colder surface of the coil. Since all of the moisture is deposited on the coil, there will be no frost on any other part of the freezer compartment or on any food packages stored in this area. Because frost cannot accumulate on the coil indefinitely, some provision must be made for its removal. This is done by means of an automatic defrost cycle.

The automatic defrost cycle for the freezer compartment is controlled by a timer which has a 12-hour electrically-driven movement. A revolving cam in the timer operates the defrosting contacts. The timer operation is governed by the compressor since the timer motor operates only when the compressor is running. The only time that the timer motor operates independently of the compressor is during a defrost cycle (when the compressor is not running). If it is assumed that the compressor will run 50% of the time, there will be at least one automatic defrost a day in the freezer compartment. If the compressor should run 100% of the time (under extreme ambient conditions), there would be a maximum of two defrost cycles during a 24-hour period.

The Defrost Sequence

The defrost sequence is as follows:

1. The timer turns off the compressor and the freezer compartment fan. At the same time, it energizes the heater on the coil and the heater on the drain trough. At this point, the compressor and fans are not operating, but the heaters are activated.
2. The heater melts the frost on the coil, and the resulting water drains down into the trough through a tube in the bottom of the freezer compartment.
3. When the temperature at the bimetal thermodisc reaches approximately 65°, the coil heater is turned off. The coil should be completely defrosted by this time. Approximately 10 minutes will elapse from the start of the defrost cycle to the time when the coil heater is cut off. If, for any reason, the bimetal disc should fail to function when the interior evaporator temperature reaches 65°, the timer will cut off the coil heater after approximately 18 minutes from the start of the defrost cycle. The food will not be affected by the defrost cycle because the fan is off and warm air is not being circulated. The food packages are protected by the insulating cold air in the evaporator and the insulation afforded by the plate covering the cold coil.
4. The drain heater stays on for approximately 8 minutes after the coil heater has been shut off by the thermodisc, and is cut off by the action of the timer. This 8-minute delay in cutting off the drain-trough heater is to insure that all the water is drained from the trough before the refrigeration cycle begins again.
5. Since the defrost cycle interrupted an *on* cycle, the compressor is activated immediately to bring the coil down to the proper temperature as quickly as possible. There is, however, an additional 5-minute delay before the fan operation is resumed. This is to prevent the circulation of warm moist air.

Defrost Timer Operation

Figs. 4, 5, and 6 show the action of a typical defrost timer during normal operation, as well as the variation in the electrical circuit during the defrost cycle. Fig. 4 indicates the circuit conditions during normal operation. One side of the power line goes to the timer motor and to contact terminal #1. The other side of the power line runs through the cold control in the fresh-food compartment and to the other side of the timer motor. This same side of the power line also goes to one side of the compressor, and then to one side of the fan motor. Contact terminal #1 makes contact with terminal #4, which goes to the other side of the compressor. Terminal #4 also makes contact with terminal #5 which goes to one side of the fan switch. When the fan-switch plunger is depressed, and terminals #4 and #5 are making contact, the fan motor is energized.

Courtesy Hotpoint Co.

Fig. 4. Electric circuit during normal operation.

The condition shown in Fig. 5 is the one that exists at the beginning of the defrost cycle. Terminals #1 and #4 have separated, which interrupts the power to the compressor. Terminals #4 and #5 are also separated, which interrupts the power to the fan switch, de-energizing the fan motor. Timer terminals #1 and #2 make contact, sending power to one side of the drain heater and to the cold coil thermostat. The other side of the drain heater and one side of the coil heater connect to one side of the power line. This connection energizes the drain-trough heater. The coil heater is energized when the thermodisc closes.

Fig. 6 shows the electrical situation after approximately 18 minutes of the defrost cycle have expired. Terminals #1 and #2 break contact and

Courtesy Hotpoint Co.

Fig. 5. Electric circuit at start of defrost cycle.

Courtesy Hotpoint Co.

Fig. 6. Electric circuit near end of defrost cycle.

de-energize both the coil heater and the drain heater. Remember that in normal operation, the cold coil thermostat will de-energize the coil heater after approximately 10 minutes of the defrost cycle. Terminals #1 and #4 make contact and energize the compressor, while terminals #4 and #5 are still open to allow the 5-minute delay before the evaporator fan starts again. After the 5-minute delay, terminals #4 and #5 make contact and the evaporator fan is energized once again. The action of the timer in breaking the circuit of the cold coil heater is purely a mechanical one. If the cold coil thermostat is operating properly, it will remove the coil heater from the circuit after approximately 10 minutes of the defrost cycle, or when the temperature at the point of the cold coil thermostat reaches either 55 or 65 degrees.

The schematic wiring diagrams and pictorial wiring diagrams, just illustrated are representative of other makes of refrigeration. However, if in doubt and there is no schematic diagram pasted on the refrigerator, write the company that made the refrigerator for a schematic diagram of same.

DEFROST THERMOSTAT TESTING

The freezer compartment defrost thermostat contacts are usually set to a temperature change of about 15 degrees. Since a test might be made just after a freezer defrost cycle, it could appear that the defrost thermostat was inoperative because the contacts were still open. In order to make a more positive check the following steps are recommended.

1. Disconnect the power cord.

2. Remove the breaker strip from the fresh-food compartment.

3. Connect the probes of a test lamp to the ends of the defrost thermostat leads. Refer to the wiring diagram on the cabinet for the correct color code. (Needle-type probes are best for this test, since they can be inserted into the connectors.)

4. Plug the test lamp into the wall receptacle. If the test lamp has a bright glow, the contacts are closed, but in case the lamp has a dull glow, the contacts are open and additional testing is necessary.

5. It is important that the refrigerator line cord be disconnected, otherwise an erroneous reading will be made, or the defrost and drain-trough heater could be energized. Points A and B in Fig. 7 show the circuit being tested and the points at which the test lamp is connected. If the initial test has shown the freezer thermostat contacts open, the serviceman should use a can of refrigerant for

the next part of the test. Leave test lamp connected to points A and B.

TOTAL RUNNING WATTS - MIN. 150 - MAX. 190

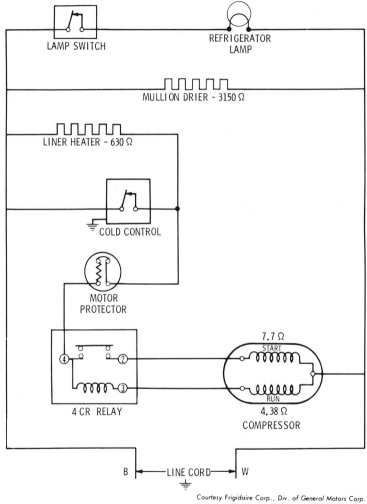

LAMP SWITCH

REFRIGERATOR LAMP

MULLION DRIER - 3150 Ω

LINER HEATER - 630 Ω

COLD CONTROL

MOTOR PROTECTOR

④ ②
③
4 CR RELAY

7.7 Ω
START
RUN
4.38 Ω
COMPRESSOR

B ← LINE CORD → W

Courtesy Frigidaire Corp., Div. of General Motors Corp.
Fig. 7. Schematic wiring diagram of Model FCD-123T.

6. Remove the food from the freezer compartment and take off the freezer cover plate. Spray the defrost thermostat with a can of

refrigerant for about 15 seconds. This should reduce the temperature to the vicinity of 15°, thus closing the contacts. When the contacts close, a click will be heard and the test lamp will glow brightly. *If the contacts do not close after approximately three sprayings of refrigerant, and the lamp retains its dull glow, then the serviceman has made a positive check and should then replace the thermostat.* The serviceman having tested the thermostat in the freezer and finding it operating normally, should test the freezer heater.

If you are accustomed to using a volt-ohmmeter and a clamp-on ammeter, these instruments are far superior to a test lamp. Note on the schematic wiring diagrams, that at the top there are wattages given. These are highly important in troubleshooting domestic refrigerators as they indicate to you whether the unit is drawing the right amount of power. This can be expressed as

$$\text{Volts} \times \text{Amperes} = \text{Watts}$$

The ohmmeter can be used for testing continuity, but you must be certain that the cord to the power supply is unplugged, or you might burn up the ohmmeter. If you use the voltmeter instead of the test lamp mentioned, you will obtain the same results with less danger to yourself or the components of the refrigerator.

In using the volt-ohmmeter, always check to be sure that you have it set for ohms or the maximum voltage that you expect to encounter.

COLOR	SYM
BLACK	B
WHITE	W
RED	R
YELLOW	Y
GREEN	G
TAN	T
BLUE	BU
BROWN	BR
PINK	PK
ORANGE	OR

Courtesy Frigidaire Corp., Div. of General Motors Corp.

Fig. 8. Pictorial wiring diagram of Model FCD-123T.

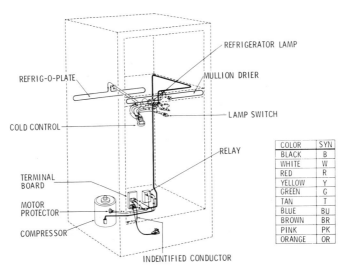

COLOR	SYN
BLACK	B
WHITE	W
RED	R
YELLOW	Y
GREEN	G
TAN	T
BLUE	BU
BROWN	BR
PINK	PK
ORANGE	OR

Courtesy Frigidaire Corp., Div. of General Motors Corp.

Fig. 9. Pictorial wiring diagram of Model FCD-150T.

Fig. 10. Schematic wiring diagram of Model FPI-152T.

Courtesy Frigidaire Corp., Div. of General Motors Corp.

Fig. 10. Schematic wiring diagram of Model FPI-152T.

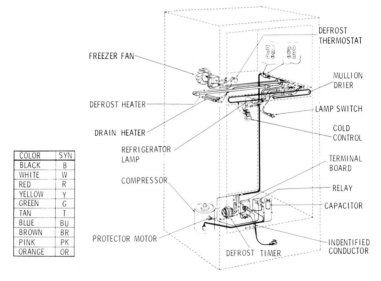

COLOR	SYN
BLACK	B
WHITE	W
RED	R
YELLOW	Y
GREEN	G
TAN	T
BLUE	BU
BROWN	BR
PINK	PK
ORANGE	OR

Courtesy Frigidaire Corp., Div. of General Motors Corp.

Fig. 11. Pictorial wiring diagram of Model FPI-152T.

TOTAL RUNNING WATTS - MIN. 245 - MAX. 300

Fig. 12. Schematic wiring diagram of Model FPCI-152T-7.

SOLENOID - 400 Ω
WATER VALVE

SEE ICEMAKER LABEL
FOR INFORMATION

ICE MAKER

DRAIN HEATER - 8820 Ω

REFRIGERATOR LAMP

LAMP SWITCH

MULLION DRIER - 1310 Ω

DEFROST HEATER
25 Ω

DEFROST
47000Ω THERMOSTAT

COLD
CONTROL

FREEZER
FAN

MOTOR
PROTECTOR

CAPACITOR

COMPRESSOR
10.7 Ω

1000Ω

RELAY

2.8 Ω

DEFROST TIMER

LINE CORD

Courtesy Frigidaire Corp., Div. of General Motors Corp.

WATER VALVE
(IF USED)

ICE MAKER

DEFROST
THERMOSTAT

ICE MAKER WIRING HARNESS
(IF USED)

FREEZER FAN

MULLION
DRIER

DEFROST HEATER

LAMP
SWITCH

DRAIN HEATER

COLD
CONTROL

REFRIGERATOR LAMP

TERMINAL
BOARD

RELAY

COLOR	SYN
BLACK	B
WHITE	W
RED	R
YELLOW	Y
GREEN	G
TAN	T
BLUE	BU
BROWN	BR
PINK	PK
ORANGE	OR

COMPRESSOR

CAPACITOR

IDENTIFIED
CONDUCTOR

MOTOR
PROTECTOR

DEFROST TIMER

Courtesy Frigidaire Corp., Div. of General Motors Corp.

Fig. 13. Pictorial wiring diagram of Model FPCI-152T-7.

255

TOTAL RUNNING WATTS - MIN. 300 - MAX. 360

Courtesy Frigidaire Corp., Div. of General Motors Corp.

Fig. 14. Schematic wiring diagram of Model FPF-200TI.

Fig. 15. Pictorial wiring diagram of Model FPF-200TI.

Fig. 16. Schematic diagram illustrating the automatic defrost thermostat connections.

Courtesy Hotpoint Co.

CHAPTER 10

Cabinet Maintenance And Repairs

Proper fit of the door is of the utmost importance in any refrigerator. The door seal should always be tested before a refrigerator is placed in operation or when servicing a refrigerator in the home. A poor door seal usually results in cabinet sweating inside, excessive frosting of the cooling unit, high percentage of running time, high power consumption, slow ice freezing, high cabinet air temperature, and possible sweating on the outside front of the cabinet. A typical door construction and seal assembly is illustrated in Fig. 1.

A good method of testing the tightness of a door seal is to place a light inside the cabinet; this should be a strong flashlight. The light will be visible at any point where the door gasket does not make sufficient contact with the cabinet. Another method to locate imperfect door seals is by the use of a 0.003-in thick metal feeler. Locate the point of a poor seal by inserting the feeler at various points around the door between the gasket and the cabinet front with the door closed. A magnet-enclosed gasket seal is shown in Fig. 2.

If a poor seal is located, check the gasket first to see that it is not excessively worn. Check the hardware to see that it is not sprung or worn, and that the screws are tight. A poor seal can be corrected by rehanging the door, by replacing the gasket, or by properly adjusting or replacing the hardware. If, however, a poor seal is caused by the door being sprung

Fig. 1. Cross-section of typical refrigerator door; (A) Represents inside door panel; (B) Door seal; (C) Outside door panel; (D) Wooden frame; (E) Wooden frame; (F) Insulation.

MAGNET

Fig. 2. Cross-section of a typical magnetic door gasket.

out of line or by the front of the cabinet being out of line, the poor seal can be corrected as follows:

1.   If the leak is on the hinge side of the door, open the door and hold with the left hand. Then with the heel of the right hand strike all

259

along the hinge edge until the door is sprung in sufficiently to tighten the seal.

2. If leakage is on the top-latch side, spring the door in at the top until contact is made along this side by opening the door, holding it at the bottom, and pushing in at the top. Reverse this process if the leakage is at the bottom on the latch side.

3. To stop a leakage at the top or bottom of the door, spring the door in at the top or bottom by opening the door, holding one end, and pushing on the end to be sprung in.

4. Place friction tape in back of the door gasket at any point where the gasket does not make a tight fit after the above procedure has been followed. Be sure to fold the tape so that it does not show above the gasket.

DOOR REPLACEMENT

Doors are obviously replaceable by removing the hardware and transferring it to the new door, as shown in Fig. 3. Carefully center the door in the opening and draw the hinge screws tight in rotation so that unequal pressure of the screws will not throw the door out of line.

Hinge Replacement

Hinges usually have a metal cover that conceals the screws. To remove this cover, insert a knife blade in the slot and pry out. The screws are now exposed and can be removed, and the hinge can be replaced. The hinge screws are embedded in heavy screw plates attached to the inside of the steel case. This prevents any sagging or distortion of the cabinet.

Latch Replacement

First, remove the escutcheon that usually encloses the base of the handle by inserting a small screwdriver in the slot under the handle at the bottom of the escutcheon. Pry it off at the bottom, then lift it up and off. One type of escutcheon is removed by squeezing it in the middle of its plate to elongate it, then simply slipping it off. The defective latch can now be removed and replaced with a new one.

Latch Strike Adjustment

Three latch-tongue adjustments may be made to obtain the correct door closure and gasket seal. To make *up and down* adjustments, pencil a line

Courtesy Whirlpool Corporation

Fig. 3. Exploded view of typical door parts.

on the cabinet flange just opposite the front edge of the latch mechanism nylon well (Fig. 5). Loosen the mounting screws and adjust the latch tongue up or down until the lower point of the tongue is approximately 1/16″ above the penciled line. To make *in and out* adjustments, slowly close the door until a slight resistance is felt as the latch tongue contacts the rear edge of the mechanism nylon well. The distance between the door gasket and the cabinet flange should be approximately ¼″ when the tongue touches the well. Only one distinct resistance should be noted as the door is closed.

261

Courtesy Whirlpool Corporation

Fig. 4. Replacement of food liner in a typical refrigerator door.

If two contacts occur, the latch tongue is positioned too low on the door. Shims may be added to or removed from the latch tongue for in and out adjustments. *To make sideways adjustments,* hold the door in a partially open position and measure the distance between the cabinet flange and the side on the latch tongue. Loosen the mounting screws and adjust the tongue sideways until this distance is approximately ¼''.

NYLON WELL

PENCIL LINE

LATCH TONGUE

1/16"

Fig. 5. Typical latch tongue adjustment.

Interior Light Arrangement

Refrigerators are equipped with an automatic interior electric light, as shown in Fig. 6. The switch is operated by the door and is designed and wired to complete the lamp circuit and light the lamp when the door is opened. When the door is closed, the switch contacts open, and the light is extinguished.

Due to careful design of the switch and simplicity of the circuit, service complaints dealing with this part of the refrigerator are very infrequent. Trouble is most often found to be a burned-out lamp. Thus, if the lamp will not light, check for a burned-out or loosened lamp. If no trouble is found here, check the light-switch plunger; it may be loose or stuck. Check all terminals of the light circuit for loose connections. Sometimes a broken wire may be responsible for the failure, although this is a rather remote possibility.

263

Fig. 6. Pictorial diagram of an electrical system including interior light arrangement in a typical refrigerator.

CHAPTER 11

Household Freezers

Household freezers are manufactured in various sizes to suit family requirements. The purpose of household freezers is to facilitate maximum food storage over a period of time. They are usually termed as horizontal or vertical units, depending on individual requirements and installation space available. Household freezers are usually manufactured in capacities of from 13 to 20 cubic feet, and are normally equipped with hermetically sealed compressor units.

CYCLE OF OPERATION

The operation of the refrigeration system can best be described with reference to the refrigerant flow diagram shown in Fig. 1. The temperature of the freezer is regulated by a conventional temperature control, which starts the motor-compressor when the evaporator requires refrigeration, and stops the motor-compressor when the evaporator has been sufficiently cooled. When the temperature control contacts close, the motor-compressor starts operating and reduces the pressure in the evaporator. The refrigerant evaporates, absorbing heat in the process, thus cooling the evaporator. The vapor is drawn through the suction line by the compressor, which compresses the refrigerant gas and forces it into the condenser under high pressure. The compression of the vapor causes the temperature to rise permitting the condenser to transfer the heat to the cabinet shell and into the air surrounding the cabinet as shown in Fig. 2. As the compressed vapor gives up its heat, it condenses to a liquid in the condenser.

Fig. 1. Refrigerant-flow diagram of horizontal freezer.

Fig. 2. The routing of the tubing in a typical horizontal freezer.

The capillary tube is made small in diameter in order to control the amount of liquid refrigerant forced through it from the condenser to the evaporator. For a short period of time after the motor-compressor shuts off, the pressure in the condenser continues to force refrigerant through the capillary tube until the pressure is substantially equalized throughout the system. If the motor-compressor attempts to start immediately after it shuts off (before equalizing has occurred), the motor-compressor may stall and trip the overload protector, thereby breaking the electrical circuit. This prevents the motor windings from burning out.

CHARACTERISTICS OF A CAPILLARY-TUBE SYSTEM

A refrigeration system that uses a *capillary tube* as the refrigerant control has certain characteristics and advantages. A capillary tube allows the liquid refrigerant to continue its flow from the condenser into the evaporator after the temperature control has shut off the motor-compressor. This effects a reduction in the condenser pressure during the *off* cycle, and permits the motor-compressor to start easily against a relatively low pressure when refrigeration is again required.

The amount of refrigerant in a capillary-tube system is small but critical. There should be just enough refrigerant so that, during normal operation, evaporation will take place for the entire length of the freezing coil, but not beyond. If there is *too little* refrigerant, there will be insufficient cooling of the freezing coils. If there is *too much* refrigerant, the excess liquid will flow into the suction line. Refrigerant in this line will cause it to frost during the *on* cycle, and the frost will melt and drip off during the *off* cycle.

ELECTRICAL SYSTEM

The hermetically sealed motor-compressor requires a 60-cycle, 115-volt ac current, and has a capacity of from 1/12 to ⅓ hp, depending on the size of the total food-storing capacity of the freezer cabinet.

Fan Motor

The fan for cooling the motor-compressor in some models is mounted on the compressor housing. It is connected directly across the power supply to the motor-compressor after the temperature control and before the overload protector, as shown in Fig. 3. Connected in this manner, the fan will continue to cool the motor-compressor even if the overload protector has opened because of overheating. The fan will stop only when the temperature controls cut out.

267

Fig. 3. Wiring diagram of electrical connections in a horizontal freezer.

Temperature Control and Signal Light

The temperature control, in addition to its conventional function, also controls the operation of the signal light, as noted in Fig. 3. The temperature control is an electrical switch controlled by the temperature of the control feeler tube. As the feeler tube gets warmer the temperature control switches on and the motor-compressor starts. When the temperature of the feeler tube has been brought down to a predetermined temperature, the switch contacts open and the motor-compressor stops.

The temperature at which the control opens (turns off the motor-compressor) is called the *cut-out point,* and the temperature at which the control closes (turns on the motor-compressor) is called the *cut-in point.*

COMMON
TO MOTOR-COMPRESSOR
COMMON
TO SIGNAL LIGHT

D
C
B
A

PUSH
ROD

BELLOWS

CONTROL ARM

PIVOT

SPRING

CONTROL
FEELER
TUBE

CAM

CONTROL KNOB

MOTOR-COMPRESSOR NOT RUNNING;
SIGNAL LIGHT ON
(FOOD STORAGE TEMPERATURES ARE SAFE)

Fig. 4. Cutaway view of temperature control during normal off cycle.

Adjusting the temperature control knob changes both the *cut-in* and *cut-out* temperatures, by varying the spring tension against which the diaphragm of the control exerts pressure. The design of the control provides adequate adjustment to meet all normal requirements. For average use, the temperature control knob should be set to the normal position, which is the warmest operating position. Since there is no *off* position, this makes it impossible to shut off the freezer accidentally.

The temperature control also operates the signal warning light. The signal circuit (inside the temperature control) will be closed and the signal light located in the top part of the temperature-control escutcheon will be *on* when safe temperatures are being maintained in the food compartment. The expanding action of the control bellows (when the temperature in the food compartment rises 15° F or more above the cut-in setting of the control knob) opens the signal-circuit contacts and causes the signal light to go out when food-storage temperatures are unsafe.

269

The signal light will be *on* at all times, except as follows:

1. *Temperature control knob turned colder.* If the control knob is turned from a position near normal to a much colder position, so that the cut-in point of the new control knob position is 15° F or more below the temperature at the control feeler tube, the signal light will go out. When the temperature is reduced to less than 15° F above the cut-in temperature of the new position, the signal light will come on.
2. *Large amounts of warm food in freezer.* If a large amount of warm food is put in the freezer, the light may go out. It is best to freeze smaller amounts of food at one time.
3. *Freezer door left open too long.* The signal light will go out if the freezer door is left open for an extended length of time.
4. *Blown fuse.* A blown fuse in the electrical supply (house circuit) will cause the signal light to go out.

MOTOR-COMPRESSOR RUNNING; SIGNAL LIGHT ON
(FOOD STORAGE TEMPERATURES ARE SAFE)

Fig. 5. Cutaway view of temperature control during normal on cycle.

COMMON
TO MOTOR
COMMON
TO SIGNAL

D
C
PUSH ROD
B
A
BELLOWS
CONTROL ARM
PIVOT
SPRING
CONTROL FEELER TUBING
CAM
CONTROL KNOB

MOTOR-COMPRESSOR RUNNING; SIGNAL LIGHT OFF
(FOOD STORAGE TEMPERATURES ARE UNSAFE)

Fig. 6. Cutaway view of temperature control when signal light is off.

5. *Service cord disconnected from wall outlet.* The signal light will go out if the service cord plug is disconnected from the wall outlet.

6. Bulb (signal light) burned out.

7. Loose wire connections at the control terminals.

8. *Defective operation of refrigeration system.* If defective operation of the refrigeration system causes the temperature in the food compartment to rise 15° F or more above the cut-in point of the control-knob position, the signal light will go out.

To obtain a clearer understanding of the temperature control which regulates the operation of the motor-compressor and signal light, refer to Figs. 4, 5, and 6. During a normal *off* cycle of the motor-compressor, the feeler tube of the temperature control is cold and the bellows is contracted, allowing the spring to pull the control arm down, as illustrated in Fig. 4. Contacts *A* and *B* are open and the motor-compressor is idle. Contacts *C* and *D* are closed and the signal light is on, indicating a safe food-storage temperature.

271

FOR PURPOSES OF CLARITY, ILLUSTRATION DOES NOT SHOW
EXACT ROUTING, NUMBER OF PASSES OR PARTS LOCATION.

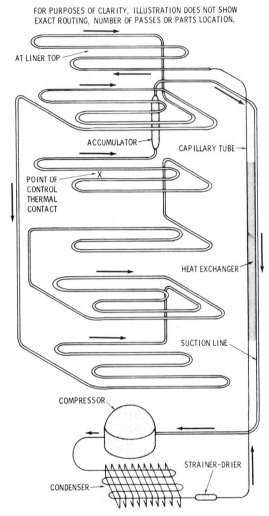

Fig. 7. Refrigerant-flow diagram of upright household freezer with fan-cooled condenser.

During a normal *on* cycle of the motor-compressor, the bellows, control arm, and spring are in an intermediate position, as shown in Fig. 5. In this position, contacts A and B are closed, and power is supplied to the motor-compressor. Contacts C and D are also closed, so that the signal light is on and a safe food-storage temperature is indicated. If, for

272

some reason, the temperature in the food compartment is 15° F or more above the cut-in temperature of the control, the bellows is expanded to a point that forces the right-hand end of the control arm upward, as shown in Fig. 6. Contacts *A* and *B* are closed, supplying power to the motor-compressor, but the push rod breaks contact at *C* and *D* and causes the signal circuit to open. The signal light will stay off, indicating an unsafe food-storage temperature, until the motor-compressor restores proper temperature in the food compartment. The bellows will then contract, and the signal-circuit contacts *C* and *D*, will close.

UPRIGHT FREEZERS

One basic refrigeration system is used on practically all upright freezers, the only major variation being the use of a fan-cooled condenser (Fig. 7) located beneath the cabinet on a few models. Other upright freezers use a static condenser (Fig. 8) located on the back of the cabinet. All units are of the conventional hermetically sealed design. For service purposes, the compressor is considered as one part, and the freezer shelf and heat-exchanger assembly as another and completely separate part.

The Refrigeration System

The first few passes of the condenser tubing (static type Fig. 8) form the oil-cooler loops that carry the partially cooled refrigerant gas back through the compressor. This lowers the operating temperature of the compressor and results in increased efficiency for the unit. The condenser releases the heat absorbed by the refrigerant and changes the hot gaseous refrigerant into a cool liquid refrigerant.

The capillary tube controls the flow of refrigerant to the freezer shelves. Part of the capillary tube is soldered to the suction line to form a heat exchanger. The heat from the capillary tube is transferred to the suction line, and the cold suction line further cools the liquid refrigerant in the capillary tube. When the refrigerant leaves the capillary tube and enters the larger tubing of the freezer shelves, the sudden increase in tubing diameter forms a low-pressure area, and the temperature of the refrigerant drops rapidly as it changes to a mixture of liquid and gas. This cold mixture passes through the top shelf (or top freezer plate in some models) then to the bottom shelf, ascending through each additional shelf until it reaches the accumulator.

In the process of passing through the shelf tubing, the refrigerant absorbs heat from the storage area and is gradually changed from a liquid

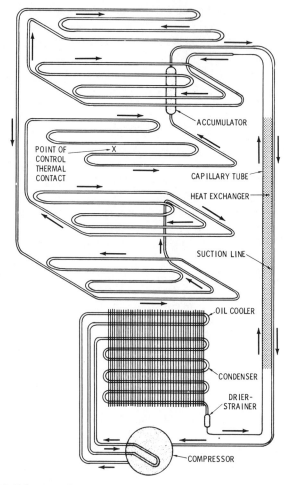

FOR PURPOSES OF CLARITY,
ILLUSTRATION DOES NOT SHOW
EXACT ROUTING, NUMBER OF PASSES OR PARTS LOCATION

ACCUMULATOR

POINT OF
CONTROL
THERMAL
CONTACT

X

CAPILLARY TUBE

HEAT EXCHANGER

SUCTION LINE

OIL COOLER

CONDENSER

DRIER-
STRAINER

COMPRESSOR

Fig. 8. Refrigerant flow diagram of upright freezer with static condenser.

and gas mixture to a gas. The accumulator traps and evaporates any small amount of liquid refrigerant remaining in the low side, thereby preventing the liquid from entering the suction line to the compressor.

The refrigerant gas passes through the suction line to the compressor to be compressed and then repeats the cycle. The temperature-control and signal-light circuits are basically the same as those previously discussed in the section on horizontal freezers.

Electrical System of Automatic Defrost Freezer

The electrical systems of automatic defrost freezers are quite similar to the electrical systems of automatic defrost refrigerators. Fig. 9, shows a schematic wiring diagram and Fig. 10 shows the same wiring in pictorial form.

TOTAL RUNNING WATTS - MIN. 215 - MAX. 315

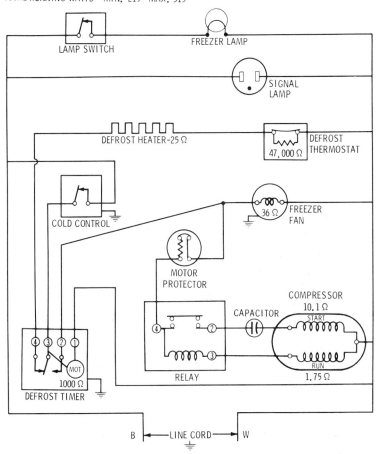

Courtesy Frigidaire Corp., Div. of General Motors Corp.

Fig. 9. Schematic wiring diagram of Model FP-156V.

COLOR	SYN
BLACK	B
WHITE	W
RED	R
YELLOW	Y
GREEN	G
TAN	T
BLUE	BU
BROWN	BR
PINK	PK
ORANGE	OR

Courtesy Frigidaire Corp., Div. of General Motors Corp.

Fig. 10. Pictorial wiring diagram of Model **FP-156V.**

Service instruction charts have been worked out which will be of help in locating and repairing faults likely to be encountered in day-to-day service work.

CAPACITOR

RECEIVER

FAN AND MOTOR

CONDENSER

CONDENSER

COMPRESSOR

Fig. 11. An air-cooled condensing unit as employed in the larger home freezers.

HOUSEHOLD FREEZER TROUBLE CHART

Symptom and Possible Cause	Possible Remedy
Compressor Does Not Run	
(a) Low voltage	(a) Check with voltmeter. Voltage should be within 10% of that stamped on the compressor-motor.
(b) Defective compressor, relay or overload protector	(b) Test the compressor. To test the compressor without any cabinet wiring in

HOUSEHOLD FREEZER TROUBLE CHART (Con't)

Symptom and Possible Cause

Possible Remedy

the circuit, unplug the compressor lead cord from the junction assembly and connect it directly to the electrical outlet (use a short extension cord if necessary). If the compressor starts, the trouble is not in the compressor, starting relay or overload protector. Plug the lead cord into junction assembly and check the temperature control and all cabinet wiring for defects or poor connections. If the compressor does not start, check the overload protector, starting relay and motor windings. To check an overload protector short across its terminals. If the compressor starts, replace the overload with one having the same part number. If the compressor does not start, look for other trouble (low line voltage, defective starting relay, defective compressor). To check a starting relay, pull it off of the compressor terminals. Remove the line cord lead from the terminal on the relay and connect it directly to the run terminal of the compressor. Plug the line cord into wall outlet, and at the same time, momentarily short between the run terminal and the start terminal. If the compressor starts, the relay is defective and must be replaced. If the compressor does not start, it is defective and must be replaced.

Signal Light Off

(a) Burned out bulb.

(b) Line cord unplugged.

(c) Loose wire connection at control.

(a) Replace.

(b) Plug in to proper outlet.

(c) Reconnect loose wire at control.

Compressor Cycles On Overload Protection

(a) Low voltage.

(b) Defective starting relay or overload protector.

(a) See previous note in this section on low voltage.

(b) See previous note in this section for proper check.

HOUSEHOLD FREEZER TROUBLE CHART (Con't)

Symptom and Possible Cause

Possible Remedy

(c) Poor air circulation around freezer.

(c) Defective freezer or compressor-motor fan. Check and replace if necessary.

(d) Overcharged or restricted system.

(d) An overcharged system may have a frostback condition on the suction line. Trouble of this kind is usually recognized by water drippings when the compressor stops. Excessively long running time, a warmer than normal freezer compartment are other signs of an overcharged system. If a system is seriously overcharged, it must be evacuated and recharged.

Compressor Runs But Freezer Is Too Warm Or Too Cold

(a) Improperly calibrated control.

(a) The temperature control may be re-calibrated by varying the cut-in and cut-out temperature. Turn the calibration screw right to "raise" and left to "lower" the cut-in or cut-out temperature.

(b) Poor gasket seal.

(b) Check and repair as required.

Compressor Runs Too Much Or All Of The Time

(a) Incorrect control knob setting.

(a) Adjust temperature control calibration screw.

(b) Defective or improperly calibrated control.

(b) Replace defective control. Recalibrate control as noted in the foregoing.

(c) Poor gasket seal.

(c) Repair or replace as required.

Compressor Runs But Freezer Is Too Warm Or Too Cold, Or The Compressor Runs Too Much Or All Of The Time

(a) Undercharged or partially restricted system.

(a) If the system is undercharged the freezer will be colder than normal. An undercharged system must be purged, evacuated, and recharged with the proper amount of *Freon*. Before recharging, however, test for refrigerant leaks as instructed elsewhere in this book. Bent tubing, foreign matter or moisture in the system may cause a partial restriction in the low side tubing. This is usually indicated by frost-free tubing between the restriction and the capillary tube and by frost-

279

HOUSEHOLD FREEZER TROUBLE CHART (Con't)

Symptom and Possible Cause

Possible Remedy

covered tubing between the restriction and the suction line. The restriction acts like a second capillary tube, increasing the pressure behind it (warming) and decreasing the pressure beyond it (cooling). Replace the component part if there is a partial restriction in the refrigerant tubing.

(b) Compressor not pumping adequately.

(b) A compressor which is not pumping adequately will produce very little cooling effect. All cooling surfaces may be covered with a thin film of frost, but the temperature will not descend to the cut-out temperature of the control, even with continuous running of the compressor. Place hand on the accumulator surface for 2 or 3 seconds, then examine the surface. If all frost is melted where touched, install gauges and check operating pressures. If high side pressures are lower than normal, and low side pressures are higher than normal, suspicions of defective compressors will be confirmed and the compressors must be replaced.

Noisy Operation

(a) Tubing vibrating against compressor cabinet.

(a) Adjust or fasten tubing.

(b) Loose compressor mounting. Unlevel cabinet or weak floor.

(b) Fasten compressor securely. Make sure the freezer is level by checking, both from front to rear, and from side to side, with a carpenter's level. An adjustable leveler is provided at each corner of the cabinet. Always level the freezer and check the gasket seal whenever the freezer is installed, and again if the freezer is moved.

Compressor With Capacitor in the Starting Winding Circuit Keeps Tripping Overload Switch

(a) Capacitor defective

(a) Test with capacitor checker or a new capacitor. Replace if defective

(b) Overload relay defective

(b) Replace overload relay

CHAPTER 12

Styles of Domestic Refrigerators and Food Arrangement

Refrigerators for household use come in a great number of styles and sizes. Some of the variations involve different cubit foot interior dimensions, single door refrigerators with an interior door for the freezer compartment, double door units with one door for the refrigerator compartment and a separate door for the freezer compartment. They also come double wide, with a door on one side for the refrigerator compartment and another door on the other side for the freezer compartment. This last style, sometimes has cold water storage built into the door, with automatic fill and automatic ice cube dispenser. These features are usually optional.

Listed below are some comparative figures of operation on different types of refrigerators.

	Average Wattage Est.	kWh Annual
Refrigerator (12cu.ft.)	241	728
Refrigerator (Frostless— 12 cu. ft.)	321	1217
Refrigerator/Freezer (14 cu. ft.)	326	1137
Refrigerator/Freezer (Frostless—14 cu. ft.)	615	1829

These are average figures, intended only to give some comparison as to operating costs. The frostless refrigerator, covered in Chapter 7 will of course cost more to operate, since heater coils are used in defrosting.

Food arrangement is of importance, since there must at all times be air circulation throughout the refrigerator, in order to keep the food at the proper temperature to prevent early spoilage.

There are some very interesting facts concerning market trends, which have been published by ASHRAE. This information contained within the quotation marks, are taken and has been reprinted with permission from ASHRAE, 1975 Equipment Handbook.

MARKET TRENDS

"The household refrigeration industry began its real growth in the 1920's, and before 1960 there was an electric refrigerator in nearly 100% of the wired homes in the United States. As a result, this large market is now primarily dependent upon the establishment of new homes and upon replacement sales."

"On the other hand, only one in five homes has a freezer, though the need is often satisfied by the freezer storage volume in the combination refrigerator-freezer."

"During 1970, 2-door combination refrigerator-freezers accounted for nearly 80% of the refrigeration market with the side-by-side type having grown rapidly in the last five years to about 19% of this quantity and the bottom-mounted freezer type diminishing to about 1%. The no-frost type of system accounted for nearly 63% of those sold."

"This increase in size and complexity of refrigeration appliances has caused a modest increase in the average selling price, though refinements in design and the developments of new materials have resulted in the customer receiving more for his money. Figures derived from data published by the U. S. Department of Labor show that in 1970 a typical refrigerator or freezer sold, at wholesale, for about 84% of the 1957-59 average price of an equivalent model having about the same storage and volume figures."

Ice Service

"Through a variety of manual, automatic, and semi-automatic means, refrigerators and an increasing number of freezers perform the important service of providing ice for use in beverages."

"The manual ice tray is made of aluminum or plastic and is filled at the kitchen sink and emptied outside the cabinet or into a container in the cabinet. Ice trays are frozen by placing them in the stream of air that is substantially below 32° F or in contact with directly refrigerated evaporator surfaces. Generally, the water freezes faster when the direct contact method is used."

"The use of automatic ice making equipment in household refrigerators is increasing annually. During 1972, 12% of the refrigerators produced in the U. S. were equipped with factory-installed automatic ice makers, and another 34% were produced with field installed ice makers. The ice makers used vary in design, but all provide the user with a convenient supply of ice cubes without the inconvenience and mess normally associated with manually filling and handling ice cube trays."

"The household ice maker mechanism is located in the freezer section of the refrigerator and requires an attachment to a water line. The ice freezing rate is primarily a function of the system design. The ice mold has been attached directly to a refrigerated surface, but today most ice makers are applied to no-frost type refrigerators, and the water is frozen by passing refrigerated air over the ice mold. Since the ice maker has to share the available refrigerator capacity with the freezer and food compartments, the ice production rate is usually limited by design to 4 to 6 lb/24 hr. An ice production rate of approximately 4 lb/24 hr., coupled with an ice storage container capacity of 7 to 10 lbs., is adequate for most users."

Figs. 13 and 14 of Chapter 7, illustrate the electrical system of a typical ice making attachment for a refrigerator.

FOOD ARRANGEMENT

The proper arrangement of food is of the utmost importance if the best result of refrigeration is to be obtained. Generally, foods should be stored according to the delicacy of flavor, as well as to their keeping qualities. Reference to Fig. 1 will indicate the proper food arrangement in a single-door refrigerator. The same relative arrangement should be maintained in a two-door refrigerator (Fig. 2). Food arrangement in apartment-type refrigerators is shown in Figs. 3 and 4. As noted in the illustrations, modern refrigerating units include a storage bin in the base of the cabinet to be used for dry vegetables which do not require refrigeration, such as potatoes, squash, dry onions, and turnips. The inner door panel is used for the storage of eggs, bottled foods, liquids, etc.

Courtesy Gibson Refrigerator Corporation

Fig. 1. Food arrangement in a typical one-door type refrigerator.

In this connection it should be noted that good air circulation is vital to the efficient refrigeration and preservation of food. If the air is restricted from circulating to all parts of the cabinet, the food in the lower area will not be refrigerated sufficiently. It is therefore important that packages and other items of food be placed in the refrigerator in such a way as to allow sufficient air circulation. Large bulky items, especially square packages that are pushed against the rear wall of the cabinet, prevent the cold air from circulating downward to the lower shelves and vegetable drawers. Space should be left between packages, and nothing should be placed too close to the cabinet walls. It is especially important that the refrigerator cabinet never be overloaded.

Placing hot foods or hot liquids in the refrigerator imposes an excessive load on the unit. Foods and liquids should be cooled to room temperature before they are placed in the refrigerator. Doing this will reduce the operating time of the unit.

Courtesy Gibson Refrigerator Corporation

Fig. 2. Food arrangement in a typical two-door refrigerator with special freezer compartment.

DOOR OPENINGS

The user should be instructed that door openings should be kept at an absolute minimum. One way to do this is to remove all of the items from the refrigerator at one time when a meal is being prepared, instead of opening the door each time something is needed. Hot, humid weather imposes a greater load on the refrigerator unit. During humid weather, moisture will accumulate on the cold inner walls of the refrigerator cabinet. This is quite normal, and is similar to the formation of moisture on the outside of a glass which is filled with a cold liquid. Although this condition is normal during periods of high humidity, it is annoying, and can be minimized by opening the refrigerator door less frequently.

285

Courtesy Gibson Refrigerator Corporation

Fig. 3. Method of connection for water cooled refrigerator.
single-door refrigerator.

WARM WEATHER EFFECT

Warm weather operation, particularly during the summer months and in locations where air contains a considerable amount of moisture, will impose an additional effect on the refrigeration unit, known as *sweating*. When the warm moisture-laden air comes in contact with the cold surfaces of the refrigerator, the moisture condenses. The result of this is a condition which affects the inside surfaces of the refrigerator cabinet. This is a completely normal condition during warm, humid weather. The extent to which this sweating condition will take place is contingent upon the number of door openings, the length of time which the door is left open, the temperature of the air outside the refrigerator, and the relative humidity.

A loose door seal will allow warm air to leak into the cabinet, and cause excessive sweating during warm weather. If the door seal is good, the user must be informed that a certain amount of sweating is normal. Also,

Courtesy Gibson Refrigerator Corporation

Fig. 4. Food arrangement in typical hotel-apartment refrigerator installation with an independent freezer unit.

he should be told that the condition can be kept to a minimum by following these recommended procedures:

1. Keep all liquids and moist food covered. This will prevent the moisture from settling on the interior surfaces of the cabinet.
2. Keep the number of door openings to a minimum—*this is extremely important.*
3. Defrost often. On models which do not have the automatic defrosting mechanism, it might be necessary to defrost as often as twice a week during warm, humid weather.

COOLING COIL FROSTING

The refrigerator unit will run longer during very hot humid weather, and will cycle more frequently than in cooler weather. If the control knob is set *too high* under these conditions, the *off* cycle (on automatic defrost units) may not be long enough to allow the cooling coil to defrost completely. In this event, the entire cooling coil will become heavily frosted within a very short period of time. This condition is a result of the inability of heat from the heating wires to penetrate the thick frost surface. This will tend to increase the temperature in the fresh-food area.

The solution to this problem is to turn the thermostat to a low setting for a period of 12 hours, or until the cooling coil has defrosted completely. Then the control knob should be kept at a lower setting to allow the cooling coil to defrost completely during any *off* cycle that might occur. Because each *on* cycle will begin with a cooling coil free from frost, better fresh-food temperatures will be obtained.

CLEANING

Regular cleaning of the refrigerator interior is of the utmost importance. A warm baking-soda solution is very good for cleaning the interior and will keep it fresh and sweet smelling. In the absence of baking soda, warm sudsy water may be used. Plastic interior pans should also be cleaned in order to keep the food fresh and free from odors. Use only mild sudsy water (the same as that used for hand dishwashing) or a baking-soda solution. Exteriors should be washed regularly with clean sudsy water.

ODOR PREVENTION

In addition to the regular cleaning, prompt attention to spillovers (especially milk), or an orange or lemon that has been overlooked, is most important to maintaining a sweet smelling refrigerator. When cleaning both the refrigerator and freezer sections, pay particular attention to the underside of the shelves and the moisture pan, and examine the crevices in the door seal, especially in humid climates.

Package any foods to be frozen in moisture and vapor proof wrapping paper. This is especially important in case of fish, onions, and foods containing garlic. It is also a good idea to rinse the hands in lemon water after handling fish and before handling the outside wrappings of fish packages. Charcoal canisters (available from hardware and department stores) can be placed in the refrigerator or food freezer to help in odor removal. Whenever a refrigerator is to be turned off for a few weeks (or when it is to be moved), it is important that it be thoroughly cleaned and dried. The door should be left ajar for air circulation, except during the actual move. A closed-up odor takes time to disappear after the refrigerator is cooled down and in use again. Swinging the door wide each time it is opened will permit a greater change of air inside the refrigerator, and will help speed up this process.

CHAPTER 13

Installation Methods

Years ago there were home refrigeration systems that required the cabinet to be located in one location and the refrigeration unit in another such as in the basement. Now, practically all home refrigeration is in a self-contained unit. A few instructions for the installation of the self-contained units are appropriate.

All that is required is to place and level the cabinet, remove the shipping bolts, test for leaks, and put the unit into service. The service is complete when the temperature control dial is set in the desired position.

A method practiced by numerous distributors of household refrigerators is to uncrate the cabinet at the service shop. This is an excellent plan to follow, inasmuch as it permits the unit to be operated and checked through its cycle. It permits the delivery of a tested unit and also eliminates uncrating, with the accompanying disorder on the customer's premises. However, when this method is impractical, the following procedure is recommended:

1. Determine where the customer wants to locate the refrigerator, then make an inspection with the customer, bearing in mind the following important factors that are necessary for an efficient and smoothly running refrigerator.

2. Perform the uncrating operations carefully so that the finish of the cabinet will not be scratched or marred.

3. Remove the shipping bolts and wood blocks that hold the condens-

ing unit secure during shipment. Some cabinets are shipped with mounting springs or suspension parts in a separate package, usually accompanied by detailed instructions for this part of the installation. Regardless of the style of mounting, be certain that the unit floats freely on the spring or suspension mounting so it will be able to absorb any vibration caused by the operation. Unpack and install the shelves and defrosting trays.

4. The back of the cabinet should be at least 3'' from the wall, as illustrated in Fig. 1.

5. If the refrigerator is to be placed in a recessed wall or built-in cupboard, there must be an opening of at least 3'' at the top and sides of the cabinet for proper ventilation, as shown in Fig. 2.

6. Be sure that the floor is solid. If not, it may be necessary to brace the floor directly beneath where the refrigerator is to be located.

Fig. 1. Minimum distance to wall when installing self-contained refrigerator.

Leak Test

As a general rule, most self-contained units are shipped with the refrigerant confined in the condenser or liquid receiver. During shipment, it is possible that, due to strains at one or more joints, a leak develops.

Fig. 2. Method of placing self-contained refrigerator in recessed wall space underneath a built-in cupboard. Dimensions shown are minimum for proper ventilation.

INSTALLATION OF ABSORPTION-TYPE REFRIGERATORS

Household refrigerators employing the absorption cycle do not differ in any important respect (with regard to location of units) from that of the previously discussed units in which the compression method of refrigeration was used. All units of this type require heat energy instead of mechanical energy to make the change in condition required in the refrigeration cycle. The heat energy required is usually obtained from a gas flame, with the gas piped from a convenient place (usually in the basement of the building).

There are two principal types of units used, namely:

1. Water-cooled units.
2. Air-cooled units.

A typical refrigerator employing water-cooled condensers is shown in Fig. 3. This type of unit usually necessitates tapping the water system of the house or apartment in which the installation is to be made. Although this is a rather simple problem, it should be done with care and in accordance with the Underwriter's Code or any local rules that affect such work. In addition, a pipe must be installed from the gas supply line; in most locations this will require a permit from the gas company.

Cooling Water Line Installation

The purpose of the cooling water line is to furnish an uninterrupted supply of cool water to the condenser. After it is determined where the customer wants to locate the refrigerator, ascertain by means of a careful inspection that the location is suitable with respect to ventilation, accessibility, maintenance, and repairs, etc., then proceed. Make certain that the connections to the unit are made at a point in the water supply line where full pressure and flow will be constantly maintained. Close the main valve in the water supply line.

Make the extension from the house line with a ½'' diameter pipe to a point that is within approximately 10 ft. of the unit. At this point, a shut-off valve, water filter, and pressure regulator should be installed, as shown in Fig. 3. The connection between the water pressure regulator and the unit should be made with copper tubing. When securing the lines, only brass fittings and nipples should be employed. The water line should be separated from the gas line to prevent condensation in the gas line.

Drain Line Installation

The drain line is that part of the piping required to drain the heated water out of the condenser so that a constant flow of water in the system is maintained. This line is connected to the water control valve outlet of the unit and to a convenient drain outlet, or waste line. When this latter method is used, a check valve must be installed in the drain line to prevent water from backing up into the unit.

Fig. 3. Food arrangement in a typical hotel-apartment single-door
refrigerator.

It is customary to install all connections as close to the unit as possible;
this facilitates the inspection and adjustment of the water control valve.
Vertical drops in the drain line should not exceed 5 to 6 ft., since it may
result in a partial vacuum in the water control and increases water
consumption. An elimination of this tendency is sometimes accomplished by the installation of a larger pipe (½'' or ⅝'') instead of the
⅜'' standard tubing.

295

Gas Line Installation

When installing a gas line, first turn off the main valve at the gas meter. The gas in the line is consumed by lighting a gas burner connected to the line. It is considered good practice to tap the refrigeration service main directly after the meter and ahead of any other gas appliances to ensure a steady and uninterrupted supply of gas.

The gas line usually consists of ½'' black-iron pipe; this pipe should be extended to within approximately 10 ft. of the unit, at which point the gas shut-off valve is installed in a closed position, and the main gas valve is opened at the meter. All pilot lights that became extinguished when the gas was shut off should be lighted, and the gas jets that were opened to exhaust the gas from the line should be closed.

Gas Pressure Regulator

The gas regulator is now installed in the line next to the gas shut-off valve as shown in Fig. 4. The function of this gas regulator is to automatically regulate the gas so it will maintain a constant burner pressure independent of any gas-line pressure fluctuations, thus ensuring a uniform rate of gas supply to the burner. The gas regulator should be installed at least one foot from the gas thermostat in a horizontal position and with the arrow on the regulator pointing in the direction of gas flow. The installation of the gas line is completed with copper tubing connected to the thermostatic gas control on the unit.

Traps in the line should be avoided, since they have a tendency to become filled with condensation, thus interfering with the free flow of gas. The gas line should be insulated whenever it is in direct contact with cold walls or where it runs through cold rooms or refrigerated compartments. Before the installation is completed, a careful check for gas leaks as well as a thorough inspection of all joints should be made.

Fig. 4. Method of connection for air cooled refrigerator.

CHAPTER 14

Compressor Lubrication System

Some mention was made in Chapter 3: Compressors, about lubrication. At the present time, the oil refining industry, by continuous research, is able to advise and supply the potential user with the type of oil best suited to his purpose.

MOTOR LUBRICATION

In a hermetically sealed compressor unit, the motor and compressor are lubricated by the splash system. This will apply to smaller commercial units as well as domestic units.

Lubrication of motor and fan bearings is usually accomplished by continuously feeding small quantities of oil through a piece of wicking or felt from a reservoir directly to the bearings. This makes it necessary to use an oil that is light enough in body to pass through the wick readily, but heavy enough to form a proper film on the bearing surface at operating temperatures. The oil must also have a low ''pour point'' so that it will flow freely to the bearings when starting up cold. The best oil will resist deterioration due to oxidation and polymerization, which develops stickiness and decreased lubrication. The use of properly refined oil will eliminate most of these difficulties. When one considers the small amount required, and the costly damage that may result from using an inferior oil, it can be seen that the best is the cheapest.

Grade 20W, non-detergent automobile oil serves well for lubricating motor bearings of the bronze sleeve type, or poured babbitt bearings. The small condenser fan motors are usually supposed to be oiled for the life of the unit. They are usually single-bearing motors, with the bearing on the end opposite from the fan. If one of these fans has dry bearings and sticks, it will generally not burn the winding as they are high impedance type motors.

Should this happen, one may in many cases drill an ⅛" hole in the bearing cap, and with an oil can relubricate the felt wick, which is used to hold the oil for the bearing lubrication, and then turn the shaft back and forth until oil gets into the bearing and frees the shaft. Use a self tapping metal screw to close the drilled hole.

In oiling motor bearing, do not over oil them, as it will just run out and possibly get into the motor windings.

Many motors now come with ball bearings. Some of these, usually in the smaller size motors, are sealed on both sides, and cannot be greased. If they get noisy, replace them with a direct replacement bearing.

On larger motors, the bearings are usually greasable. You may readily tell, as the bearing enclosure on the end bells will have two ⅛" pipe plugs. A ball bearing may be quickly ruined, by using the wrong type of grease, or by over greasing them as they will heat up. Remove one plug and replace it with an Alemite fitting. Use the high grade ball bearing grease. Ball bearing grease is a short fiber grease. When you are ready to grease the bearings, remove the ⅛" pipe plug, which is usually about opposite to where you install the greasing fitting. Now pump grease in until it starts to come out the plug hole. Stop adding grease, start the motor and let it run for a little while until the grease quits coming out the plug hole. Replace the ⅛" pipe plug and you have thus removed grease pressure on the ball bearing.

Set specific times for checking the motor lubrication, if an operator is in attendance, good times to grease are New Years and the 4th of July.

COMPRESSOR LUBRICATION

There are so many types of refrigeration compressors that it is impossible to write specifically what will fit all requirements. Even when pressures and machines are quite similar in design, they are apt to require individual consideration for proper lubrication. Speed, clearance, temperature conditions, etc., have an important bearing, and the system of operation and the refrigerant used are the determining factors. Most oils are chosen for a given job as a result of actual tests; therefore, servicemen in the field will profit by always using the oil specified by the manufacturer of the unit being serviced, or by using an oil known to have the same specifications. Buy oils from reputable refiners only, preferably those who have had real experience in refrigeration problems.

LUBRICATION METHODS

Compressors are lubricated in various ways, depending on their size and other factors. Compressors serving commercial refrigeration are lubricated by various methods, such as:

1. Splash
2. Semiforced feed
3. Forced feed

Enclosed booster compressors are usually equipped with an internal forced-feed lubrication system which supplies oil under pressure to the main bearings, connecting rod bearings, and the compressors shaft seal. With reference to Fig. 1, the lubrication system consists of a rotary gear pump driven from the compressor crankshaft, the oil filter, and internal oil passages. The normal crankcase oil level covers the oil filter and the oil-pump suction to assure that the pump is always primed and to minimize the possibility of gas binding. The oil pump takes suction from the crankcase through the oil filter and discharges this oil vertically into a passage in the pump housing. From the pump, the flow of discharge oil divides, part flowing to the pump end and part flowing through an internal pipe to the seal end of the compressor.

On two-cylinder compressors, part of the oil which flows to the pump end lubricates the pump end main bearing, and the remainder flows into the crankshaft through drilled and piped passages to lubricate the connecting rod bearings, crankpin, and wristpin. Oil passing through the bearings falls back into the crankcase. A relief valve and a pressure gauge are connected to the oil pump discharge above the pump end main bearing. The oil which flows to the seal end passes into the seal end main bearings near the center. The oil which flows through the bearing toward the pump end falls directly into the crankcase. That oil passing through the bearing toward the seal end flows into the shaft seal space to flood the space and lubricate the seal surfaces, and returns to the crankcase through an external vent line above the shaft seal housing.

Courtesy the Frick Company

Fig. 1. Sectional view of a typical compressor showing lubrication details.

Oil Failure Switch

On large commercial compressors, to avoid operation at oil pressures below safe minimum, an oil failure switch is usually furnished. The switch is activated by pressure difference between the crankcase pressure and the oil pump discharge pressure. This switch opens and interrupts the

301

control circuit to the compressor motor if the differential oil pressure falls below a safe limit.

NECESSARY LUBRICATION PRECAUTIONS

Halocarbon refrigerants are miscible with oil and, during operation, a certain amount of oil is carried through the system with the refrigerant. This is a normal condition, and the oil which leaves the crankcase when an idle system is started will soon begin to return as the system reaches normal operating temperatures and pressures. During shutdown periods, a certain amount of refrigerant is absorbed by the oil in the compressor crankcase. The actual concentration of refrigerant in the oil depends on the duration of the shutdown period, the temperature of the oil, and the pressure in the crankcase, the concentration increasing with lower oil temperatures and higher standby crankcase pressures.

When the compressor is started, the crankcase pressure is rapidly lowered to the suction pressure, permitting the refrigerant to boil out of the oil. This causes the oil in the crankcase to foam, and, if the concentration of refrigerant is great enough, it is possible that sufficient oil may be carried out of the crankcase to cause temporary loss of oil.

There are a number of methods of perventing too much loss of oil from the compressor, after a shut down.

To avoid excessive concentrations of refrigerant in the compressor crankcase oil during shutdown periods, a crankcase oil heater is usually furnished. This heater is immersed in the crankcase oil and should be wired into the compressor motor control circuit so that the heater is de-energized when the compressor is running and energized at all times while the compressor is idle.

Crankcase oil heaters cannot prevent liquid entering the crankcase during off cycles as the result of poor piping arrangements or leaking liquid feed devices. They will, however, maintain the crankcase oil at a temperature higher than other parts of the system, thus minimizing the absorption of the refrigerant by the oil.

Operation with this arrangement is as follows: Whenever the temperature control device opens the circuit, or the manual control switch is opened for shutdown purposes, the crankcase heater is energized and the compressor keeps running until it cuts off on the low-pressure switch. The crankcase heater remains energized during the complete off cycle, and it is thus important that a continuous live circuit be made available to

302

the heater during the off time. The compressor cannot start again until the temperature control device or manual control switch closes, regardless of the position of the low-pressure switch.

The use of a crankcase heater with single pump-out at the end of each operating cycle requires:

1. A liquid-line solenoid valve in the main line or in the branch to each evaporator.
2. Use of a relay or the compressor motor starter auxiliary maintaining contact for obtaining a single pump-out operation before stopping the compressor.
3. A relay or auxiliary starter contact for energizing the crankcase heater during the compressor off cycle and de-energizing it during the compressor on cycle.
4. Electrical interlock of the refrigerant solenoid valve with the evaporator fan or chilled-water pump so that the refrigerant will be stopped when either the fan or pump is out of operation.
5. Electrical interlock of the refrigerant solenoid valve with the safety devices, such as the high-pressure cutout, oil safety switch, and motor overloads, so that the refrigerant flow valve will close when the compressor stops through the action of one of these safety devices.

Domestic refrigerators and freezers are often placed in garages that are not heated. This is fine in the hot months, but when the temperature drops below the boiling point of the refrigerant, the refrigerant is absorbed into the oil, and upon start-up, the oil will flow with the refrigerant to a degree, and possibly lower the level of oil in the compressor. If the home refrigerator or freezer is in such a cold spot, heat must be applied to the compressor so that the above results do not damage the compressor. This may be accomplished by a small lamp bulb, placed so that it will not cause a fire. The proper method is to not place these refrigeration items where they will be subjected to extremely cold temperatures.

The crankcase heater is used primarily on larger types of compressors. With the smaller compressors, the unit should be pumped down if it is to be shutoff for a considerable length of time. This is accomplished by the installation of a vacuum gauge on the suction side of the compressor, shutting off the compressor valve from the rest of the system, and pumping the compressor crankcase down to about one pound of pressure.

After accomplishing this, place tags on the equipment to explain what has been done, so whoever starts it again will know exactly what to do to the unit. This will also save loss of refrigerant through the shaft seal.

Compressors are sometimes equipped with flooded type evaporators to prevent oil loss after a shutdown. It is recognized that neither automatic pump-down control nor a single pump-out operation are practical in systems employing flooded evaporators, unless suction-line solenoid valves are added to the system. Therefore, with flooded type evaporators, the following control arrangements are considered satisfactory:

1. Manual operation as described in the following. No crankcase heaters required.
2. Automatic control from temperature controllers or other devices, provided crankcase heaters are used and energized on the off cycles and the liquid solenoid valve closes whenever the compressor is stopped. Where water cooling of the compressor is employed, a solenoid valve in the water supply line should close whenever the compressor is stopped.
3. Same as the foregoing, except with the added precaution of a single pump-down of the compressor for night or weekend shutdowns. This can be accomplished by manually closing the compressor suction stop valve. If the single pump-out procedure is used, the compressor will automatically pump down once, then shut off on low-pressure cutout and stay off until manually restarted. This feature cannot be used if the low-pressure cutout also acts as the last capacity control step.

SERVICING DATA

Before adding oil to a compressor, moisture must be reduced to an absolute minimum to avoid undesirable chemical reactions as well as the possible formation of ice and the resulting check of refrigerant flow. The oil should be dry when it is purchased, and it should be so handled that it will not have a chance to absorb moisture before it is put to work in a machine. Large users sometimes remove moisture by centrifugal or filter presses. Also, they may buy oils of lower quality (and, incidentally, lower price) and refine and filter this oil until it meets their specifications. This should not be attempted by anyone who does not have a well equipped laboratory and competent chemical control.

The percentage of moisture in oil is related to its dielectric strength which may be determined with suitable equipment. The oil should pass at least a 25, and preferable 30, kilovolt (kV) test. The latter is equivalent to approximately 0.01 per cent, the former to about 0.03 per cent, by weight of moisture. Do not purchase refrigeration oil in too large containers. When adding oil, pour what you feel that you will need into a thoroughly cleaned and dried jar, and place the suction tube into the jar to draw the oil into the compressor. If you remove a filler plug to pour the oil directly into the compressor, thoroughly clean the area around the filler plug.

Acidity

An oil should not only be neutral when purchased, but should remain so during storage and after being charged into a machine. In use, an oil may become acidic due to acids formed by the action of moisture on the refrigerant, as the result of oxidation of certain constituents of the oil, or through formation of organic acids from saponifiable matter due to the action of oxygen and(or) moisture. Obviously, prevention of development of acidity during operation may be accomplished by keeping the system dry and using neutral oil having a low oxidation number and free from saponifiable matter.

Wax Separation

When methyl chloride, *Freon-12* (and other *Freons*), and *Carrene* (methylene chloride) are mixed with oils in the amounts usually circulated (less than 1 to 10 per cent or higher) in refrigerating systems, wax will separate upon cooling sufficiently. This phenomenon is not to be confused with the possible separation of wax from the pure oil. An oil may test wax free and still separate wax on being mixed with the refrigerant. Due to the limited solubility of oil in liquid sulfur dioxide, wax separation does not present a problem.

Removal of Oil Sludges and Solids

The effects of the formation of an oil sludge in a refrigerating machine are obvious, but a few of the results are:

1. Blocks oil passages and prevents proper lubrication.
2. Packs behind piston rings, reducing compressor efficiency.
3. Causes sluggish valve action, also reducing efficiency.
4. Carries over to the expansion valve or float valve, causing sticking and failure to function properly.

305

5. Remains in condenser coils and reduces cooling efficiency.
6. Carries into the expansion coil with detrimental insulating effect.

According to the *Ansul Chemical Company*, solids removed from refrigeration systems have always had an air of mystery about them. However, mystery always surrounds something not understood. Through continuing research, much has been clarified in the past few years. It is important that all in the industry realize the simplicity of the problem and its solution.

The various kinds of solids are easily classified both as to type and origin. They are:

1. Free metals.
2. Corrosion.
3. Metal oxides.
4. Oil sludges.
5. Ice in capillary tubes, expansion valves, or coils.
6. All other extraneous solids.

For convenience, these six types of solids, their origins and remedies, are listed in Table 1 at the end of this chapter. The remedies for cleanup and recurrence of these refrigeration solids are simple and can be summed up this way. All are a case of filtration of solids, absorption by a desiccant of water and acid, and absorption of oil decomposition products by the desiccant from the oil in the beginning stages of breakdown. In many cases, the oil breakdown progresses far enough to impair lubricity. It is difficult to tell when the breakdown has reached this point and, because of this, an oil change is usually recommended when oil breakdown is involved.

Much money is spent to put controls and protective devices on equipment for the safety of the compressor. In many cases these devices are never used. Often no means of protection against moisture and other contaminant troubles is installed. Yet moisture problems are far more prevalent than any of the problems which the common protective devices seek to overcome. This fact is proved by surveys which show that approximately 80 per cent of all service calls are either directly due to moisture or traceable to it. Thus, a good filter-drier is the best insurance.

In a clean, dry, acid-free system, there is nothing to filter. Moisture is the key to corrosion solids and oil sludges. If moisture levels are low enough, the acidity remains low and these troublesome solids do not form. A factory-clean system fitted with the proper filter drier at the factory and installed with care should operate indefinitely. The filter-drier is a means of prevention for the manufacturer and an all-around remedy for the service contractor.

SUMMARY

The problem of lubrication is an important factor in successful refrigerant-plant operation and has been fully covered in this chapter. Lubrication and control methods, together with crankcase oil heaters, and servicing data, are also included. Suggested methods for removal of refrigeration solids have been tabulated to assist in refrigeration-plant operation and maintenance.

Table 1. Suggested Methods for Removal of Refrigeration Solids

Type of Solid	Possible Cause	Possible Remedy
Free Metals Powdered iron, bronze, or copper. Chips of copper or solder.	Compressor wear or incomplete cleaning of compressor body. Chips of copper may result from sawing copper lines or from raw edges broken from flared joints. Origin of solder is obvious. Check for metallic iron with magnet. Corrosion solids do not respond.	Remove from system with filter drier or plain filter.
Corrosion Solids Chlorides of iron, copper, and aluminum. Occasionally fluorides of those metals.	Action of acid on these metals. Acid may result from moisture, oil breakdown, hermetic motor burn-out, use of carbon tetrachloride or similar chlorinated solvents, or from heat breakdown of the refrigerant by soldering on lines containing refrigerant vapor.	Filter out solids with filter drier. Remove acid and water with proper type of filter drier.
Metal Oxides Iron oxide and copper oxide.	Iron oxide (iron rust) results either from allowing equipment to stand open to moist air or, most commonly, from liquid water circulating in the system along with the refrigerant. Copper oxide results from heating copper lines without passing an inlet gas, such as nitrogen, through the lines.	Remove from system with a good filter drier. Water must be removed along with the iron oxide to prevent formation of more iron oxides and corrosion solids. Remove copper oxides with filter or filter drier.

307

Table 1.   Suggested Methods for Removal of
Refrigeration Solids (Continued)

Type of Solid	Possible Cause	Possible Remedy
Oil Sludges Vary in consistency from blown oil to dry powder or tarry mass.	Result of heat in the presence of moisture and acid. May occur even if system is dry and acid free if compression ratio is too high— substantially above 10 : 1.	If oil is not too bad, it can be cleaned up by the proper activated alumina drier.
Ice in Capillary Tubes Expan.ion Valves, or Coils	System contains more moisture than the refrigerant can hold in solution at the expansion valve, capillary tube, or coil tempera- ture.	Dry system with good drier below freeze-up and corrosion range.
All other Extraneous Solids	Usually carelessness. Occasionally find desiccant or core sand, but this is very unusual.	Good filter or filter drier.

REVIEW QUESTIONS

1.   Name various compressor lubrication methods.

2.   What is the function of an oil-pressure switch?

3.   State the operation of various compressor controls.

4.   What precautions should be observed prior to adding oil to a compressor?

5.   How does acidity affect refrigeration compressor lubrication?

6.   What are the various methods used for removal of refrigeration solids?

CHAPTER 15

Refrigeration Control Devices

In order to be of service, it is necessary that the refrigeration unit function without attention; that is, it must be made fully automatic in its operation. This automatic operation is obtained by the use of certain control devices whose function it is to connect and disconnect the compressor at periodic intervals as required to keep the stored food at the low temperatures required.

Refrigerators are usually equipped with the conventional capillary tube for refrigerant control and a temperature control switch for motor-compressor control. The motor and compressor in such systems are located in a common enclosure and are therefore termed a *sealed system*. Among the several advantages of the sealed system is its compactness, with an accompanying saving of space in the unit, in addition to the simplicity of the necessary control components.

Because of the numerous instances where older household refrigerators are still in use, it is important that the serviceman have a thorough knowledge in the operation of a few common types of these older systems. This will enable him to understand others, since they all work on the same basic principles. In general, these controls may be divided into two classes, namely:

1. Those employing a mechanical means for their operation.
2. Those employing an electrical arrangement for their operation.

Often, however, the two may be combined and the system made to function as a result of the combination. An automatic expansion valve (Fig. 1), is an example of a control which is purely mechanical, while a thermostatic type of control is a device for motor-compressor control and is an example of a control in the electrical class.

While the function of a refrigerating system is the removal of heat and the consequent governing or maintaining of desired temperatures, many types of controls are employed other than merely temperature controls to complete the system. A great many controls have an indirect bearing on the temperature, although their primary purpose is to definitely operate some part of the system. The compression method of refrigeration has two main types of controls. They are:

1. Refrigerant controls (liquid, vapor).
2. Motor controls.

Refrigerant controls are usually exercised by employment of the following:

1. Automatic expansion valve.
2. Thermostatic expansion valves.
3. Capillary or choke tube.

Motor controls are usually exercised by the use of:

1. Pressure control.
2. Thermostatic control.

AUTOMATIC EXPANSION VALVE

The automatic expansion valve is used on household evaporators of the dry type. It is located at the inlet to the evaporator and is used to control the flow of refrigerant from the receiver to the evaporator. The purpose of the expansion valve is to reduce the high-pressure liquid to a low-pressure liquid as it enters the cooling coil. When the pressure has been reduced by the compressor suction to a predetermined value depending on the valve setting, the valve opens, admitting refrigerant to the evaporator. The admitted refrigerant evaporates almost immediately at a low temperature and is pumped away before admitting additional refrigerant to the evaporator.

LOW PRESS. REFRIG. VAPOR
HIGH PRESS. REFRIG. VAPOR (F-12)
LOW PRESS. REFRIG. LIQUID (F-12)
HIGH PRESS. REFRIG. LIQUID (F-12)
OIL (150 VISCOSITY)

Fig. 1. Illustrating the refrigerant circuit and component arrangement when using an automatic expansion valve as a means of controlling the refrigerant.

In reference to Fig. 2, which shows a typical automatic expansion valve, the operation is as follows: The condensing unit starts and reduces the evaporator back pressure to a point where the spring pressure forces the valve bellows, yoke, and needle downward until the valve is opened. This condition exists until sufficient refrigerant is allowed to pass through

311

the valve to raise the back pressure enough to overcome the spring tension, at which time the valve tends to close. Thus, during the time the condensing unit is in operation, a nearly constant pressure is maintained within the evaporator by the action of this valve. During the idle period, the rising evaporator pressure exerts a closing force which holds the valve needle securely on its seat, thereby reducing the flow of refrigerant.

Courtesy American Standard Control Division

Fig. 2. Typical automatic expansion valve.

THERMOSTATIC EXPANSION VALVES

The thermostatic expansion valve differs from the automatic expansion valve in that its action is governed by the temperature, whereas the automatic expansion valve is governed entirely by pressure. In construction, it is similar to the automatic expansion valve except that it has, in addition, a thermostatic element. This thermostatic element is charged with a volatile substance (usually a refrigerant that is the same as that used in the refrigeration system in which the valve is connected). It functions according to the pressure-temperature relationship and is connected (clamped) to the evaporator suction line of the evaporator and to the bellows operating diaphragm of the valve by a small sealed tube. A cutaway view of a thermostatic expansion valve is shown in Fig. 3.

Courtesy Frigidaire Corp., Div. of General Motors Corp.

Fig. 3 A typical thermostatic expansion valve.

Its operation is governed by the temperature, thus a rise in the evaporator temperature will increase the temperature of the evaporated gas passing through the suction line to which the thermostatic expansion valve bulb is clamped. The bulb absorbs heat, and since its charge reacts in accordance with the pressure-temperature relations, the pressure tending to open the valve needle is increased and it opens proportionally.

Briefly, the greater the evaporator gas temperature rise in the evaporator, the wider the valve opens, and vice versa. The wider the valve opens, the greater is the percentage of coil flooding. This improves the heat transfer. It also causes the compressor to operate at a higher average suction pressure. Hence, an increase compressor capacity will also increase overall system capacity.

For the best performance of a thermostatic expansion valve, the power element bulb of the valve must be fastened to the suction line or drier coil as close to the evaporator outlet as possible. It is EXTREMELY IMPORTANT that the bulb be securely clamped to the suction line with metal clamps.

After the unit has stabilized, the adjustment of the valve shall be such that the frost line from the evaporator stops at or slightly beyond the bulb of the thermostatic expansion valve. This keeps liquid refrigerant from going back to the suction side of the compressor.

The adjustment of this valve can be very touchy. A small turn of the adjustment valve either way, can make quite a change in the refrigerator operation. A case in point; an automatic ice flaking machine in a drugstore, was not making and releasing the ice flakes properly. The service man added refrigerant and checked it a couple of times, but it still did not do the job. Another man was called in and with about one-half turn of the thermostatic expansion valve adjustment, the machine operated perfectly. It did, however, take several tries at the valve setting and a little watching afterwards to be assured that the expansion valve was operating properly.

VALVE SELECTION

A thermostatic expansion valve selected or specified for any installation should be matched to the individual application. To do this is far more complex today than in years past. Valve selection involves several important factors that must be given careful consideration. They include:

1. Refrigerant used in system.
2. Refrigeration load.
3. Suction temperature.
4. Liquid inlet temperature.
5. Type of evaporator.
6. Operating conditions.

Many different refrigerants are used today. A valve to be selected for any given application should be designed specifically for the refrigerant in the particular system. Capacity of the valve must be matched to that of the evaporator. Manufacturers' literature indicates valve capacities covering a broad application range. The type of evaporator installed in the system dictates the type of expansion valve selected. If the evaporator has a considerable pressure drop, or is equipped with a distributor, an external equalizer valve must be used. Equally important, the type of evaporator and the type of installation should be considered in order to properly apply the various types of power element charged valves that are available.

OPERATION

A diagrammatical arrangement illustrating the operation of a typical thermostatic expansion valve is given in Fig. 4. Here, the illustration shows that part of the refrigeration system in which the temperature and pressure are being analyzed. Also shown is the cycle of action indicating the changes in the values of temperature and pressure at the various points of the cycle as well as the direction of movement of the valve needle.

By following the oval cycle of action in a clockwise direction, it will be observed that when the temperature (T_1) decreases at the thermostatic power element bulb at the evaporator outlet, the pressure (P_1) decreases in the bulb and in the thermostatic power bellows. The bellows then contract, thus closing the needle valve (N) and decreasing the refrigerant pressure (P_2) at the suction side of the valve and the inlet of the evaporator. The temperature (T_2) then decreases, due to the throttled action of (N), causing the temperature (T_1) at the outlet to the compressor to increase.

Just as soon as the temperature (T_1) increases at the outlet and at the thermostatic bulb, the pressure (P_1) increases in the bulb and the thermostatic power bellows. This causes the bellows to expand and the needle valve (N) to open. The refrigerant pressure (P_2) and the temperature (T_2)

315

increase at the suction side of the valve and at the inlet of the evaporator as a new supply of liquid refrigerant is admitted to the evaporator. As the refrigerant evaporates, the temperature (T_1) decreases and the cycle of action repeats itself.

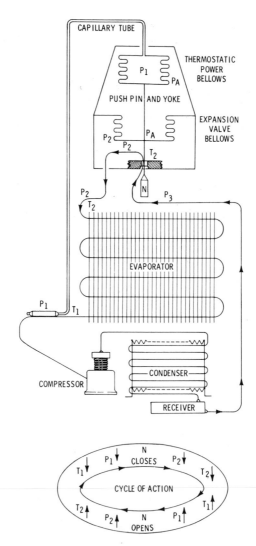

Fig. 4. Diagram illustrating the working principles of a thermostatic expansion valve.

316

External Equalizer Application

In order to compensate for an excessive pressure drop through an evaporator, the thermostatic expansion valve must be of the external equalizer type, with the equalizer line connected either into the evaporator at a point beyond the greatest pressure drop or into the suction line at a point on the compressor side of the remote-bulb location. In general, and as a rule of thumb, the equalizer line should be connected to the suction line at the evaporator outlet. By using the external equalizer type of thermostatic expansion valve with the equalizer line connected to the suction line, the true evaporator outlet pressure is imposed under the thermostatic expansion valve diaphragm. The operating pressures on the valve diaphragm are now free from any effect of the pressure drop through the evaporator, and the thermostatic expansion valve will respond to the superheat of the refrigerant gas leaving the evaporator.

In general, for best evaporator performance, the thermostatic expansion valve should be applied as close to the evaporator as possible and in such a location as to make it easily accessible for adjustment and servicing. On pressure-drop and centrifugal-type distributors, apply the valve as close to the distributor as possible, or as shown in Fig. 5.

In multiple-valve applications, the external equalizing tubes should not be connected directly into the suction line or manifold. To secure maximum efficiency from the evaporator, connect the equalizing tube of each valve into the evaporator it controls, just beyond the point of excessive pressure drop. This will insure the suction pressure of only that evaporator being transmitted to the equalizing chamber of the valve and not the average suction pressure of all the evaporators.

Courtesy Alco Valve Company

Fig. 5. Thermostatic expansion valve application on a pressure-drop distributor feeding an evaporator.

317

Valve Selection

Thermostatic expansion valve size is determined by the Btu/hr or tons load requirement, the pressure drop across the valve, and the evaporator temperature. It should not be assumed that the pressure drop across the thermostatic valve is equal to the difference between discharge and suction pressures at the compressor. This assumption will lead to incorrect sizing of the valve.

The pressure at the thermostatic valve outlet will be higher than the suction pressure indicated at the compressor, due to frictional losses through the distribution header, evaporator tubes, suction line fittings, and hand valves. The pressure at the thermostatic valve inlet will be lower than the discharge pressure indicated at the compressor due to frictional losses created by the length of the liquid line, valves, fittings, and possible vertical lift. The only exception to this is where the valve is located considerably below the receiver and the static head built up is more than enough to offset frictional losses. The liquid line should be properly sized, giving due consideration to its length, plus the additional equivalent length of line due to the use of fittings and hand valves. When a vertical lift in the liquid line is necessary, an additional pressure drop due to loss in static head must be included.

Sizing of Expansion Valves

The selection and sizing of thermostatic expansion valves is of utmost importance for the best coil performance. Valve capacity must be at least equal to the coil-load rating, and never more than twice that value. Any valve which is substantially oversized will tend to be erratic in operation, and this will penalize both coil performance and rated capacity output. Liquid-line strainers should always be installed ahead of all thermostatic expansion valves.

Thermostatic expansion valves are normally rated with *Freon-12* refrigerant at 40° F evaporator temperature, 10° F superheat, and 60 psi differential (pressure at valve inlet minus pressure at valve outlet). For capacities of other differentials, or when used with other refrigerants such as *Freon-22* and *Freon-502*, the valve manufacturer's ratings must be consulted and closely followed in reference to capacity correction factors.

Although it is frequently assumed that, when thermostatic expansion valves are used in low-temperature applications, some increased capacity results due to a higher pressure differential, this is not always true because of variations in valve design. It is always advisable under wide-range conditions to secure the valve manufacturer's recommendations. As a further precautionary note, the power-element charges of all thermostatic expansion valves must be properly selected for the operating temperature ranges and the type of refrigerant used in the system.

Thermostatic expansion valves should be located as close as possible to the evaporator inlet, and the bulb attached or inserted at a point where the refrigerant will not trap in the suction line. Keep the bulb away from the tees in common suction lines so that one valve will not affect any other valve. Externally equalized valves should be used on all multicircuited evaporators. In general, internally equalized valves are applied with single-circuited coils, except when excessive drops are encountered.

HOW TO TEST A THERMOSTATIC EXPANSION VALVE

It is quite a simple matter to make a complete and accurate test of these valves in the field. In most cases, the regular service kit contains all of the necessary equipment. The equipment required is as follows:

1. Service drum full of *Freon* or methyl chloride (in the shop, a supply of clean dry air at 75 to 100 pounds pressure can be used in place of the service drum). The service drum is merely for the purpoe of supplying pressure and, for this reason, the refrigerant used does not have to conform with the valve being tested; in other words, a drum of *Freon* would be perfectly satisfactory for testing with sulfur dioxide, methyl chloride, or *Freon* valves.

2. A high-pressure and a low-pressure gauge. The low-pressure gauge should be accurate and should be in good condition so that the pointer does not have too much lost motion. The high-pressure gauge is not absolutely necessary, but is recommended so it will show the pressure on the inlet of the valve.

3. Fittings and connections are required to complete the hook up, as shown in Fig. 5.

4. Some finely crushed ice is necessary, and one of the most convenient ways of carrying this around is to keep it in a thermos bottle. Otherwise, a milk bottle or other container is satisfactory. Whatever the container is, it should be completely filled with crushed

319

ice. Do not attempt to make this test with the container full of water and a little crushed ice floating around on top.

Fig. 6. Illustrating the necessary equipment to test a typical thermostatic expansion valve.

Procedure

Connect the valve, as shown, with the low-pressure gauge screwed loosely into the adapter on the expansion valve outlet. The gauge is screwed up loosely so as to provide a small amount of leakage through the threads. Proceed as follows:

1. Insert the bulb in the crushed ice.

2. Open the valve on the service drum and be sure that the drum is warm enough to build up a pressure of at least 70 pounds on the high-pressure gauge connected in the line to the valve inlet.

3. The expansion valve can now be adjusted. The pressure on the outlet gauge should be different for various refrigerants as follows:

Freon ....22 pounds
Methyl chloride ....15 pounds
Sulfur dioxide .... 3 pounds

Fig. 7. Illustrating refrigerant circuit and component arrangement in a typical capillary and sealed household refrigeration system.

Note: *Be sure to have a small amount of leakage through the gauge connection while making this adjustment.*

4. Tap the body of the valve lightly with a small wrench in order to determine if the valve is in operation. The needle of the gauge should not jump more than one pound.

5. Now screw the gauge up tight so as to stop the leakage through the

threads and determine if the expansion valve closes off tightly. With a good valve, the pressure will increase a few pounds and then either stop or build up very slowly. With a leaking valve, the pressure will build up rapidly until it equals the inlet pressure.

6. Again loosen the gauge so as to permit leakage through the threads, and then remove the feeler bulb from the crushed ice and warm it up with the hand or by putting it in water at about room temperature. The pressure should increase rapidly, showing that the power element has not lost its charge. If the pressure does not increase when this is done, it is a sign that the power element is dead.

Note: With the new gas charged expansion valves, the amount of charge in the power element is limited and the pressure will not build up above the specified pressure. This pressure is always marked on the power element and must be considered when testing gas charged valves.

7. With high pressure showing on both gauges, as outlined in the preceding paragraph, the valve can be tested to determine if the body bellows leaks. This should be done by loosening the packing nut and using a halide leak detector or soap suds to detect the escape of gas. When making this test, it is important that the gauge and other fittings are screwed up tight so as to eliminate leakage at other points.

Precautions

Be sure that the service drum has liquid in it and is warm enough to build up sufficient pressure. The high-pressure gauge used as shown in Fig. 6 will save a lot of trouble because it will show when there is not enough pressure on the inlet side of the valve. During the winter time especially, the service drum may become cold and develop insufficient pressure to make a satisfactory test. Be sure that the thermos bottle or other container is full of finely crushed ice and does not have merely a little ice floating on top of the water.

CAPILLARY TUBE SYSTEMS

The capillary tube or restrictor system is presently in universal use on household refrigerators because of the simplicity in operation, since it contains no valves or adjustments. Because of its nature, this type of system requires more accurate design to meet the particular requirements. The restrictor, in a sense, is a fixed control having no movable element responsive to load variations. Its element of variable control lies only in the natural variation of the factors affecting the rate of flow of the refrigerant. The most important of these are the pressure difference, the volume, and the density of the refrigerant. The refrigerant circuit and component arrangement are shown in Fig. 7.

The positive force to push the refrigerant through the restrictor is the pressure differential between the inlet and outlet—the inlet being the condenser pressure and the outlet the evaporator pressure. Acting against this positive force is the resistance offered by the friction within the restrictor. The temperature of the condensing medium has a greater effect on the pressure difference than the load change. The friction loss or pressure drop in the restrictor depends on the velocity, density, volume, and viscosity of the refrigerant, and on the diameter and length of the restriction.

The velocity depends on the volume and density of the refrigerant delivered by the compressor. The volume and density depends on the temperature and pressure of the refrigerant and whether it is in a liquid or vapor state. The viscosity of the pure refrigerant changes very little. However, the presence of oil will affect the viscosity by increasing the pressure drop. The diameter and length of the restrictor are fixed quantities in any unit.

The restrictor or capillary tube system accomplishes the reduction in pressure from the condenser to the evaporator without the use of a needle valve or pressure reducing valve, but by using the pressure drop or friction loss through a long small opening. With this system there is no valve to separate the high-pressure zone of the condensing unit from the low-pressure zone of the unit and evaporator. Therefore, the pressures through the system tend to equalize during the off cycle, being retarded only by the time required for the gas to pass through the small passage of the restrictor.

Capillary tubes, although not used to any great extent in commercial refrigeration applications, are used extensively in domestic refrigeration and also in special refrigeration units such as water and milk coolers, ice-cream cabinets, etc. The capillary tube (Fig. 8) consists of a very small diameter tube, the length of which depends on the size of the unit to be served, the refrigerant used, etc. This tube is usually soldered to the suction line between the condenser and the evaporator to effect the necessary heat exchange. This device acts as a constant throttle or restrictor on the refrigerant, and due to its length and small diameter, offers sufficient frictional resistance to the flow of refrigerant to build up a head pressure of an amount needed to produce condensation of the gas.

Fig. 8. Capillary tube showing the strainer connection.

If the two units (condenser and evaporator) were simply connected by a large tube, the pressure would rapidly adjust itself to the same value in both of them. Just as a small diameter water pipe holds back water, allowing a pressure to be built up behind the water column with but a small rate of flow, similarly the small diameter capillary tube holds back the liquid refrigerant, thus enabling a high pressure to be built up in the condenser during the operation of the compressor, and at the same time, permitting the refrigerant to flow slowly into the evaporator.

Due to the danger of foreign substances clogging the narrow passages in the tube, thus preventing the flow of refrigerant to the evaporator, a filter drier is usually provided and connected between the condenser outlet and the capillary tube inlet.

Capillary-Tube Restrictions

Restrictions of the capillary tube may be caused by foreign particles lodged in the tube, moisture freeze-up, or by a bend or kink. A restriction or stoppage in the capillary will cause excess liquid refrigerant to collect in the condenser, resulting in excessive head pressure. This condition may be recognized by the fact that the frost level on the evaporator will be low while, at the same time, the condenser will be cool; the compressor will operate for a short period of time and then cycle on the overload. A starting and stopping of the unit several times will indicate if the stoppage is of a temporary or permanent nature. If the condition is of a permanent character, the indications are that the capillary tube is clogged and should be cleaned or, if necessary, replaced.

To clean the capillary tube, remove it from the system carefully, noting the direction in which the refrigerant flows through the coil. Then secure a drum of carbon dioxide and attach the coil to it in such a way that pressure will flow through it in the reverse direction. This will usually succeed in clearing the obstacle. If pressure is applied from and in the same direction as the refrigerant, the plug will probably be forced tightly into place and will be unremovable.

In many cases where the capillary hole in the tube is not very small, it is possible to straighten the tubing and run a piece of fine piano wire through it to clear the obstacle. The tube must be straightened, otherwise the wire cannot be pushed through. When moisture freeze-up causes a restriction, it occurs at the outlet end of the capillary tube. Normally, a frost build-up can be noticed in this area. However, insulation wrapped around the tubing may conceal or limit the amount of frost formation. Expose the discharge end of the capillary and apply heat at this point. If there is enough head pressure, and if the restriction is caused by a moisture freeze-up, a gurgling noise will be heard as the heat releases the refrigerant through the tubing. It is possible that this moisture will be picked up by the drier and cause no further trouble. However, if the freeze-up happens again, replace the heat exchanger or the drier, and purge, evacuate, and recharge. If the trouble remains, change the heat exchanger. A kink in the capillary tube will have about the same symptoms as a moisture freeze-up except for the frost accumulation. Check the capillary tube along its entire length and, if possible, straighten the tube enough to relieve the restriction. Check the unit operation to see if the situation has been corrected. If the trouble still exists, replace the defective part.

Replacement of Capillary Tubes

Replacement should be performed in the shop after discharging the unit. In replacing the capillary tube, make sure that the same length is employed and that the bore (inner diameter) is exactly the same as that of the old tube. To check the bore of the old tube, cut several sections from it and carefully fit small drill bits into the ends until one is found that fits snugly. The drill bit can then be measured with a micrometer so that the internal diameter of the tubing can be determined.

The capillary tube is usually soldered to the evaporator and filter connections with ordinary soft solder and may be removed with a heavy-duty soldering iron or a torch. The capillary tube must also be unsoldered from the suction line. Care must be taken when installing the new tube to see that the tube is clean where it is to be soldered. Care must also be taken to prevent flowing too much solder into the connection which would possibly close the end of the tube. After cleaning the ends of the tube with sandpaper or emery cloth and removing the pins from the ends of the tube, apply soldering acid and tin the tube at the point where the solder will eventually make the joint. Do not tin the tube all the way to the end as it would be practically impossible to install the tube and solder the connection without closing the ends with solder. Insert the tube 1-½ inches into the evaporator fitting. Insert the other end of the tube as far as possible into the filter connection, then pull the tube back out ⅛ in. The tube may then be soldered to both connections, and clamped and soldered to the suction line in the same way as the original tube.

After changing the capillary tube, the unit should be baked out thoroughly and recharged. In case of breakage of the tube, a satisfactory field repair can be made by cutting off the old capillary tube several inches from where it enters the evaporator and joining the ends with a new capillary tube of exactly the same bore by means of a special capillary-tube union furnished by the manufacturer of the unit. This consists of a union with threaded openings at each end, into which capillary tube nuts are fitted. The finished assembly should of course have the same length as the old capillary tube. This method makes it unnecessary to remove the evaporator plug when making repairs to the capillary tube. The expansion system is sometimes referred to as being the dry system while the low-side float system is the flooded system. This is sometimes true, but not necessarily so, for the expansion valve may be used on a flooded system where the evaporator is so designed as to take the low-pressure vapor out of the top with the evaporator practically filled with liquid refrigerant.

DOMESTIC ELECTRICAL SYSTEMS

The electrical wiring may differ considerably from unit to unit, depending on the protective feature of the motor-compressor as well as the defrosting circuit or circuits. All systems are made up of a cabinet wiring harness which includes the line cord, compressor leads, light switch, control leads, and warmer wire leads (where used). In its simplest form, the wiring components (Fig. 9) consist of the following:

1. Motor-compressor.
2. Light switch.
3. Temperature control.

Courtesy Frigidaire Corp., Div. of General Motors Corp.

Fig. 9. Schematic wiring diagram of Model D-100.

Some refrigerator units use natural draft for condenser cooling, whereas others use forced draft provided by a cooling fan usually located underneath the cabinet.

The functional requirements of the motor and compressor devices are:

1. Starting of the unit.
2. Stopping of the unit.
3. To protect the unit against overload caused either by high refrigerant pressure or excessive current drawn by the motor.

The motor control mechanism in modern household refrigerators is actuated by a conventional thermostatic control switch. The length of the operating period of the compressor determines the quantity of the refrigerant pumped and the amount of cooling produced. In order to function properly, the compressor must be started when the temperature of the evaporator has risen to a predetermined point and stopped when the temperature has dropped to another predetermined point, both of which must be at such levels as to maintain the desired storage space temperature. This temperature in most household refrigerators is approximately 38° F. The start-stop controls for the motor-compressor are operated directly by the temperature of the refrigerant in the evaporator. As the temperature of the refrigerant in the evaporator rises due to absorption of heat from the food load in the unit or from the water to be frozen, the pressure of the refrigerant rises. The rise in pressure caused by the rise in the refrigerant temperature is used to act through a small syphon bellows on a switch to start and stop the compressor. It is in this manner that the temperature in modern refrigerating units is automatically controlled to provide the desired cooling.

Compressor Wiring

Compressor motor sizes vary with the cabinet capacity and are from 1/12 to about a ⅓ horsepower. The compressor motor employs a starting and running winding. The running winding is energized during the complete cycle of operation, whereas the starting winding is energized only during the starting period, when additional torque is required to overcome inertia of the unit.

Starting Relay

At the moment of starting, when the temperature control closes the electrical circuit, a surge of electric current passes through the running winding of the motor and through the relay coil. This energizes the starting relay coil and pulls up the starting relay armature, closing the starting winding contacts. The current through the starting winding introduces a second out-of-phase, magnetic field in the stator and starts the motor. As the motor speed increases, the running winding current is reduced. At a predetermined condition, the running winding current, which is also the current through the starting relay coil, drops to a value below that necessary to hold up the starting relay armature. The armature drops and opens the starting winding contacts and takes the starting winding out of the circuit. The motor then continues to run on the running winding as an induction motor. A schematic diagram of a household refrigerator circuit is shown in Figure 9, and a pictorial wiring diagram of the same circuit is shown in Figure 10. These are the simplest circuits you will probably come across.

In the next chapter more complex diagrams will illustrate the later models with more equipment connected.

Overload Protector

In series with the motor windings is a bimetallic overload protector. Should the current in the motor windings increase to a dangerous value, the heat developed by passage of current through the bimetallic overload protector will cause it to deflect and open the controls. This breaks the circuit to the motor windings and stops the motor before any damage can occur.

After an overload or a temperature rise has caused the bimetallic overload protector to break the circuit, the bimetallic strip cools and returns the contact to the closed position. The time required for the overload switch to reset varies with room temperature.

Temperature Control

The temperature control or thermostatic switch as the device is sometimes called, is the connecting link between the evaporator and the electrical system. Briefly, this control mechanism (Fig. 11) consists essentially of a thermostatic bulb (which is clamped to the evaporator), together with a capillary tube and bellows. These three interconnected elements are assembled in one unit and charged with a few drops of refrigerant, such as sulfur dioxide, or a similar highly volatile fluid.

As the temperature of the bulb increases, gas pressure in the bulb-bellows assembly increases, and the bellows pushes the operating shaft upward against the two spring pressures. The shaft operates the toggle or snap mechanism. Consequently, the upward travel of the shaft finally pushes the toggle mechanism off center and the switch snaps closed, starting the compressor.

COLOR	SYN
BLACK	B
WHITE	W
RED	R
YELLOW	Y
GREEN	G
TAN	T
BLUE	BU
BROWN	BR
PINK	PK
ORANGE	OR

Courtesy Frigidaire Corp., Div. of General Motors Corp.

Fig. 10. Pictorial wiring diagram of Model D-100.

As the compressor runs, the control bulb is cooled, gradually reducing the pressure in the bulb-bellows system. This reduction of bellows pressure allows the spring to push the shaft slowly downward until it has finally traveled downward far enough to push the toggle mechanism off center in the opposite direction, snapping the switch open and stopping the compressor. The control bulb then slowly warms up until the motor again starts and the cycle repeats itslef. The thermostatic control is set to

330

close the compressor circuit at a certain evaporator temperature and open the circuit at another predetermined temperature, and in this manner any desired cabinet temperature may be maintained.

Fig. 11. Simplified illustration showing the working principles in a bulb bellows system of temperature control.

TEMPERATURE CONTROL BY SUCTION PRESSURE

As previously mentioned, the temperature may be controlled for commercial refrigerating units by means of thermostatic switches. These are similar to domestic-type temperature control switches, but usually operate at a lower voltage than that at which the motor operates. This is accomplished by the thermostat controlling a low voltage relay to actuate the control switch for the compressor unit.

Pressure motor control switches are most common in multiple and large commercial units. Those on the larger commercial units operate at a lower voltage than that supplied to the motor, and actuate a relay to control the motor starting switch. Since the size of the motors used are variable, the control switches must operate at a lower voltage and current rating than the motor. The motor starting switch control pulls little current, so these switches can operate the magnetic coil that pulls the starting switch into start position.

The low pressure or suction side of the refrigeration unit is an indicator of the temperature of the evaporator coil, and therefore one control works well regardless of the number of coils connected to it.

This pressure control, installed in the suction line at the compressor compound valve, is actuated by a bellows whose pressure is that of the suction pressure, even though it may be at a minus or plus pressure. As the pressure builds up, the bellows pressure increases and the toggle switch is closed to start the compressor. As the pressure drops, the bellows pressure of the switch also drops, and at a preset point the contacts open and the compressor stops.

Another important control to have is a high pressure cutout switch, which should be installed on compressors, especially the larger ones. This high-pressure cutout switch may be a separate unit from the low pressure control, or may be combined in the same enclosure.

Recording chart mechanism, are quite often placed on refrigeration walk-in boxes, and more especially on walk-in freezer boxes, so that the temperature variation for a 24 hour period will be recorded. This eliminates someone watching the temperature at frequent periods. A thermostat bulb is installed in the box, and a capillary tube leads through the wall to a clock mechanism operated chart. The recorder has a pen that will draw a line showing temperatures at all times during the 24 hour period.

SERVICING VALVES

If the valve does not appear to feed enough refrigerant to the evaporator, as evidenced by too low a suction pressure or too high a superheat, the trouble may be due to:

1. A dirty liquid screen.
2. Adjusting stem screwed in too far (improper superheat adjustment).
3. Too high a pressure drop through the evaporator.
4. Lack of refrigerant in the system. (This is a very common source of trouble.)
5. Vaporization of the refrigerant in the liquid line due to excessive pressure drop. This condition, which is not uncommon on *Freon-*

12 installations, materially reduces the capacity of the valve. The remedy is to be sure the liquid line is of proper inside diameter in relation to its length, or else cool the liquid and gas mixture to recondense the gas by the use of an interchanger or subcooler.

6. The power-element bulb being located in a liquid trap in the suction line or evaporator.
7. Too low a head pressure.

If a valve appears to be feeding too much refrigerant, it may be caused by:

1. Adjustment stem screwed out too far.
2. Power-element bulb not making good contact with the suction line.
3. Dirt or scale on the valve seat.
4. Bulb placed in a location warmer than the refrigerated fixture.

If the suction pressure is lower than anticipated, this is no fault of the valve, provided the superheat of the vapor at the end of the evaporator is normal. The low suction pressure may be due to the load being light at the time, or the condensing unit having more capacity than necessary. If a refrigerating system is not functioning properly, do not arbitrarily blame the valve. Many refrigerator control valves are needlessly replaced when the difficulty is due to improper installation of the valve or to some irregularity in another part of the entire system. Always thoroughly investigate the entire system and make a definite analysis of the difficulty before attempting to rectify it. Be sure the functions of a thermostatic refrigerant control valve, its possibilities, and limitations are understood, and always be sure it is properly installed and adjusted before deciding to replace it.

SUMMARY

To obtain the necessary control of the refrigerant flow, various automatic control devices have been introduced. The purpose of these devices is to provide means by which the refrigerator machine is automatically connected and disconnected at certain periods to keep the food in storage at a required temperature. This freedom from attention has been obtained by the development of automatic controls to perform the various operations necessary for continuous production of low temperatures. As pointed out, the control devices may operate on mechanical or electrical principles or may function as a combination of both. It should be noted, however, that since it is not possible to include all the different control types in this chapter, the construction, operation, and adjustment of typical designs only are provided. Thus, the common types of liquid refrigerating metering devices, such as float valves, capillaries, automatic and thermostatic expansion valves, including testing and adjustment methods, are fully described. Refrigeration control devices operating on electrical principles are fully explained in the following chapter.

REVIEW QUESTIONS

1. What is the purpose of refrigeration control devices?

2. State the difference between mechanical and electrical control devices.

3. How do automatic controls operate?

4. What is the function of float valves in a refrigeration system?

5. Describe capillary tubes and their operation.

6. What is an automatic expansion valve?

7. How does an automatic expansion valve differ from that of a thermostatic expansion valve?

8. What is the function of an external equalizer?

9. Describe the selection and sizing of thermostatic expansion valves.

10. How may a thermostatic expansion valve be tested?

CHAPTER 16

The Electrical System

Because of the dependence upon an unfailing source of electricity for control and motor service, it is important that the serviceman or operator know the laws of electricity as well as the principles of motor operation. In this connection, it should be noted that the equipment for motor starting, control and protection varies with the type and service application.

While motors are classified either ac or dc, the direct current type will not be covered as the chances of finding dc motors used for refrigerators are quite remote.

It is not the intent of this one lesson to give a complete electrical course, as this is a subject unto itself. A rough coverage will be given, so that you will have an idea of the electrical operation of refrigeration units and control devices.

UNITS OF ELECTRICITY

In a direct current system, the products of *volts* multiplied by *amperes* equals *watts*. Thus one watt is produced when one ampere flows under a pressure of one volt. This relationship is written:

$$\text{Ampere} \times \text{volts} = \text{watts, or}$$
$$IE = W$$

The relationship between current flow, resistance and pressure is written:

$$\text{Amperes} \times \text{resistance} = \text{volts, or}$$
$$IR = E$$

where,

I = current in *amperes*,
R = resistance in *Ohms*,
E = pressure in *volts*.

The unit of power is usually expressed in kilowatts or horsepower. The relationship between horsepower, kilowatt, and watt is:

$$1 \text{ kilowatt} = 1000 \text{ watts}$$

One horsepower is equal to 746 watts. Therefore, to reduce *kilowatts* to *horsepower*, it is necessary to divide 1000 by 746, or

$$1 \text{ kW} = \frac{1000}{746} = 1.34 \text{ HP}$$

From this we know that *kilowatts* may be converted into *horsepower* by multiplying by 1.34. The formula given are fundamental only and may be rewritten to suit any particular condition or requirement, as shown in Table 1.

Table 1.   Variations of Ohms Law

Amperes	$=$	$\dfrac{\text{Volts}}{\text{Ohms}}$	$=$	$\dfrac{\text{Watts}}{\text{Volts}}$	$=$	$\sqrt{\dfrac{\text{Watts}}{\text{Ohms}}}$
Resistance	$=$	$\dfrac{\text{Volts}}{\text{Amperes}}$	$=$	$\dfrac{\text{Watts}}{\text{Amperes}^2}$	$=$	$\dfrac{\text{Volts}^2}{\text{Watts}}$
Volts	$=$	Amperes x Ohms	$=$	$\dfrac{\text{Watts}}{\text{Amperes}}$	$=$	$\sqrt{\text{Ohms x Watts}}$
Watts	$=$	$\dfrac{\text{Volts}^2}{\text{Ohms}}$	$=$	Amperes^2 x Ohms	$=$	Amperes x Volts

THE ELECTRIC CIRCUIT

A simple way to consider electricity is to use the example of water flowing through a pipe. In a manner similar to the flow of water in a pipe, electrical conductors are arranged to form a path for the flow of an electric current. The paths in which electric current flows are called *electric circuits*. Current, which is measured in amperes, is the movement of electrons (water) through a conductor (pipe).

In Fig. 1, a simple electric circuit is shown diagrammatically. This consists of a generator, motor, switch, and a series of wire conductors. Fig. 2 shows a water system consisting of a pump, water turbine, pipes, and valve which corresponds to this electric circuit. The amount of water flowing through the pipe expressed in gallons per minute would correspond to the flow of current through the wire expressed in amperes. The pressure of water in the pipe expressed in pounds per square inch would correspond to the pressure of electricity, or the voltage of the circuit as it is usually expressed.

Fig. 1. A simple electric circuit consisting of a generator, motor, switch and a series of wire conductors.

DIRECT CURRENT (dc)

The flow of electricity in a wire is often compared to the flow of water in a pipe. In a *direct current* circuit a voltmeter connected across the line and an ammeter connected in series with it (Fig. 3), will read the true voltage and current. The product of the two readings will be the power in *watts* used.

Direct current flows in one direction, as indicated in Fig. 3. It flows from negative to positive (− to +). The only place that you will come across direct current (dc) in refrigeration, will be in automobile air-conditioning. Here the automobile uses a dc source.

Fig. 2. A water system consisting of a pump, turbine, pipes and valves which correspond to the electric circuit.

Fig. 3. Illustrating the method of connecting instruments in a circuit to measure voltage and current.

SINGLE PHASE AND THREE PHASE CURRENT

A short explanation of single phase and three phase power is in order. One hertz, formerly known as c/s, is one complete sine wave in one second (Fig. 4) and is 360 electrical degrees. The ordinary ac current that we use is 60 cycles per second, and therefore is said to have a frequency of 60 hertz. There are two alternations in one cycle, so 60 hertz current alternates 120 times per second. This means that the voltage and current start at zero, go to maximum voltage in a positive direction, return to zero, go to maximum voltage in the negative direction and then back to zero. This takes place 60 times each second. Note that the direction of flow has alternated twice in one cycle. The values that you read on a voltmeter or an ammeter will be 0.707 or 70.7% of the peak voltage or current, and is the effectual value of the voltage or current.

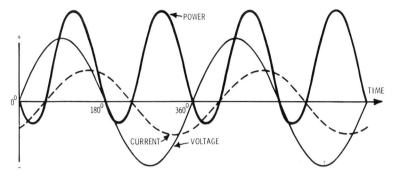

Fig. 4. Illustrating the relationship between current, voltage, and power in an alternating current circuit.

With dc, the voltage and current are always in step with each other, so the reading of a voltmeter multiplied by the reading of an ammeter, will give the power in the dc circuit in watts. This is not always true with ac. Figure 4, illustrates a case where the load might be a resistive load, such as incandescent lighting. When this is true, the voltage and current will be *in phase*. In this case voltmeter and ammeter readings could be used to arrive at the power in watts.

Reactive loads (loads made up of winding, such as motors, ballasts for flourescent lamps, etc.) cause self-inductance to set up in the coils, causing a voltage in the opposite direction to the voltage from the source. This causes the current in the windings to lag behind the applied voltage. Thus, the current reaches its maximum after the voltage has past its maximum, and this creates a *lagging power factor*. Capacitors will cause the current to lead the applied voltage, causing a *leading power factor*. This is shown in Fig. 5.

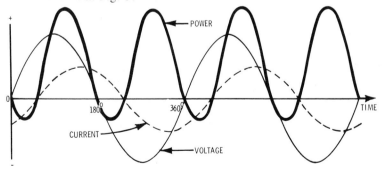

Fig. 5. Illustrating the current and voltage out of phase.

In an ac circuit, if you use a voltmeter and ammeter and multiply the results, you will not get watts, but will get volt-amperes (VA). If at the same time you put a wattmeter in the circuit, it will read watts, and on a motor load the VA will be larger than the watts. This is due to the current lagging the voltage in time, resulting in a lagging power factor. The voltage will be the applied voltage but the current will be larger than if the same load was on dc, since with the voltmeter and ammeter, you are reading apparent power, which in VA will be larger than the reading in watts on the wattmeter.

Thus, in figuring the size of conductors needed to supply a motor load, you have to use the full load amperes on an ac circuit, which as just stated will be more than the full load amperes if it was a dc motor and fed from dc. The difference between the apparent power (VA's) and the real power (watts) is called wattless power and accomplishes nothing in pulling the load.

For induction motors, it is a lagging power factor, meaning current is lagging voltage. The power factor of a motor varies for different designs. It is a common error for the uninformed to assume that some motors are more efficient than others because the name plate for one motor shows a lower current value than that of another motor of equal horsepower. Actually, the amount of power really drawn from the line in kilowatts can be the same. The motor having the higher current merely has a lower value of *power factor*. This will easily be found by the assistance of the following example:

Two single-phase ac motors, A and B (Fig. 6), each having a capacity of one horsepower, has the data shown in the illustration on their respective nameplates.

The power taken from the line by the respective motor is:

Motor A—True Power = $110 \times 10 \times 0.80 = 880$ watts
Motor B—True Power = $110 \times 12.5 \times 0.64 = 880$ watts

The two motors using the same amount of power will cost the same to operate, even though motor B takes more current than motor A. There is one thing to notice, however. The motor with the lowest power factor will have more current flowing, which will require a change in size of thermal element in the starter to provide equal protection. Also, larger wiring has to be used to carry the heavier current.

340

MOTOR A

MOTOR B

Fig. 6. Illustrating the typical data given on motor nameplate.

FREQUENCY AND SLIP

As previously explained, frequency is the number of hertz (cycles per second) and alternations is half the number of alternations ac makes per second. The frequency and the number of poles in the stator of a motor will determine the speed of the motor on ac. The only motors that run pole for pole, or at synchronous speed, are synchronous motors and hysteresis motors such as clock motors. With the motors used in home refrigeration and air conditioning, the motors run at a little less than synchronous speed. This is called slip, and regulates the power output of the motor. Thus a single-phase or three-phase motor at no load will run faster than at full load. The reason for this is that as the load increases, there is more slip, causing the motor to pull more current to compensate for the additional load. The details of this action are explained in Audel's *"Electrical Course for Journeymen and Apprentices."*

To find the synchronous speed of any ac motor, use the equation below:

$$\text{Motor rpm} = \frac{120 \times 60}{4}$$

Thus, for example:

$$\text{Motor rpm} = \frac{120 \times \text{frequency}}{\text{number of motor poles}} = 1800 \text{ rpm}.$$

Thus, all 4-pole 60-hertz motors that operate at a maximum of 1800 rpm will usually be about 3% below this speed, due to slippage, or 1750 rpm. A 4-pole 50-hertz motor that operates at a maximum of 1500 rpm will usually be about 3% below this speed, due to slippage, or 1450 rpm.

341

$$\text{Motor rpm} = \frac{120 \times 50}{4} = 1500 \text{ rpm.}$$

Other examples at other frequencies can be worked out, remembering that the minimum number of poles is 2. An odd number of poles like 3, 5, 7, 9, etc., cannot be used. It must be an even number or any number that can be divided by 2. The only ac motors that run pole for pole, or in other words at synchronous speed, are synchronous motors.

Single-phase ac current is used mainly on domestic refrigeration and on smaller commercial refrigeration.

THREE-PHASE CURRENT

Most motors on larger commercial refrigerators are of the three-phase type. Fig. 7, illustrates three-phase voltage wave patterns. Notice that no two reach maximum at the same time, but that they are spaced 120 electrical degrees apart. On say a motor, if you take an ammeter reading on a motor on one phase, this figure will have to be multiplied by 1.732, to get the total current that the motor is pulling, if you are determining horsepower. However, the ampere listing on the nameplate is the current per phase.

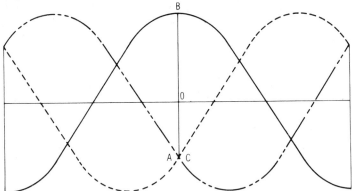

Fig. 7. Three emf's 120° apart in phase.

TRANSFORMER CONNECTIONS

The following are schematic diagrams of transformer windings, to show how single-phase and three-phase current is derived for service to buildings. Only the secondary side of the transformer will be illustrated, as this is where the power for the refrigeration motors will be derived.

Fig. 8 illustrates the type service that most homes will receive.

Fig. 9 shows a closed Delta-bank transformer, which can give us 120/240 volt single-phase, as well as 240 volt three-phase service. Many commercial establishments receive this type of service.

Fig. 10, shown a Wye three-phase transformer supply with available 120/208 volt single phase available, also 208 volt three phase and, if necessary, 120 volt three-phase. This type of service is very popular with commercial establishments. By using 277 volt transformers instead of the 120 volt transformers that are shown, we can get 277/480 volt single-phase or 480 volt three-phase. This system is necessary for motors of large horsepower.

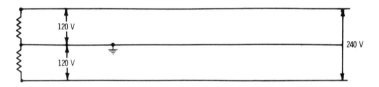

Figure 8. Single-phase, 120/240 volt supply

Fig. 9. Voltage and current in a delta transformer.

ENERGY

Energy is the ability to do work. Energy can neither be created or destroyed. It may, however be converted from one form to another. In the conversion of one form of energy to another, the conversion is accom-

Fig. 10. Voltage and current in a wye transformer.

panied by some form of losses. *Potential energy* is stored energy, such as in a battery. *Kinetic energy* is energy in motion. Since the rate of doing work is *power*, energy is equal to the product of power and time. For example, a 5 horsepower motor requires 2 hours to pump the water necessary to fill an elevated storage tank; the energy represented by this work is five times two hours, or ten horsepower-hours. Now, a ten-horsepower motor could do the same work in half the time, but the energy used would still be ten horsepower-hours. Mechanical energy is measured in horsepower-hours. Electrical energy is measured in *watt-hours,* or *kilowatt-hours.* A *kilowatt-hour* equals a *thousand watt-hours.*

If power is known to remain constant on a circuit for a particular time interval, the energy can be determined by measuring the power with a wattmeter and multiplying this reading by the time in hours. For instance, if you measure the amount of energy consumed by one of our previously described motors during a period of 24 hours, assuming that the motors are running only 50% of the time mentioned, and assuming a constant current, we obtain:

Energy consumed = 880 × 12 = 10,560 watt hours

(or 10.56 kilowatt hours)

In power systems, however, it is an exception for both the load and voltage on a circuit to remain constant for any great length of time.

Consequently, the method of measuring energy just described is not practical. To be practical, the measuring instrument must be sensitive to all changes in power and its recording element must sum up the power demand continuously for small intervals of time. An instrument that does this may be called an integrating watt meter, but is generally known as a *watt-hour meter*, because its dials record the watt-hours (more commonly kilowatt hours) of energy passed through it. The *watt-hour meters* used in residences are smaller and, in most cases, have only one single-phase element.

MOTOR EFFICIENCY

The ratio of the actual power delivered at the pulley or shaft and the power input at the terminals of a motor (computed in the same units) is called the *motor efficiency*.

It is written:

$$\text{Efficiency} = \frac{\text{Output}}{\text{Input}}$$

An example in calculating the efficiency of a single-phase ac motor may be had by considering a motor whose nameplate has the following data:

$$\text{1 HP. 10 AMPS, 110 VOLTS, PF 80\%.}$$

Since one horsepower equals 746 watts, a substitution in our equation gives:

$$\text{Efficiency} = \frac{746}{110 \times 10 \times 0.8} = 0.848 \text{ or } 85\% \text{ approx.}$$

It will readily be observed that the electrical power or the input to the motor may conveniently be measured at the motor terminals and reduced to horsepower by the usual method of calculation. The mechanical power developed by the motor is not readily available since the nameplate gives only the average power which the motor will deliver under normal conditions. The usual method adopted for calculating the mechanical power developed by a motor, and from which the efficiency may be obtained as previously described, is by the means of a *Prony brake test*.

As shown in Fig. 11, this brake consists of two blocks shaped to fit around the pulley of the motor. By means of two long bolts with handwheels at the top, the pressure of the blocks against the pulley can be varied. Two bars fastened to the top and bottom of the brake form an extension arm by which the torque of the motor is exerted upon the platform scale. The force factor, F, of the torque is measured as the net reading on the scale in pounds; the radius factor, R, is measured in feet from the center of the shaft perpendicular to the bearing point on the scale.

Fig. 11. Diagrammatic representation of arrangement for Prony brake test applied to the pulley of an electric motor.

The mechanical power of the motor equals its torque times the distance through which it operates per minute, or:

$$P = \frac{T \times 2 \times \text{rpm}}{33,000}$$

from which,

$$T = \frac{P \times 5250}{\text{rpm}}$$

where,

P = power in terms of horsepower.
T = torque (= F × R) in pound feet.
2π = 6.2832 (the ratio of the circumference of a circle to the radius).
rpm = speed in revolutions per minute.
33,000 = number of foot pounds per minute equivalent to one horse-power.

As an example, we may make a test of a 4-horsepower motor which has a force of 5.5 pounds, the radius 2.5 feet, and the speed 1440 rpm. The power delivered will be:

$$P = \frac{5.5 \times 2.5 \times 6.2832 \times 1440}{33,000} = 3.77 \text{ horsepower}$$

To find the efficiency, we must measure (by means of a wattmeter or other suitable method) the input to the motor at the same time that the output is determined. Both output and input are then converted to the same units, preferably watts, and the efficiency found by the formula.

$$P = \frac{T \times 2 \times \text{rpm}}{33,000}$$

Other methods based upon the same principles, but somewhat more convenient in measuring the torque (or mechanical output of fractional-horsepower motors), are by means of a specially designed Prony brake as shown in Fig. 11.

The Prony brake, in this case, differs from that previously described only in construction. It requires a pulley, brake arm, and scale (may be either platform scale or spring balance). If a platform scale is used, be sure that the load is applied to the center of the platform. If a spring balance scale is used, the pull must always be at right angles to the brake arm, and in either case, the scale must have small enough graduations to accurately read torque on smaller rated motors. The brake arm should be made so that the distance between the center of the pulley and contact point where the load is measured is exactly 12 inches. The scale reading will then be in pound-feet. Before starting the test, make sure that the direction of rotation is such that the brake arm will be moved *against* the balance.

In order to measure starting torque, clamp the arm to the pulley tight enough to allow the pulley to turn very slowly, then read the scale. To measure pull-in torque, release the clamp until the motor is just able to throw off the brushes and pull up to speed. Read the scale just as the brushes are leaving the commutator. The true pull-in torque is the highest scale reading at which time the motor brushes will throw-off and stay clear of the commutator.

PULLEY SPEED AND SIZES

On a motor, the driving pulley is called the *driver,* and the driven compressor pulley or flywheel is called the *driven.* In some cases a gear drive is encountered in which the total number of teeth must be counted. For gears, therefore, the number of teeth should be substituted whenever diameter occurs. By the use of a formula, the unknown can be easily found if the other three factors are known. For instance, if one diameter and the two speeds are known, the other diameter can be ascertained. In order to illustrate the formula, the diagrammatic layout in Fig. 12 will be of assistance. The diameter and speed of both the driver and driven are given as a typical example.

The one important item to consider is that the load on the motor does not increase in direct proportion to the changes in speed of the compressor by changing pulley sizes. It is very easy to overload the motor by increasing its pulley size. Load checks on the current drawn by the motor should be taken after any pulley changes from those of the manufacturer's design.

CALCULATION OF DRIVER DIAMETER

Where the rpm of the driver, the rpm of the driven, and the diameter of the driven are known, the diameter of the driver may be found by multiplying the diameter of the driven by its rpm and dividing the product by the rpm of the driver. Thus, in accordance with the previous data, the problem may be written:

$$\text{dia. of driver} = \frac{\text{dia. of driven} \times \text{rpm of driven}}{\text{rpm of driver}} \text{ or}$$

$$Dr = \frac{Dn \times Rn}{Rr}$$

where,

Dn = Diameter of driven pulley.
Dr = Diameter of drive pulley.
Rn = Revolutions per minute of driven pulley.
Rr = Revolutions per minute of driver pulley.

Fig. 12. Illustrating the Prony and Rope brake methods of testing torque for fractional horsepower motors.

Example:

$$\frac{17 \times 200}{1700} = 2'' \text{ diameter}$$

CALCULATION OF THE DRIVEN DIAMETER

Where the rpm of the driven, the rpm of the driver, and the diameter of the driver are known, the diameter of the driven can be determined by multiplying the diameter of the driver by its rpm and dividing the product by the rpm of the driven. In accordance with this, we may write:

$$\text{dia. of driven} = \frac{\text{dia. of driver} \times \text{rpm of driver,}}{\text{rpm of driven}} \text{ or}$$

$$Dn = \frac{Dr \times Rr}{Rn}$$

Example:

$$\frac{2 \times 1700}{200} = 17'' \text{ diameter}$$

CALCULATION OF DRIVER RPM

Where the diameter of the driver, the diameter of the driven, and the rpm of the driven are known, the rpm of the driver can be found by multiplying the diameter of the driven by its rpm and dividing the product by the diameter of the driver. This may be written:

$$\text{rpm of driven} = \frac{\text{dia. of driver} \times \text{rpm of driver}}{\text{dia. of driven}} \text{ or}$$

$$Rn = \frac{Dr \times Rr}{Dn}$$

Example:

$$\frac{2 \times 1700}{17} = 200 \text{ rpm}$$

Where the diameter of the driven, the diameter of the driver and its rpm are known, the rpm of the driven can be found by multiplying the diameter of the driver by its rpm and dividing the product by the diameter of the driven. Therefore, we obtain:

$$\text{rpm of driven} = \frac{\text{dia. of driven} \times \text{rpm of driver,}}{\text{dia. of driver}} \text{ or}$$

$$Rr = \frac{Dn \times Rn}{Dr}$$

Example:

$$\frac{17 \times 200}{2} = 1700 \text{ rpm}$$

In the preceding calculations, no attention has been given to belt slip which may take place. Actual tests bear out that the slip factor may amount up to 3%, depending upon the size of pulleys, type of belt used, etc.

Fig. 13. Illustrating a typical belt arrangement for a motor and compressor drive.

ALTERNATING-CURRENT MOTORS

Alternating-current motors are divided into two main classifications (depending on the type of power used) as:

1. Polyphase.
2. Single phase.

Polyphase motors are further divided into three classes, as:

1. Squirrel-cage induction.
2. Wound-rotor induction.
3. Synchronous.

Single-phase motors, although somewhat less efficient than the polyphase type, are used mainly in applications where the power requirements are small. Thus, fractional-horsepower motors are normally of the single-phase type. In integral-horsepower sizes, however, polyphase motors are commonly used.

Squirrel-Cage Motors

Squirrel-cage motors (Fig. 14) provide a variety of speed and torque characteristics. The speed of a squirrel-cage induction motor is nearly constant under normal load and voltage conditions, but is dependent on the number of poles and the frequency of the ac source. In the case of a squirrel-cage motor, however, the motor slows down when loaded an amount just sufficient to produce the increased current to meet the required torque.

The difference between synchronous speed and load speed is called the *slip* of the motor. This slip is usually expressed as a percentage of the synchronous speed. Since the amount of slip is dependent on the load, the greater the load the greater the slip will be; i.e., the slower the motor will run. This slowing down of the motor, however, is very slight even at full load, and amounts from 1 per cent to 4 per cent of synchronous speed. Thus, the squirrel-cage motor is considered to be a constant-speed type. Certain designs of the squirrel-cage motor provide a high starting torque and these types are used on refrigerator compressors and on reciprocating pumps. Motors of this type used on highly pulsating loads are normally provided with flywheels.

Wound-Rotor Motors

Wound-rotor motors (Fig. 15) are used for applications requiring high starting torque at a low starting current. A wound-rotor motor, along with its controller and resistance, can develop full load when starting with little more than full-load current. Motors of this type are provided with insulated phase windings on the rotor, and the terminal of each phase is connected to a collector ring. Stationary brushes riding on the collector rings are connected to adjustable resistors to provide a secondary winding by means of which the motor characteristics can be varied to suit various operation conditions.

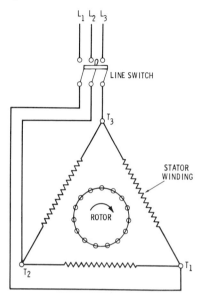

Fig. 14. Schematic diagram of an across-the-line nonreversible single-speed, manually-operated, three-phase, squirrel-cage induction motor.

Fig. 15. Schematic diagram showing the connection arrangement using a face-plate starter and speed regulator for a three-phase, wound-rotor motor.

Variations in operating speed are essential on many applications. It is often desirable to vary the operating speed of conveyors, compressors, pulverizers, stokers, etc., in order to meet varying production requirements. Wound-rotor motors, because of their adjustable and varying

353

speeds, are ideally suited for such applications. However, if the torque required does not remain constant, the speed of the wound-rotor motor will vary over wide limits, a characteristic constituting one of the serious objections to the use of wound-rotor motors for obtaining reduced speeds.

Synchronous Motors

Synchronous motors are used for continuous-duty applications at a constant speed, such as fans, blowers, pumps, and compressors. The outstanding advantage of the synchronous motor is that its power factor can be changed to compensate for the low-power factor of other drives in the same location. When synchronous motors are employed exclusively as power-factor correction devices, they are termed *synchronous capacitors,* because the effect on the power system is the same as that of a static capacitor.

The power factor of a synchronous motor may be varied by adjustment of the field current. Thus, by adjusting the field current, the power factor may be varied in small steps from low *lagging* power factor to a low *leading* power factor. This makes it possible to vary the power factor over a considerable range, and places the characteristics of the motor under the immediate control of the operator at all times.

Because of the higher efficiency possible with synchronous motors, they can be used advantageously on most loads where constant speed is required. Typical applications of high-speed synchronous motors (above 500 rpm) are such drives as centrifugal pumps. Dc generators, belt-driven reciprocating compressors, fans, blowers, line shafts, centrifugal compressors, rubber and paper mills, etc.

The field of applications of low-speed synchronous motors (below 500 rpm) includes drives such as reciprocating compressors (largest field of use), Jordan engines, centrifugal and screw-type pumps, ball and tube mills, vacuum pumps, electroplating generators, line shafts, rubber and band mills, chippers, metal-rolling mills, etc. Synchronous motors are rarely used in sizes below 10 HP. The principal field of application is in sizes of 100 HP and larger.

Starting and Controlling Polyphase Motors

Polyphase induction motors may be started and controlled by several methods, as:

1. Directly across the line.
2. By means of autotransformers.

3. By means of resistors or reactors in series with the stator winding.
4. By means of the star-delta connection.

Before discussing the various methods which are available for reducing starting current and improving line-voltage conditions, it is important to have a thorough knowledge of the effects of reduced-voltage starting on the motor as well as on the power system. Any method which reduces the starting current to the motor is accompanied by a reduction in starting torque. Therefore, it is essential to know something about the load-torque characteristics in determining if a given current limitation can be met. In other words, there are boundary conditions in which the permissible current to be taken from the line would not provide the needed output torque at the motor shaft necessary for the successful acceleration of its connected load. With all starting methods, the torque of a squirrel-cage motor varies as the square of the applied voltage at the motor terminals.

Across-the-line Method of Starting—This is generally the most economical method of starting, but is usually limited to motors up to 5 HP on account of the large starting current required. With this method of starting, the motor is connected directly to the full line voltage by means of a manually-operated switch or a magnetic contactor.

Starting with Autotransformer—In the case of the autotransformer type of starting, the current taken from the line varies as the square of the voltage applied to the motor terminals, and it is convenient to remember that the torque and line currents are reduced at the same rate. Thus, an autotransformer starter designed to apply 80 per cent of the line voltage to the motor terminals will produce 64 per cent of the torque that would have been developed if the motor had been started on full voltage. At the same time, the motor will draw 64 per cent as much of the current from the line as would have been required for full voltage starting.

Starting with Resistors or Reactors—With resistor or reactor starting, the starting current varies directly with the voltage at the motor terminals because the resistor or reactor is in series with each line to the motor and must carry the same current that flows in each motor terminal. It is evident, therefore, that the resistor and reactor type of reduced-voltage starting requires more line current in amperes per unit of torque in foot pounds than does the autotransformer type. Thus, if a motor connected to a loaded centrifugal pump, for example, is started with the 80 per cent tap on the autotransformer, the initial torque is 64 per cent. If, on the other hand, the motor were started with a primary resistor that limits the starting voltage to 65 per cent of line voltage, the initial torque would only be 42.25 per cent.

On some power systems, it is necessary to meet a restriction on the rate of current increase in starting. The rate of increase of current is determined to meet the conditions as they exist at that particular point on the system where the motor is started.

Starting with the Star-Delta Method—Some three-phase induction motors of the squirrel-cage type may be started by the star-delta method. This starting method is associated only with motors designed for their full power to be by a delta-connected three-phase winding. There must also be provided additional leads from the motor which, when regrouped, will result in a star arrangement of the three-phase winding. Six main leads from the motor are required to accomplish the switch from a start-across-the-line (star connection) to a run-across-the-line (delta connection). The starting connection is always star, since the voltage is $1/\sqrt{3}$ or 57.8 per cent of the delta or line voltage.

From the foregoing data, it follows that this type of reduced-voltage starter (which is limited to 57.8 per cent of the line voltage at starting) can be employed only where the motor has a light starting load. In all other applications, higher starting voltages are obtained through the use of resistors, reactors, or autotransformer, as previously outlined.

SINGLE-PHASE MOTORS

Single-phase motors, as their name implies, operate on only one phase. They are used extensively in fractional-horsepower sizes in most domestic and commercial applications. The advantages of using single-phase motors in small sizes are that they are less expensive to manufacture than other types, and they eliminate the need for three-phase ac lines. Single-phase motors are used in fans, refrigerators, portable drills, grinders, etc.

A single-phase induction motor with only one stator winding and a cage rotor is like a three-phase induction motor with a cage rotor, except that the single-phase motor has no revolving magnetic field at the start, and hence no starting torque. However, if the rotor is brought up to speed by external means, the induced currents in the rotor will cooperate with the stator currents to produce a revolving field, causing the rotor to continue to run in the direction in which it was started.

Several methods are used to provide the single-phase induction motor with a starting torque. These methods identify the motor as split-phase, capacitor, shaded-pole, repulsion, etc. Another class of single phase motors is the ac series-wound universal type. The ac series motors have electrical characteristics like dc series motors. Only the more commonly used types of single-phase motors are described. These include:

1. Split phase.
2. Capacitor.
3. Shaded pole.
4. Repulsion-start.
5. Ac series motor.

Split-Phase Motors

Split-phase motors (Fig. 16) have a high-resistance auxiliary winding which remains in the circuit during the starting period, but is disconnected through the action of a centrifugal switch as the motor comes up to speed. Under running conditions, it operates as a single-phase induction motor with only one winding in the circuit. The main winding is connected across the supply lines in the usual manner, and has a low resistance and a high inductance. The starting or auxiliary winding, which is physically displaced in the stator from the main winding, has a high resistance and a low inductance. This physical displacement, in addition to the electrical phase displacement produced by the relative electrical resistance values in the two windings, produce a weak rotating field which is sufficient to provide a low starting torque.

Fig. 16. Winding arrangement of a split-phase motor.

After the motor has accelerated to 75 per cent or 80 per cent of its synchronous speed, a starting switch (usually centrifugally operated) opens its contacts to disconnect the starting winding. The function of the starting switch is to prevent the motor from drawing excessive current from the line and also to protect the starting winding from damage due to

357

heating. The motor may be started in either direction by reversing either the main or auxiliary winding.

Split-phase motors of the fractional-horsepower size are used in a variety of equipments, such as ventilating fans, oil burners, washers, and numerous other applications.

Capacitor Motors

The capacitor motor (Fig. 17) is a modified form of a split-phase motor, having a capacitor in series with the starting winding. The capacitor motor is manufactured in fractional-horsepower sizes and develops a high starting torque. They are used for constant-drive applications, such as fans, blowers, centrifugal pumps, refrigerator compressors, etc. During the starting period, a winding with a capacitor in series is connected to the motor circuit, and when the motor comes up to speed, a centrifugal switch cuts the capacitor and second winding out of the circuit.

Fig. 17. Capacitor-start motor winding diagram.

Capacitor motors are practically always dual voltage motors, and can be either 120 or 240 volts. They have replaced the repulsion-induction motor practically 100%. Fig. 18, shows the connections for both 120 and 240 volts.

Shaded-Pole Motors

The shaded-pole motor (Fig. 19) employs a salient-pole stator and a cage rotor. The projecting poles on the stator resemble those of a direct-current motor, except that the entire magnetic circuit is laminated and a portion of each pole is split to accommodate a short-circuited copper strap called a *shading coil*. The shaded-pole motor is similar in operating characteristics to the split-phase motor, and has the advantage of simple construction and low cost. It has no sliding electrical contacts and is reliable in operation. However, it has a low starting torque and low efficiency. It is manufactured in fractional-horsepower sizes and has

358

found employment in a variety of household applicances, such as fans, blowers, electric clocks, hair dryers, and other applications requiring a low starting torque.

Fig. 18. Dual-voltage capacitor-start motor.

Most shaded-pole motors have only one edge of the pole split, and therefore the direction of rotation is not reversible. However, some shaded-pole motors have both leading and trailing pole tips split to accommodate shading coils. The leading pole tip shading coils form one series group and the trailing pole tip shading coils form another series group. Only the shading coils in one group are simultaneously active, while those in the other group are on open circuit.

Fig. 19. Schematic winding arrangement in a shaded-pole motor.

Repulsion-Start Motors

The repulsion-start motor (Fig. 20) has a form-wound rotor with commutator and brushes. The stator is laminated and contains a distributed single-phase motor. In its simplest form, the stator resembles that of a single-phase motor. The motor is equipped with a centrifugal device, the function of which is to remove the brushes from the commutator and place a short-circuiting ring around the commutator as the motor comes up to speed. After short-circuiting the commutator, the motor operates with the characteristics of a single-phase induction motor.

Fig. 20. Schematic arrangement of the windings in a repulsion-start induction motor.

MOTOR-COMPRESSOR CONTROLS

There are two types of motor controls used in the operation of commercial refrigeration plants, namely:

1. Manual.
2. Automatic.

The manual control may be the ordinary snap and knife switches which must be operated by hand when the actuating mechanism is started or stopped. This type of control, while providing temperature control for a given set of conditions, will not provide automatic regulation. Also, with manual control, constant attention is required by the operator to maintain as near as possible the predetermined condition. In addition, the cost of operation is usually found to be excessive.

Automatically operated controls, on the other hand, provide all the features that are considered most desirable, both for temperature and for economy of operation. Here the operator or owner is relieved of all responsibility for maintaining the pre-determined conditions, and consequently the operating costs will be kept at a minimum.

Contactors, Starters, and Relays

Most condensing units with motors larger than 1-½ HP have a starter or contactor. As a rule, these are furnished with the unit.

Relays are a necessary part of many control and pilot circuits. These automatic switches are similar in design to contactors, but are generally lighter in construction and carry a smaller amount of current.

Magnetic contactors are normally used for starting polyphase squirrel-cage and single-phase motors. Contactors may be connected at any convenient point in the main circuit between the fuses and the motor, and small control wires may be run between the contactor and the point of control.

Protection of the motor against prolonged overload is usually accomplished by time-limit overload relays which are operative both during the starting period and during the running period. The action of the relays is delayed long enough to take care of heavy starting loads and momentary overloads without tripping.

The operation of a contactor, as applied to any individual installation, depends on two main parts—the operating coil and the thermal elements of the overload relay. While the contactor proper is generally "universal," it is necessary to have the proper coil and the proper thermal elements installed, depending on the line voltage, frequency, and the protection desired.

Relays are furnished in a large number of types, the employment of which depends on the size of the condensing unit, the power-source voltage, and other factors. As used on small single-phase motors, a relay is a necessary part of the motor-starting equipment. It consists essentially of a set of normally-closed contacts which are in series with the motor-starting winding. The electromagnetic coil is in series with the auxiliary winding of the motor. When the control contacts close, the motor starting and running windings are both energized. A fraction of a second later the motor comes up to speed and sufficient voltage is induced in the auxiliary winding to cause current to flow through the relay coil. The magnetic force is sufficient to attract the spring-loaded armature which mechanically opens the relay starting contacts. With the starting contacts open, the starting winding is out of the circuit, and the motor continues to run on only the running winding. When the control contacts open, power to the motor is interrupted. This power interruption causes the relay armature to close the starting contacts, placing the motor in a condition to start a new cycle when the control contacts again close.

Motor Overload Protector

Motors for commercial condensing units are normally protected by a bimetallic switch operating on the thermo or heating principle. This is a built-in motor overload protective device which limits the motor winding temperature to a safe value. In its simplest form, the switch or motor protector consists essentially of a bimetal switch mechanism which is permanently connected in series with the motor circuit, as shown in Fig. 21.

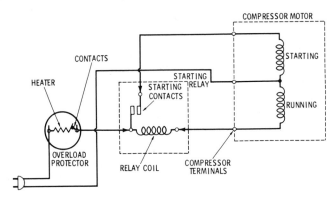

Fig. 21. Circuit arrangement of a typical domestic refrigerator.

When the motor becomes overloaded or stalled, excessive heat is generated in the motor winding due to the heavy current produced by this condition. The protector located inside the motor is controlled by the motor current passing through it and the motor temperature. The bimetal element is calibrated to open the motor circuit when the temperature, as a result of an excessive current, rises above a predetermined value. When the temperature decreases, the protector automatically resets and restores the motor circuit.

The advantage in the overload device obviously will reduce service calls due to temporary overloads. When overload conditions occur, the protector will stop the motor for a short period of time to allow it to cool off, after which it will reset itself and allow the motor to run. This cycling may continue for some time until the load has been reduced to a point where the motor can safely handle it.

In this connection, it should be noted that the servicing of motors equipped with built-in overload protectors must be handled with care, since the compressor may be idle due to an overload and hence will start as soon as the motor cools off. This could result in a serious mishap to the operator or serviceman. To avoid such difficulties, open the electrical circuit by pulling the line plug or switch prior to any repair or servicing operation.

Motor Winding Relays

A motor winding relay is usually incorporated in single-phase motor-compressor units. This relay is an electromagnetic device for making and breaking the electrical circuit to the starting winding. A set of normally-closed contacts are in series with the motor starting winding (Fig. 21).

The electromagnetic coil is in series with the auxiliary winding of the motor. When the control contacts close, the motor starting and running windings are both energized. A fraction of a second later the motor comes up to speed and sufficient voltage is induced in the auxiliary winding to cause current to flow through the relay coil. The magnetic force is sufficient to attract the spring-loaded armature which mechanically opens the relay starting contacts. With the starting contacts open, the starting winding is out of the circuit, and the motor continues to run on only the running winding. When the control contacts open, power to the motor is interrupted. This allows the relay armature to close the starting contacts. The motor is now ready to start a new cycle when the control contacts again close.

OIL-PRESSURE CONTROLS

Oil-pressure controls (Fig. 22) are designed to provide protection against breakdowns due to low lubrication oil pressure. Total oil pressure is the combination of crankcase pressure and the pressure generated by the oil pump. The net oil pressure available to circulate the oil to the bearing surfaces is the difference between the total oil pressure and the refrigerant pressure in the crankcase. This control measures that difference in pressures, which will be referred to as *net oil pressure*. A built-in time-delay switch, accurately compensated for ambient temperature, allows for pressure pickup on start and avoids nuisance shut-downs on pressure drops of short duration during the running cycle.

When the compressor is started, the time-delay switch is energized. If the net oil pressure does not build up to the cut-in point of the control within the required time limit, the time-delay switch trips to stop the compressor. If the net oil pressure does rise to the cut-in point within the required time after the compressor starts, the time-delay switch is automatically de-energized and the compressor continues to operate normally.

Courtesy Penn Controls, Inc.

Fig. 22. Exterior and interior views of an oil-pressure failure control switch with a time-delay feature.

If the net oil pressure should drop below the cut-out setting (scale pointer) during the running cycle, the time-delay switch is energized and, unless the net oil pressure returns to the cut-in point within the time-delay period, the compressor will be shut down. The compressor can never run more than the predetermined time on subnormal oil pressure. Fig. 23 shows the schematic diagram of a typical compressor with an oil-pressure control connected.

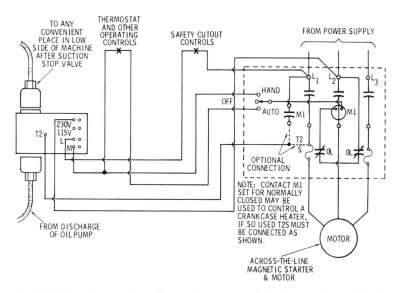

Fig. 23. Wiring diagram of an oil-protection control switch when used on a 230-volt power system.

Solenoid Valves

Solenoid valves are used on multiple installations, and are electrically operated. A solenoid valve, when connected as shown in Fig. 24, remains open when current is supplied to it, and closes when the current is shut off. In general, solenoid valves are used to control the liquid refrigerant flow into the expansion valve, or the refrigerant gas flow from the evaporator when the evaporator or fixture has reached the desired temperature. The most common application of the solenoid valve is in the liquid line and operates with a thermostat. With this hookup, the thermostat may be set for the desired temperature in the fixture. When this temperature is reached, the thermostat will open the electrical circuit and shut off the current to the solenoid valve. The solenoid valve then closes and shuts off the refrigerant supply to the expansion valve. The condensing unit operation should be controlled by the low-pressure switch. In some other applications, where the evaporator is to be in operation for only a few hours each day, a manually-operated snap switch may be used to open and close the solenoid valve, at the convenience of the operator.

365

Fig. 24. Solenoid valve when connected in suction and liquid evaporator lines.

TEMPERATURE CONTROLS

In modern condensing units, low-pressure control switches are being largely superseded by thermostatic control switches. Briefly, a thermostatic control consists of three main parts—a bulb, a capillary tube, and a power element (switch). The bulb is attached to the evaporator in such a manner as to assure contact with the evaporator, and contains a volatile liquid, usually a refrigerant such as sulfur dioxide or methyl chloride. The bulb is connected to the power element by means of a small capillary tube.

Referring to Fig. 25, as the evaporator temperature increases, the bulb temperature also increases, which raises the pressure of the thermostatic liquid vapor, causing the bellows to expand, thus actuating an electrical contact which closes the motor circuit starting the motor and compressor. As the evaporator temperature decreases, the bulb becomes colder and the pressure decreases to the point where the bellows contracts sufficiently to open the electrical contacts controlling the motor circuit. In this manner, the condensing unit is entirely automatic, producing just exactly the proper amount of refrigeration to meet any normal operating condition. A typical thermostatic control is illustrated in Fig. 26.

An automatic temperature-control system is generally operated by making and breaking an electric circuit or by opening and closing a compressed-air line. When using the electric thermostat, the temperature is regulated by controlling the operation of an electric motor or valve, and, when using the compressed-air thermostat, temperature regulation is obtained by actuating a compressed-air operated motor or valve. Electrically-operated temperature-control systems are used generally by manufacturers for practically all installations. However, compressed-air temperature control systems have applications in extremely large central and multiple installations in close temperature work, and where a large amount of power is required for small control devices.

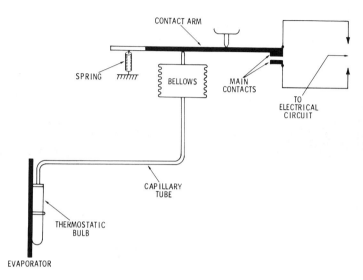

Fig. 25. Schematic diagram illustrating the working principles of a thermostatic control switch.

Bimetallic Thermostats

The so-called bimetallic type of thermostat operates as a function of expansion or contraction of metals due to temperature changes. Such thermostats are designed for control of heating and cooling in air-conditioning units, refrigeration storage rooms, greenhouses, fan coils, blast coils, etc.

The working principles of such a thermostat is shown in Fig. 27. As noted, two metals, each having a different coefficient of expansion, are welded together to form a bimetallic unit or blade. With the blade securely anchored at one end, a circuit is formed and the two contact points are closed to the passage of an electric current. Because of the fact that an electric current produces heat in its passage through the bimetallic blade, the metals in the blade begin to expand, but at a different rate. The metals are so arranged that the one which has a greater coefficient of expansion is placed at the bottom of the unit. After a certain time interval,

Courtesy Penn Controls, Inc.

Fig. 26. Exterior and interior views of a typical heavy-duty refrigeration pressure control with a visible calibrated scale indicating cut-in and cut-out pressure settings. A control of this type is normally used with manual-starting compressor motors not integrally protected, thus making line starters unnecessary.

Fig. 27. Working principles of a simple bimetallic thermostat.

the operating temperature is reached and the contact points become separated, thus disconnecting the appliance from its source of power. After a short period, the contact blade will again become sufficiently cooled off to cause the contact points to join, thus re-establishing the circuit and permitting the current to again actuate the circuit leading to the

368

appliance. The foregoing cycle is repeated over and over again and, in this manner, prevents the temperature from rising too high or from dropping too low. Fig. 28 shows the construction features of this type of control.

COVER RETAINER SPRING

DIAL LOCK SCREW

CLOSES ON HEAT RISE

h

R

Y

Courtesy Penn Controls, Inc.

Fig. 28. Bimetallic thermostat with wiring hookup (low-voltage type).

DEFROST CONTROLS

Automatic defrosting is accomplished in several ways, the control method used depending on the type of refrigeration system, size and number of condensing units, and other factors.

Defrost Timer Operation

In small and medium-size domestic applications, an automatic defrost-control clock may be set for a defrost cycle once every 24 hours or as often as deemed necessary. The defrosting is usually accomplished by providing one or more electric heaters which, when energized by the action of the electric clock, provide the heating action necessary for complete defrosting.

The defrost controls, as usually employed, are essentially single-pole double-throw switching devices in which the switch arm is moved to the defrost position by an electric clock. The switch arm is returned to the normal position by a power element which is responsive to changes in temperature. As the evaporator is warmed during a defrost period, the feeler tube of the defrost control is also warmed until it reaches the defrost cut-out point of approximately 45° F. The defrost-control bellows then forces the switch arm to snap from the defrost position to the normal position (Fig. 29), thus starting the motor compressor.

Another common method of automatic defrosting is the so-called *defrost-cycle* method in which the defrost cycle occurs during each compressor "off" cycle. In a defrost system of this type, the defrost heaters are connected across the thermostat switch terminals. When the thermostat switch is closed, the heaters are shunted out of the circuit. When the thermostat opens, the heaters are energized, completing the circuit through the overload relay and compressor.

Fig. 29. Schematic arrangement of an electric clock motor as employed in a 24-hour defrost control.

Hot-Gas Defrosting

In any low-temperature room where the air is to be maintained below freezing, some adequate means for removing the accumulated frost from the cooling surface should be provided. As a result of recent developments, an improved hot-gas method of quick defrosting for direct-expansion low-temperature evaporators is available. In order to apply this method, it is necessary to have two or more evaporators in the system.

This is because the hot gas required to defrost part of the system must be provided by the heat absorbed from the other cooling surface in a given system. The process of defrosting a plate bank with hot gas can be accomplished automatically by installing the proper controls.

ELECTRICAL SYSTEM MAINTENANCE

The first thing to check when the unit will not run is the voltage at the compressor terminals. Be sure the contact points on the pressure control are closed. The voltage at the terminals should be within plus or minus 10% of the rated voltage shown on the compressor nameplate. If extreme low voltage exists, the cause may be due to inadequate wiring to the unit. Do not use 230-volt units on 208-volt circuits.

Starting Capacitors

For a quick check to determine whether a starting capacitor is still good, disconnect the capacitor lead wires and touch them for a second or two to line wires having the same approximate voltage as the voltage shown on the capacitor. Check a 110-volt capacitor on a 115-volt line, and a 220-volt capacitor on a 230-volt line. *(Caution: Never check a 110-volt capacitor on a 230-volt line, and do not hold a starting capacitor on the line more than a few seconds.)* Now touch the two capacitor wires together. If they spark on several attempts, the capacitor is good. If not, the capacitor is defective and should be replaced.

A more accurate means of checking a capacitor is to connect the two lead wires momentarily across the line and measure the current flow. The microfarad rating of the capacitor may be determined as follows:

$$\text{Microfarad rating} = \frac{2650 \times \text{amps (for 60 hertz)}}{\text{volts}}$$

The reading thus obtained should be within 10% of the microfarad rating stamped on the capacitor. Running capacitors may be checked in the same manner as starting capacitors.

Overload Protectors

If there is no voltage at the compressor terminals, it is possible that the overload protector has tripped and is open. The protector can be checked by bridging the protector terminals with a jumper wire (which should contain a fuse for protection). If the compressor now starts, this indicates either a defective protector or an overload condition. Do not disconnect the overload protector and leave the machine operating without any protection.

Compressor Motors

The compressor motor may be checked by removing all wires from the compressor terminals which are marked as follows: S for start winding, R for running winding, C for common. Attach a 2-wire test cord to the common and running terminals, and plug in the test cord to the power supply. Then momentarily touch the leads of an auxiliary starting capacitor, having the same microfarad rating and voltage rating as supplied with the unit, across the start and running terminals. If the motor does not start, it is probably burned out or has an open winding, or else the compressor is seized. If the motor does start, the trouble is in the electrical accessories.

Starting Relay (Voltage Type)

This type of relay is used on single-phase compressors. If the relay contacts fail to open, the motor will draw excessive current. It will cycle on the overload protector and may blow out the starting capacitor. This failure may be due to an open circuit in the relay coil, or to welded contact points. The latter may be determined by disconnecting the power supply and manually trying to open the contact points. Sometimes a cracked relay base will prevent the contact points from opening sufficiently.

The relay holding coil may be checked by connecting a fused test cord to compressor terminals C and S, and plugging the other end of the test cord into the power supply. Due to the fact that the relay requires more than line voltage to actuate, it may still not open, but by pushing the relay actuating arm with an insulated screwdriver, the contacts should remain open as long as the test current remains on. If not, the relay holding coil is faulty and the relay should be replaced by one of the same model.

Fan Motors

The fan motor should run whenever the compressor is running. If it does not, turn the fan by hand to see if it turns freely. If it turns freely and the lead wires are connected, it is defective and must be replaced. Fan motors are sometimes a source of noise which can be caused by worn bearings, bent fan blades, etc. All fan motors having oiling cups should be oiled every six months. Sealed fan motors do not require oiling.

Wiring diagrams for various types of motors with protectors and starters are shown in Figs. 30, 31, 32, 33, 34 and 35.

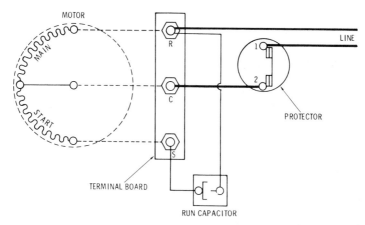

Fig. 30. Wiring diagram for a single-phase permanent split-capacitor motor with a two-terminal protector.

Fig. 31. Wiring diagram for single-phase capacitor-start motor circuit using a potential relay and a three-terminal protector.

Fig. 32. Wiring diagram for a two-phase motor with magnetic starter and thermostat.

COMPRESSOR MOTOR REPLACEMENT

If the compressor motor burns out, the following procedure must be used to clean the refrigeration system of contamination resulting from the motor burnout:

1. Close the compressor service valves and remove the burned-out compressor.
2. Install the new compressor and a filter-drier in the suction line, using an adapter at the suction valve, if available, so that gas passes through the filter-drier before entering the compressor. If an adapter is used, be sure that the hose between the filter-drier and compressor is clean. Use the largest size filter-drier with openings matching the suction line.

Fig. 33. Wiring diagram for a three-phase star-connected motor.

Fig. 34. Wiring diagram for a three-phase star-connected motor with winding protector.

375

Fig. 35. Wiring diagram for across-the-line starting of a 550-volt compressor motor with oil-protection control.

3. Evacuate the compressor and the system cleaner with a good vacuum pump, break the vacuum with refrigerant, and re-evacuate. Only the compressor and system cleaner need to be evacuated, since the service valves have been closed isolating the refrigerant in the system.

4. Open the compressor service valves, close the liquid line valve, and pump down the system. Install an oversize filter-drier (at least

one size larger than the normal selection size) in the liquid line, removing the old filter-drier if one exists.

5. Operate the system, using the usual starting procedure. Check the pressure drop across the system cleaner during the first hour of operation and change the cores if it becomes excessive.

6. Take an oil sample in 8 to 24 hours. If oil tests free of acid and is clean, remove the system cleaner.

7. When the system cleaner is removed, replace the liquid line filter-drier and install a sight glass with a moisture indicator in the liquid line.

8. In two weeks, recheck the color and acidity of the oil to see if another change of the liquid line filter-drier is necessary. Before the job is completed, it is essential that the oil be clear and acid free. The sight glass with a moisture indicator will indicate if the filter-drier should be changed to reduce the moisture content of the system.

ELECTRICAL-SYSTEM TROUBLE CHART

Symptom and Possible Cause	Possible Remedy
Compressor Does Not Run	
(a) Motor line open.	(a) Close, start or disconnect switch.
(b) Fuse blown.	(b) Replace fuse.
(c) Tripped overload.	(c) Reset overload relay.
(d) Control stuck open.	(d) Repair or replace.
(e) Piston stuck.	(e) Remove motor compressor head. Look for broken valve and jammed parts.
(f) Frozen compressor or motor bearings.	(f) Repair or replace.
(g) Control off on account of cold location.	(g) Use thermostatic control or move control to warmer location.
Unit Short Cycles	
(a) Control differential set too closely.	(a) Widen differential.
(b) Discharge valve leaking.	(b) Correct condition.

ELECTRICAL-SYSTEM TROUBLE CHART (Con't)

Symptom and Possible Cause	Possible Remedy
(c) Motor compressor overload cutting out.	(c) Check for high speed pressure, tight bearings, stuck pistons, clogged air- or water-cooled condenser, or water shut off.
(d) Shortage of gas.	(d) Repair leak and recharge.
(e) Leaky expansion valve.	(e) Replace.
(f) Refrigerant overcharge.	(f) Purge.
(g) Cycling on high-pressure cut out.	(g) Check water supply.

Compressor Will Not Start—Hums Intermittently (cycling on overload)

(a) Improperly wired.	(a) Check wiring against diagram.
(b) Low line voltage.	(b) Check main line voltage and determine location of voltage drop.
(c) Open starting capacitor.	(c) Replace starting capacitor.
(d) Relay contacts not closing.	(d) Check by operating manually. Replace relay if necessary.
(e) Open circuit in starting winding.	(e) Check stator leads. If leads OK, replace stator.
(f) Stator winding grounded.	(f) Check stator leads. If leads OK, replace stator.
(g) High discharge pressure.	(g) Eliminate cause of excessive pressure. Make sure discharge shutoff valve is open.
(h) Tight compressor.	(h) Check oil level. Correct binding.

Compressor Starts but Motor Runs On Starting Winding

(a) Low line voltage.	(a) Bring up voltage.
(b) Improperly wired.	(b) Check wiring against diagram.

ELECTRICAL-SYSTEM TROUBLE CHART (Con't)

Symptom and Possible Cause	Possible Remedy
(c) Defective relay.	(c) Check operation manually. Replace relay if defective.
(d) Running capacitor shorted.	(d) Check by disconnecting running capacitor.
(e) Starting and running windings shorted.	(e) Check resistances. Replace stator if defective.
(f) Starting capacitor weak.	(f) Check capacitance. Replace if low.
(g) High discharge pressure.	(g) Check discharge shutoff valve. Check pressure.
(h) Tight compressor.	(h) Check oil level. Check binding.

Relay Burn Out

(a) Low line voltage.	(a) Increase voltage to not less than 10% under compressor motor rating.
(b) Excessive line voltage.	(b) Reduce voltage to maximum of 10% motor rating.
(c) Incorrect running capacitor.	(c) Replace running capacitor with correct value capacitance.
(d) Short cycling.	(d) Reduce number of starts per hour.
(e) Incorrect relay mounting.	(e) Mount relay in correct position.
(f) Relay vibrating.	(f) Mount relay in rigid location.
(g) Incorrect relay.	(g) Use relay properly selected for motor characteristics.

Starting Capacitor Burn Out

(a) Short cycling.	(a) Replace starting capacitor with series arrangement or reduce number of starts per hour to 20 or less.
(b) Prolonged operation on starting winding.	(b) Reduce starting load (install suction regulating valve). Increase voltage if low.

379

ELECTRICAL-SYSTEM TROUBLE CHART (Con't)

Symptom and Possible Cause	Possible Remedy
(c) Relay contacts sticking.	(c) Clean contacts or replace relay.
(d) Improper capacitor.	(d) Check for proper capacitor rating.

Running Capacitor Burn Out

(a) High line voltage and light load.	(a) Reduce voltage if over 10% excessive. Check voltage imposed on capacitor and select one equivalent to this in voltage rating.

Unit Operates Long or Continuously

(a) Shortage of gas.	(a) Repair leak and recharge.
(b) Control contacts frozen.	(b) Clean points or replace control.
(c) Dirty condenser.	(c) Clean condenser.
(d) Location too warm.	(d) Change to cooler location.
(e) Air in system.	(e) Purge.
(f) Compressor inefficient.	(f) Check valves and pistons.
(g) Expansion valve set too high.	(g) Lower setting.
(h) Iced or dirty coil.	(h) Defrost or clean.
(i) Cooling coils too small.	(i) Add surface or replace.
(j) Expansion valve too small.	(j) Raise suction pressure with larger valve.
(k) Restricted or small gas lines.	(k) Clear restriction or increase line size.

Head Pressure Too High

(a) Refrigerant overcharge.	(a) Purge.
(b) Air in system.	(b) Purge.
(c) Dirty condenser.	(c) Clean.

ELECTRICAL-SYSTEM TROUBLE CHART (Con't)

Symptom and Possible Cause	Possible Remedy
(d) Unit location too hot.	(d) Relocate unit.
(e) Restricted water-cooled condenser.	(e) Clean or replace condenser.
(f) Water shut off.	(f) Turn on water.

Head Pressure Too Low

(a) Refrigerant shortage.	(a) Repair leak and recharge.
(b) Compressor suction or discharge valves inefficient.	(b) Clean or replace leaky valve plates.
(c) Cold room or cold water.	(c) None. Efficiency increased.

Noisy Unit

(a) Insufficient compressor oil.	(a) Add oil to proper level.
(b) Tubing rattle.	(b) Bend tubes away from contact.
(c) Mountings loose.	(c) Tighten.
(d) Oil slugging or refrigerant flooding back.	(d) Adjust oil level or refrigerant charge. Check expansion valve for leak or oversize orifice.

Compressor Loses Oil

(a) Shortage of refrigerant.	(a) Repair leak and recharge.
(b) Evaporator design.	(b) Change oil to get higher gas velocity.
(c) Gas-oil ratio low.	(c) Add 1 pt. oil for each 10 lbs. of refrigerant added to factory charge.
(d) Plugged expansion valve or strainer.	(d) Clean or replace.
(e) Oil trapping in lines.	(e) Drain tubing toward compressor.
(f) Short cycling.	(f) Refer to "Unit Short Cycles".

381

ELECTRICAL-SYSTEM TROUBLE CHART (Con't)

Symptom and Possible Cause	Possible Remedy
(g) Superheat too high at compressor suction.	(g) Change location of expansion valve bulb or adjust valve to return wet gas to compressor.

Frosted or Sweating Suction Line

(a) Expansion valve admitting excess refrigerant.	(a) Adjust expansion valve.

Hot Liquid Line

(a) Shortage of refrigerant.	(a) Repair leak and recharge.
(b) Expansion valve open too wide.	(b) Adjust expansion valve.

Frosted Liquid Line

(a) Receiver shutoff valve partially closed or restricted.	(a) Open valve or remove restriction.
(b) Restricted dehydrator or strainer.	(b) Replace restricted part.

Top Condenser Coils Cool When Unit in Operation

(a) Refrigerant undercharge.	(a) Repair leak and recharge.
(b) Compressor inefficient.	(b) Check and correct compressor.

Unit on Vacuum. Frost on Expansion Valve Only

(a) Ice plugging expansion valve orifice.	(a) Apply hot wet cloth to expansion valve. Moisture indicated by increase in suction pressure. Install drier.
(b) Plugged expansion-valve strainer.	(b) Clean strainer or replace expansion valve.

SUMMARY

In this chapter, some of the most common electrical types of mechanical refrigeration and motor control devices have been presented. Various motor types, their controls, and operation characteristics have been presented first, because the motor is the prime source of power in all compressor operations. In this connection, it should be noted that, although direct current motors are seldom used in compressor drives, their classification and characteristics are included to assist the plant operator and others in distribution districts where only direct current is available.

The various methods of motor-compressor controls, including contactors, relays, and overload devices operating on the thermal or heating principles, provide the necessary information required in plant operation. Description of temperature controls, such as thermostatic control switches and bimetallic thermostats, together with defrost methods and associated devices are also given.

Electrical system maintenance, giving details on capacitor function as used on single-phase ac motors and the checking methods to use in case of failure, has been given. The procedure for motor-compressor replacement will furnish information when a replacement due to burnout or other troubles becomes necessary. The chapter concludes by providing an electrical-system trouble chart giving possible causes and remedies when the refrigerant system becomes inoperative.

REVIEW QUESTIONS

1. State the difference between a shunt-wound and a compound-wound motor.

2. Describe the starting arrangement in a shunt-wound motor.

3. What are the speed and torque characteristics of a squirrel-cage motor?

4. Describe the applications for wound-rotor motors.

5. Why are synchronous motors employed exclusively as power-factor correction devices termed synchronous capacitors?

6. Describe control methods for polyphase motors.

7. What are the starting devices used for single-phase motors?

8. Name five types of single-phase motors.

9. What are the advantages of using single-phase motors in the small sizes rather than polyphase motors?

10. Why is the starting torque greater in a capacitor motor than in a split-phase motor?

11. What are the disadvantages of shaded-pole motors?

12. Why is the repulsion-start induction motor so called?

13. Can an ac series motor operate on either alternating or direct current?

14. Why is automatic motor-compressor control desired instead of manual control?

15. What protection methods are used against overload on ac motors?

16. What is the function of time limit overload relays in an ac motor circuit?

17. Describe and give the function of motor-winding relays.

18. What is meant by temperature control and how does it operate?

19. Describe the operation of a bimetallic thermostat.

20. How does an automatic defrost control operate?

21. What precautions should be observed in electric system maintenance?

22. What is the procedure to use in motor-compressor replacement?

CHAPTER 17

Commercial Refrigeration Principles

There are three principal methods used for producing mechanical refrigeration and consequently low temperatures. They are:

1. Use of natural ice.
2. Freezing mixtures.
3. Vaporizing liquids.

The use of natural ice as a refrigeration medium is too well known to require any explanation. The freezing mixtures, usually called brines, generally consist of natural ice with the addition of calcium chloride which lowers the freezing point of the mixture. The third method, that of vaporizing liquid, is most commonly used in mechanical refrigeration.

For any liquid, the temperature at which it boils or vaporizes is called its *boiling point*. For liquids having very low boiling points, it is not necessary that heat be supplied by any visible source, as the heat in surrounding objects alone may be sufficiently high to cause boiling or vaporization. This is true of anhydrous ammonia, for example, whose boiling point of 28° F below zero is sufficiently low to cause it to boil violently when placed in an open vessel at room temperature. The absorption of heat by the vaporization of the ammonia will cause the outside of the vessel to become heavily frosted by moisture condensed and frozen out of the air immediately surrounding it. From this it follows that a refrigeration plant could consist of only a supply of refrigerant, a

385

regulating valve, and a coil of pipe in which the refrigerant would absorb heat from the substance in which the coil is immersed, and refrigeration would be produced. Such a simple system would, however, be very uneconomical in that it would use the refrigerants only once. The ability to recover refrigerants and use them over and over again is a continuous cycle employed in the compressor system of refrigeration.

The refrigerants primarily used in commercial systems are *Freons* and methyl chloride. Anhydrous ammonia is used in large installations where an operator is in attendance. Anhydrous ammonia is a hazardous material, and should not be used where the commercial refrigeration equipment is unattended. It was used to a great extent in earlier commercial refrigeration facilities, especially in cold-storage plants and ice-making machines. Since it is classified as a hazardous material it required special treatment, even in the electrical wiring for operation of this equipment.

EFFECT OF PRESSURE ON REFRIGERANTS

As the boiling temperature of any liquid may be made to change with pressure, it is therefore an easy matter to cause the liquid refrigerant to boil at any desired temperature by placing it in a vessel where the required pressure may be obtained. The process of boiling or vaporization by which a liquid is changed to a vapor can be reversed, that is the vapor can be reconverted into a liquid by employing pressure. This is called *condensation*. An increase in pressure will, by raising the boiling point, assist in the condensation of the vapor. Liquids used as refrigerants must, by reason of their initial cost, be recovered, and the process of condensation is employed for this purpose.

In an earlier chapter of this book, the pressures and temperatures of various refrigerants were given. These will vary slightly with altitude, due to the difference in barometric pressures, but for practical purposes they are very helpful in refrigeration service.

ESSENTIAL PARTS OF A REFRIGERATION PLANT

The essential parts of a refrigeration plant are:

1. Compressor.
2. Condenser.
3. Receiver.

4. Expansion valve.
5. Evaporator coils.
6. Suitable pipe lines, shut off valves, gauges, thermostats, etc.

Since anhydrous ammonia was the most used refrigerant in earlier times, it will be covered here as it is possible that in commercial refrigeration service you may run across some of this type system still in use.

The apparatus commonly used in an ammonia compression system is shown in Fig. 1. The compressor is merely a pump which takes the ammonia vapor from the low-pressure side, compresses it, and delivers it to the high-pressure side. The *high-pressure* side of the plant comprises the compressor discharge piping, condenser, liquid receiver, and liquid piping between the liquid receiver and expansion valve. The *low-pressure* side comprises the piping leading from the expansion valve to the evaporator coil, evaporator coil, and piping leading from the evaporator coil to the compressor suction side. The pressure at the low side of the system remains practically constant and is indicated by the back-pressure gauge at the suction side of the compressor.

In a refrigeration plant of this type, ammonia vapor superheated by the work of compression passes through a *condenser* where cooling water lowers its temperature until it condenses into liquid ammonia. The liquid leaving is subcooled (82° F), which temperature is slightly below that of condensation, which is 86° F.

The *expansion valve* comprises the dividing point between the high-pressure and the low-pressure side of the system. The function of the expansion valve is to regulate the flow of ammonia from the high-pressure to the low-pressure side. The heat absorbed from the refrigerator by the liquid ammonia in the evaporating coils causes the liquid to boil until vaporization is complete and vapor is slightly superheated. After the liquid refrigerant has produced its refrigerating effect in the evaporating coils, the gas is compressed to a suitable pressure. Ammonia gas, for example, when compressed to 154.5 lbs. gauge, will condense at a temperature of 86° F. Cooling water of a lower temperature than 86° F is usually available for abstracting the latent heat from the vapor, thus converting it to a liquid (i.e., condensing it). To obtain this pressure of 154.5 lbs., this system requires the use of a compressor with some form of motive power, and also a condenser in which the gas may be liquefied.

The regulating or expansion valve releases the liquid from the condenser or its accompanying liquid container or receiver and controls the

Courtesy the Frick Company

Fig. 1. Diagram showing the piping arrangement and accessories in a direct-drive ammonia refrigerating unit.

pressure at which the evaporation and consequent refrigeration takes place. The expansion valve may be hand or automatically controlled, and its opening or closing allows more or less refrigerant to pass to the low-pressure side of the system. The liquid refrigerant in the low side is evaporated by the heat of the substance being cooled, and the resulting gas is pumped away by the compressor. To replace this, enough refrigerant is allowed to pass from the high side through the expansion valve to maintain a constant pressure and temperature in the low side. The compressor brings the low-pressure ammonia gas from the cold side, compresses it, and discharges it to the high-pressure side. This increase in pressure raises the boiling or, in this case, condensing temperature of the gas so that comparatively warm cooling water readily condenses the compressed gas again to a liquid. This liquid ammonia is usually collected in a receiver, and from there it is passed through the expansion valve. The reduction in pressure lowers the boiling point of the ammonia, causing it to evaporate.

The action of other refrigerants used in compression systems closely resemble that just described for ammonia. These systems (Fig. 2) operate fundamentally in the same manner, and are:

1. Evaporation of liquid at low pressure.
2. Compression of gas to raise its condensing temperatures.
3. Condensation of gas at high pressure.
4. Reduction of pressure on the liquid to lower its boiling temperature.

Head pressures, back pressures, and working temperatures vary in accordance with the properties of each refrigerant and with conditions encountered in individual plants.

CONDENSERS, FITTINGS, AND ACCESSORIES

It is important that a condensing unit of the proper size and loading point be selected. In making this selection, the following should be taken into consideration:

1. The capacity necessary to handle total application load without excessive running time or overload when ambient temperatures or water temperatures are at their peak.
2. Selection of type unit, whether air or water cooled.
3. Condensing unit location.
4. Maximum summer air temperature.
5. Maximum summer water temperature.
6. Quality, quantity, pressure, and cost of condensing water.

7. Total cost of operating an air-cooled unit as compared to that of a water-cooled unit.
8. Refrigerant temperature range at which the condensing unit will be operated.
9. Running time desirable.
10. Availability of electrical supply of the proper capacity and voltage.

Courtesy the Frick Company

Fig. 2. Diagram of a standard refrigerating unit using Freon-12 or Freon-22.

When necessary to locate the condensing unit above the evaporator, special care should be taken in sizing the suction line and installation of a trap at the base of the suction line to assure the return of the oil (which may circulate in the system) to the compressor.

In many instances the condenser is cooled by means of a fan on the driving motor, and the flywheel of the compressor has fan blades instead of spokes. The air is drawn through the condenser and blown across the compressor and motor of the unit. Where banks of units are installed (as in a supermarket) the separate units are mounted next to one another and in tiers, such as illustrated in Fig. 3.

Courtesy Steffen's Market

Fig. 3. Bank of air-cooled, supermarket compressors.

Where banks of refrigeration units are installed, such as shown in Fig. 3, each unit has a set of louvers (Fig. 4) which open when the unit starts, if the ambient temperature of the refrigeration unit room is too high. They will not open if the ambient temperature is cool; this is accomplished automatically and electrically.

COOLING WATER REQUIREMENTS

An unfailing supply of suitable water is of the utmost importance in all refrigeration plants. By definition, water is a chemical compound of two

391

gases; oxygen and hydrogen in the proportion of 2 parts by weight of hydrogen to 16 parts by weight of oxygen, at a normal pressure of 14.7 pounds per square inch. At 62° F, one U.S. gallon of water (231 cu. in.) weighs 8.33 lbs., or 8-⅓ lbs. for ordinary calculations.

Courtesy Steffen's Market

Fig. 4. Exterior view of on-demand louvers for air-cooled compressors.

Water cooling and refrigeration are interrelated since, without water, commercial refrigeration would not be possible. The primary considerations when planning a refrigeration system are the temperature, chemical composition, and cost of the required water supply. Condenser water may be obtained from city mains, open ponds, rivers, lakes, wells, or recirculated water. In cases where the water temperature is high due to exposure to the sun, or where it is desired to use the water over again, expediency requires artificial methods of water cooling for condenser use. Atmospheric cooling is usually obtained by one of the following methods:

1. Cooling ponds.
2. Spray cooling ponds.
3. Cooling towers.

392

Cooling Ponds

The cooling pond is usually resorted to when the location of the plant is such that a natural body of water is available nearby. Here, the water is cooled by being forced into surface contact with the air, and the amount of cooling will depend on the relative temperature and humidity of the air. Any available pond or lake may be used as a means of cooling recirculated water without the use of sprays, provided it is large enough for the purpose. The heated water should be delivered as far from the suction connections as possible. Since the cooled water tends to sink due to its greater density, the best results will be achieved when the suction connection is applied to the bottom of the pond.

Spray Cooling Ponds

The spray cooling pond differs from the ordinary cooling pond in that means are introduced to accelerate the cooling effect. This is achieved essentially by properly designed sprays or nozzles which break the water into fine drops, but not into a mist, since to be effective, the drops must fall back into the pond and not be carried away. The number of nozzles required depends on the size of the pond, i.e., the water requirement of the plant. In average practice, the nozzles are located from 3 to 8 feet above the water level and in horizontal rows from 10 to 16 inches apart. Such an arrangement is shown in Fig. 5.

The pressure at the inlet to the spray system may be from 5 to 12 pounds per square inch. Here, as in the previously discussed cooling pond, the final temperature of the water depends on air temperature and humidity. In most cases, however, a reduction of water temperature from 10° to 15° may be obtained with a single spraying.

Cooling Towers

Water cooling towers are perhaps the most common method of obtaining water cooling, and have been in general use for a considerable time. Water cooling towers are divided into two classes, depending on their design as:

1. Open towers.
2. Closed towers.

Open Towers—The open tower depends for its functioning on a number of platforms or drops for breaking up or atomizing the water, or on filming water over large surfaces exposed to the dry atmosphere. The closed tower, also called a fan tower, has much less exposed surfaces and depends for its functioning on the circulation of a large volume of dry air through falling water for its cooling effect.

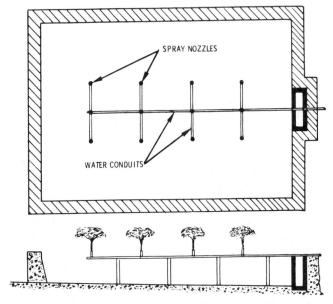

Fig. 5. Spray nozzle arrangement in a typical spray cooling pond.

The open types consist generally of drip pans installed at regular intervals, with space between to allow free access of the air, and each pan having holes for the equal distribution of the water. Shavings, boards, mineral wool, tile, and even slate, have been used with some success in the pans in this type of tower as well as in the closed type. The question is largely one of installation expense and consideration of the deterioration factor. Almost any material or device that provides satisfactory distribution and separation of the water along with adequate retardation, is satisfactory for the purpose.

Such a tower (Fig. 6) is open at the sides and depends upon the natural air circulation of the atmosphere. Its efficiency varies with the velocity of the wind, the humidity and temperature, and also upon the design of the tower for separation and retardation. The deterioration factor in this type is quite large, since the destructive effect of air and water under these conditions is considerable.

Fig. 6. Structural arrangement of an open cooling tower. In a water-cooling system of this type, the water is discharged into the top pan and, after passing through all the various pans, returns to the condenser.

Dimensions of Cooling Towers—A cooling tower for a 50-ton plant is approximately 12 × 6 feet at the bottom and 24 feet high. Sufficient air could be supplied by two 8-ft. fan blowers, which would consume about 8 HP.

Efficiency of Cooling Towers—The efficiency of a cooling tower is the number of degrees to which it can cool the water. Very seldom can the

395

water be cooled much below 75° F in the summer. The higher the temperature of the initial water, however, the more efficient the tower will be in its operation. Condenser water from steam condensers is furnished at a temperature ranging from 165° down to 80° F, and the operation of the towers under these circumstances is very efficient. Refrigerating plants have a range of temperature depending simply on the pressure maintained in the condenser and seldom rises above 120° F for the initial temperature in the cooling tower. The evaporation of the water in the cooling tower must, of course, be resupplied, and this represents a certain loss. In refrigerating plants, the loss of the water is from 5% to 15% for each circulation.

Closed Tower Construction—The closed tower is practically identical with the open type with one important modification. The walls of the tower are enclosed and air is supplied at the bottom and forced upward throughout the tower by means of a fan assembly, such as the one in Fig. 7. The air supply under these circumstances can be varied by mechanical means and the resulting cooling effect made practically independent of temperature and humidity variations of the outside atmosphere. The operation of the tower is further independent of the existence of winds for its efficient operation.

Fig. 7. A roof-top packaged cooling unit.

When considering the installation of a cooling tower, there may be no need for fans, except in cases where a larger floor space is not available. Fans require considerable expense for the installation and a continuous expense for running and repairs. In the majority of plants, there is always plenty of space either in the yard, on top of a roof, or over the condensers for a sufficiently large cooling tower without having to install a fan. If such a tower is properly constructed, it will give the same results as towers with fans, and can be installed at a low cost.

The amount of water used is an important item, unless there is an unlimited supply. If water is purchased, it should be passed over a cooling tower and used again and again. The cooling tower will pay for itself in a short time in the amount of water saved. In installations where reliability is a matter of prime importance and cost of installation a matter of minor significance, the closed type is invariably chosen. The majority of large power plants use the closed type.

Cooling Tower Control

Temperature controls for refrigerating service are designed to maintain adequate head pressure with evaporative condensers and cooling towers. Low refrigerant head pressure, caused by abnormally low cooling-water temperature, reduces the capacity of the refrigeration system.

Two systems of control for mechanical and atmospheric draft towers and evaporative condensers are shown in Figs. 8 and 9. As noted in Fig. 8, the control opens the contacts on a temperature drop. These contacts are wired in series with the fan motor (or to the pilot of a fan-motor controller), stopping the fan when the cooling water temperature falls to a predetermined minimum value corresponding to the minimum head pressure for proper operation. In the control system shown in Fig. 9, the contacts close on a temperature drop, and are wired in series with a normally-closed motorized or solenoid valve, opening the valve when low cooling temperature occurs. The cooling water then flows through a low header in the atmospheric tower, reducing its cooling effect.

COOLING SYSTEM COMPONENTS

Head-Pressure Control

In applying refrigerant condensers, it is well to take advantage of a lower than design condensing temperature whenever this is possible to

Courtesy Penn Controls, Inc.

Fig. 8. Wiring diagram of a control for a cooling tower with forced-draft cooling.

Fig. 9. Wiring diagram of a control for a cooling tower with atmospheric-draft cooling.

achieve the effects of more refrigeration capacity from the system, lower brake horsepower requirements per ton of refrigeration, and lower compressor discharge temperatures. At the same time, it is often necessary to take precautions that the condensing pressure does not drop down too low during periods of low temperature of the water or outside air. This is particularly true on jobs that have higher internal or process loads throughout the year, and also on systems having compressor capacity control. The principal problem arising from abnormally low condensing pressures is the reduction of the pressure difference available across the expansion valve to the point where it is unable to feed sufficient refrigerant to the evaporator to handle the cooling load.

Water-Cooled Condensers

With water-cooled condensers, head-pressure controls are used for the dual purpose of holding the condensing pressure up and conserving water.

Water Regulating Valves

The control commonly used with water-cooled condensers is the condenser water regulating valve. This valve is usually located in the supply water line to the condenser. It is a throttling type valve, and responds to a difference between condensing pressure and a predetermined shutoff pressure. The higher this difference, the more the valve opens up, and vice versa. A pilot connection to the top of the condenser transmits condensing pressure to the valve, and the valve assumes an opening position according to this pressure. The shutoff pressure of the valve must be set slightly higher than the highest refrigerant pressure expected in the condenser when the system is not in operation. This is to make sure the valve will close and not pass water during off cycles. This pressure will be slightly higher than the saturation pressure of the refrigerant at the highest ambient temperature expected around the condenser.

Condenser water regulating valves are usually sized to pass the design quantity of water at about 25-30 lbs. difference between design condensing pressure and valve shutoff pressure. In cooling-tower application, a simple bypass can also be used to maintain condensing temperature, employing a manual valve or an automatic valve responsive to head-pressure change. The valve can be either the two-way or three-way type.

Evaporative Condensers

When installed outdoors, the unit should be located so that there is a free flow of air around it. A minimum clearance of twice the width of the unit should be provided at the air inlet and air outlet. Care should be taken that the flow of the discharge air is not deflected back to the inlet. Short circuiting in this fashion can materially reduce the cooling capacity. Special consideration should be given to short circuiting on multiple installations. It is very desirable to position the unit so that the prevailing summer wind will blow into the inlet. It is also desirable to take precautions against a strong wind blowing into the outlet, which can reduce the capacity and might overload the fan motor. A windbreak erected about six feet from the air discharge is often useful in preventing this. In some outdoor locations, it may be desirable to place a coarse screen over the air inlet to exclude large objects, such as birds, paper, and other debris.

If the unit is installed indoors, careful consideration must be given to supplying and removing an adequate amount of air from the space. If the discharge air is allowed to recirculate, the wet-bulb temperature will rise and the capacity will be reduced. It is usually possible to omit a duct to the air inlet. The unit can draw air from the room, but provision should be made to supply an adequate amount of replacement fresh air by special blower units when required.

The piping to and from the unit should be installed in accordance with good practice, using pipe sizes at least as large as the fittings on the unit. It is desirable to use vibration eliminators on all lines to and from the units. The units should be installed so that the top cover can be lifted for maintenance. The makeup water line should be connected to the float valve. If it is desired to use the makeup line to fill the system, it may be desirable to make the pipe size larger than the float valve connection size.

In some locations, it is permissible to let the overflow drain on the roof, but if this is not desired, the overflow should be piped to a sewer. A bleed or blowdown connection should be incorporated in the piping. This should be installed in the discharge pipe from the pump, and at a high enough level so that the bleed flows only when the pump is operating. If the unit is installed outdoors and must operate in winter, the installation should be made using an indoor storage tank in a heated space.

Air-Cooled Condensers

With air-cooled condensers (Fig. 10), the pressure stabilizing method is often employed as a means of obtaining pressure control. This is achieved by backing liquid refrigerant up in the coil to reduce the effective condensing surface. The basic principle of the pressure stabilizer is simple and unique. It is a heat transfer device which transfers

400

the heat from the hot-gas discharge of the compressor to the subcooled liquid leaving the condenser. This heat exchange is controlled by the regulating valve installed between the condenser and the receiver. This valve is set at the desired operating pressure and throttles from the open position to the closed position as the head pressure drops. The throttling action of the valve forces liquid to flow through the heat-exchanger section of the pressure stabilizer. The heat picked up by the liquid raises the receiver pressure and prevents further flow of liquid from the condenser. This causes flooding of the condenser and, by reducing the effective surface, maintains satisfactory discharge and receiver pressures. This assures a solid column of liquid at the expansion valve at a pressure which insures satisfactory operation of the system.

Fig. 10. An air-cooled condenser arrangement with a hot-gas header and a pressure stabilizer.

The pressure stabilizer is provided with a spring-loaded check valve in the heat-transfer liquid section which remains closed during warm-weather operation to insure against liquid refrigerant reheating.

Driers

Moisture in a refrigeration system is one of the main causes of trouble, particularly to the compressor, resulting in the formation of acids which may attack the motor insulation, the valves, and the valve plate. Some driers are most effective when located in the refrigerated space, although there are now driers on the market which the manufacturers claim are not affected by temperature. The decision of whether or not to use a perma-

401

nent drier on the higher-temperature systems is one of judgment. In the case of package-type air conditioners and water chillers, it may not be necessary if proper dehydration techniques have been followed at the factory, no refrigerant connectións are broken in the field, and any refrigerant added is at a sufficiently low moisture content. If there is any doubt that the system is initially free of moisture, or that the system can be maintained dry in operation, then a drier is indicated for protection of the compressor and to prevent freezing at the liquid feed device.

In the case of hermetic compressors, there is an additional reason for keeping moisture out of the system. The motor windings are exposed to the refrigerant gas and excessive moisture can cause a breakdown of the motor winding insulation, possibly resulting both in burnout of the motor and the distribution of the products of decomposition throughout the refrigeration system. A full-flow drier is usually recommended in hermetic compressor systems to keep the system dry and to prevent the products of decomposition from getting into the evaporator in the event of a motor burnout.

Side-outlet driers are preferred since the drying element can be replaced without breaking any refrigerant connections. The drier is best located in a bypass of the liquid line near the liquid receiver. The drier may be mounted horizontally, as shown if Fig. 11, but should never be mounted vertically with the flange on top since any loose material would then fall into the line when the drying element is removed.

Fig. 11. A drier with piping connections.

A three-valve bypass should be used, as shown, to provide a means for isolating the drier for servicing and when using open-type compressors, to allow only partial refrigerant flow through the drier to reduce pressure drop. In systems using hermetic compressors, it is preferable to have all refrigerant going through the drier at all times for the reasons just listed in the preceding paragraphs.

Reliable moisture indicators are now on the market for installation in refrigerant liquid lines. These devices will indicate whether or not a system has excessive moisture in it. They are valuable additions to the system as a indicator of when the drier cartridge should be replaced.

Fittings and Valves

One of the most important considerations when installing a refrigerating plant is to make sure that the piping system is absolutely tight, since even the slightest leakage is dangerous. It is important to remember that the refrigeration cycle is a closed one with the compressor in series in order to obtain continuity. Tightness is important, especially with ammonia, although other refrigerants are hardly less demanding in this respect. Because pressures are almost always above the pressure of the atmosphere, the possibility of air leaking into the system is slight; but the piping and fittings must be of a material to comply with the various codes in existence, and the joints must be designed for the particular application and system performance.

Referring back to Fig. 2, of this chapter, you will see more valves shown than you will find on an ordinary commercial system. Valves D and E on the compressor are compound valves, as were explained earlier, in Figures 1 and 2 of Chapter 6. These two valves are always present, and valve B at the discharge of the receiver tank will always be found. The other valves shown in Fig. 2 of this chapter are convenient but not always installed.

Should you wish to pump the system down to change an expansion valve, you would close valve B, place a compound gauge on valve E by backing the valve stem all of the way out, and a pressure gauge on valve D, by backing the valve stem all of the way out. Then crack both valves by backing the stem out slightly. Purge the air out by loosening the nut where the line attaches to each gauge. Now start the compressor. If the compressor is controlled by a pressure actuated switch on the low side, this switch will have to be blocked closed, or it will open when the low side reaches the pre-set point. Stay in attendance with the compressor when doing this. If you hear it scrubbing oil from the crankcase, and this will be very noticeable, stop the compressor and, proceeding slowly, turn the compressor on and off by hand. This operation will draw the refrigerant back from the closed valve B on the receiver tank, back through the compressor and put it in the receiver tank.

Draw a vacuum on the low side gauge, and then shut the machine off. If the vacuum holds, crack the valve on the receiver tank slightly to allow liquid to enter the evaporator, watching the low pressure gauge. When this gauge shows about 1 pound of pressure, shut off valve B. You will want to have a very slight pressure on the system, so that when you open the system at any point, such as to change the drier-filter, replace the expansion valve, etc., there will be refrigerant pressure, however slight, to blow out and not suck air into the system. You may now replace the drier or expansion valve. Before connecting the discharge side of either of these, crack valve B, slightly allowing refrigerant to pass through and escape into the air, to purge out any air in them. Then attach the discharge side tubing to the component(s), you replaced. Open valve B all of the way, checking for leaks at the points where you were working.

Valves F, F_1, and F_2, are handy to have, but expensive. Thus, if you wish to change the drier, close valves F and F_1, which are ordinarily open, and open valve F_2, which is normally closed. Remove the drier, remembering that the high-side pressure is on it, so crack the connections slowly until the pressure is released. Remove the drier, put in a new drier, connecting the side toward valve F first, then crack valve F slightly to purge the air out, and then connect the side of the drier on the valve F_1 side. Close valve F_2, open valves F and F_1, and operate the compressor, checking the pressures at the gauges on the compressor, especially the high-side gauge for abnormal pressure. The system should be again ready to operate. However, check for leaks when under pressure.

Ordinarily, you will not have valves C and A on the condenser. These two valves are what you might call luxuries. They are nice, but expensive.

If you needed to replace the flapper valves in the compressor head, or the shaft seal, you would proceed as explained for the drier replacement.

If the flapper valves in the compressor head are to be replaced, remove the head slowly as there will be more pressure here. As you tighten the head again, crack valve B to purge the air out of the head. Close valve B, and tighten the head-bolts. Put a little pressure on the head by cracking valve B and check for leaks. You are again ready to operate the compressor.

Without valves C and A, on the condenser, if the condenser leaks, you will lose the refrigerant anyway, so you should not be too concerned.

Replace the condenser and pump down the system with a vacuum pump as previously explained, drying out the system, and recharge. A condenser problem is very seldom encountered.

Heat Exchangers

Although sometimes a subject of controversy in high-temperature application, it is generally agreed that in medium- and low-temperature refrigeration systems, heat exchangers (when properly applied) perform the following functions of utmost importance to overall system performance:

1. Subcooling the liquid refrigerant entering the thermal expansion valve reduces the flash gas load at the evaporator inlet. It likewise increases the heat content difference of the refrigerant during its evaporating phase, which produces more useful work in the evaporator.
2. In the process of subcooling, the heat extracted from the liquid refrigerant is transferred to the suction gas, thereby insuring a dry suction return to the compressor at an entering superheat level, which will produce the best possible volumetric efficiencies for the refrigerant used.
3. The increase in suction line temperature will likewise reduce the possibility of sweating.
4. The use of a heat exchanger will permit more open adjustment of the thermal expansion valve without the risk of serious floodback of liquid to the compressor under light or variable load conditions, and at the same time assuring the maximum utilization of evaporator surface.

The advantages of a heat exchanger may be cancelled out if its use increases the suction line pressure drop by an excessive amount. On a theoretical heat exchanger without pressure drop, designed to superheat the suction gas to 65° F from a normal 20° F suction temperature, it is entirely possible to obtain as high as a 16 per cent increase in the overall efficiency of the machine as compared to a system without a heat exchanger. There are many factors involved, however, to obtain such a figure and, for practical purposes, it may readily be said that the overall efficiency of a system may be increased 8 to 10 per cent by the proper adaptation of a heat exchanger.

In all low-temperature applications, it is more important to correctly size and properly apply heat exchangers. Their selection must be based on accurate performance ratings checked out against the calculated design loads involved for each evaporator or for the entire system. Care must likewise be taken to insure that both liquid and suction connections are properly sized in order to reduce entrance and exit losses to a minimum.

Balancing the System

The application of air cooling the low sides in refrigeration work is basically an air-conditioning problem. It so happens that the refrigeration industry has certain standards which make the adaptation fairly easy as compared to the adaptation of a coil to air-conditioning work. In the proper application of a coil, it is most important to make sure that there is a balance between the condensing unit and the low side. The temperature difference between the air to be passed over the coil and the refrigerant temperature in the coil is the point which determines the conditions which will be maintained in a given cooler.

For the general purpose of refrigerators involving meats, vegetables, and dairy products, it is common procedure to balance the low side to the condensing unit at a 15° temperature difference—that is, they are balanced to maintain a temperature difference of 15° between the refrigerant in the coil and the air temperature. It has been learned by experience that if this is done, one may expect to maintain 80% to 85% relative humidity in a cooler, which is a good range. A coil which is selected for a wide temperature difference will therefore maintain a lower relative humidity in service, whereas one which is selected for too close a temperature difference will produce relative humidities which are higher than required for practical operation, and surface sliming may result on stored meat products during winter periods when loads are reduced. It has been determined over years of practice that coils should be selected to balance between 12° and 15° F temperature difference for general-purpose refrigerated storage coolers.

On straight vegetable coolers where higher humidities are desired, the coil should be selected to balance the compressor at a 10° F temperature difference, as this will produce an average relative humidity of 90% within the refrigerated space. The same recommendation applies to florists display boxes and, in both cases, the maintenance of a high relative humidity in long-term storage is beneficial, whereas some exception with reference to meat products was noted previously. On low-temperature units, if one stops to consider that the amount of dehumidification is in proportion to the temperature difference, it is obvious that the closer the temperature difference, the less frost accumulation. It is recommended that coils for low-temperature work be selected to balance the condensing unit at a 10° temperature difference or less.

REFRIGERATION SYSTEM OPERATION

It is essential to thoroughly clean the compressor before placing it in service, and the best time for cleaning is just prior to the initial start of the system. The crankcase cover should be removed and the inside of the crankcase thoroughly cleaned out. Only clean lint-free rags should be used for this purpose. If waste is used, lint and threads will stick to the casting surfaces.

Before starting the compressor, make certain that the crankcase is properly filled with a good grade of machine oil as recommended by the manufacturer of the machine. The initial oil charge may be filled through the crankcase opening to a sufficiently high level for satisfactory operation. Although compressors are usually given a factory test before shipment, a no-load run of from 10 to 20 hours (depending on the size of the compressor) is recommended. Remove the head covers and discharge valves and pour oil on top of the piston as close to the cylinder walls as possible to insure initial lubrication to the cylinder walls. Rotate the compressor by hand to make sure everything is free, then start the motor and compressor and run the unit at idle to insure that every part is properly lubricated before the unit goes into actual service. It is preferable before starting the compressor to oil prime the gear pump by removing the small pipe plug on top of the strainer and pouring in oil. This oil will run back into the pump and seal the gears.

When satisfied that all parts are properly lubricated, and that there is no undue heating, the unit should be stopped and the discharge-valve and cylinder-head covers replaced, making sure that the cylinder-head cover gaskets are in place and that they are properly tightened. Before starting the plant, however, it is important to test the system for leakage at a pressure equal to the maximum discharge pressure at which the system is expected to operate. This will not apply to all commercial refrigeration, as many of the compressor units do not have oil pumps.

Starting the Plant

Some refrigeration units come with a refrigerant charge in the receiver tank and others do not. The system may come with a slight charge of an inert gas, such as nitrogen to ward off moisture-laden air. Irrespective, the complete system; i.e. new lines, evaporator coils, etc., have to be pumped down with a vacuum pump to get rid of all air and moisture. As the general rule the compressor will not be charged with refrigereant in the receiver tank, so we will proceed from that angle:

407

1. Open all valves, installing a compound gauge system on the high and low-sides. Connect the vaccuum pump as was described earlier and pump the system down, until no more bubbles are coming out through the oil container, into which the discharge of the vacuum pump was inserted. Continue pumping after this for about one hour.

2. Back the suction valve all the way out, shutting off the opening to the compressor and also to the discharge line of the compressor (suction line). This closes the connection that you had connected the vacuum pump to.

3. Shut off the vacuum pump and disconnect it. Connect a cylinder of refrigerant to where you disconnected the vacuum pump from the compressor, and crack the refrigerant cylinder valve slightly to purge the line from it to the compressor.

4. You are now ready to charge the system.

5. With the refrigerant cylinder upright, start the compressor, and screw the compressor valve in slightly, then open the refrigerant cylinder valve. Control the gas going in by the valve on the compressor, being sure that the discharge valve on the compressure is screwed in just slightly so that the high-side pressure may be checked.

6. The vacuum pump should have indicated any leaks, but do not rely on this. When you have a few pounds pressure on the entire system, stop this operation, backing out the suction valve to shut off the gas to the system from the refrigerant tank.

7. Test for leaks, as was previously explained. If no leaks are found, continue charging the system by cracking the suction valve, turning it in a turn or two.

8. Do not turn the refrigerant tank upside down, as you will scrub oil from the compressor. Leave this tank upright, so that you receive only gas into the system.

9. There are two methods used to tell when the proper charge is reached. One, is that the receiver tank has a sight glass on its end, to tell you the liquid level in the receiver. The other is a sight glass in the liquid line from the receiver tank.

10. If there is a sight glass approximately ½ of the way up on the end of the receiver tank, shut off the refrigerant when this point is reached.

11. If the sight glass in the liquid line is used, you will see bubbles in

the sight glass until pure liquid refrigerant is flowing, at which time you will have a clear sight glass and know that the system is receiving liquid refrigerant into the evaporator.

12. At this point stop adding gas, and let the system operate for a considerable time, to be sure that you have enough refrigerant.

13. If after a time bubbles appear in the sight glass, you know that you do not have quite enough refrigerant in the system.

14. All of this time watch the high-side gauge, if the pressure is abnormally high, you have not removed all of the air from the system. If this is the case, shut the system off, and in a little while bleed some refrigerant from the high-side valve. If the system was properly cleared of air by the vacuum pump, this problem will not occur.

15. If all systems seem to be operating properly, watch the evaporator for freeze-over. It should soon freeze over completely.

16. If the frost goes back beyond the bulb attachment to the suction line, adjust the expansion valve to bring the frost back to the bulb. This is a slow process, and you must not hurry it.

Stopping the Plant

If the plant is to be out of operation for some period of time, it is well to pump the system down, this is accomplished as follows:

1. Shut valve B at the receiver tank in Fig. 2.

2. Install a low-pressure gauge that reads inches of mercury on the low-side of the compressor.

3. Start the compressor. You may have to block the low pressure cut-off switch, to keep the compressor running. Pump a slight vacuum, and then stop the compressor for a little while. If there is still a vacuum on the system, crack the valve B on the receiver tank (Fig. 2) slightly, until a pound or two of refrigerant is in the system, then shut off all valves. Be sure to tag the compressor plainly, so that someone else will know what has been done.

The pressures on the low-side gauge, for turn-on and shut-off, will materially aid in setting the temperatures desired, as most commercial refrigeration systems are controlled by a low-side pressure switch.

COMPRESSOR TROUBLE CHART

Symptoms and Possible Cause   Possible Remedy

High Head Pressure at Discharge

(a) Air or noncondensable gas in system.

(a) Purge from condenser.
Increase water flow.
Check pump and flow lines.

(b) Inlet water warm.

(b) Increase water flow.

(c) Insufficient water.

(c) Clogged water pump strainer or pipe line; clean out.

(d) Condenser clogged.

(d) Clean condenser tubes.

(e) Too much liquid in condenser-receiver.

(e) Remove excess into service drum.

Low Head Pressure at Discharge

(a) Charge of refrigerant in system too low.

(a) Add refrigerant from service drum.

(b) Discharge valve leaks slightly.

(b) Repair or replace.

(c) Too much cold water.

(c) Throttle outlet water valve to condenser.

(d) Faulty thermal expansion valve.

(d) Adjust or replace expansion valve.

High Suction Pressure

(a) Over feeding of expansion valves.

(a) Adjust valve to higher superheat.

(b) Leaking compressor suction valves.

(b) Remove compressor head and examine valves.

Low Suction Pressure

(a) Insufficient refrigerant in system.

(a) Add from service drum.

(b) Restriction in liquid lines.

(b) Remove and clean strainer.

(c) Stopped suction strainer.

(c) Remove and clean carefully.

(d) Underfeeding of expansion valves.

(d) Adjust valve to decrease superheat.

410

COMPRESSOR TROUBLE CHART (Con't)

Symptom and Possible Cause   Possible Remedy

Compressor Stops From High Pressure Cutout

(a) High-pressure cutout setting incorrect.

    (a) Make correct setting.

(b) Overcharge of refrigerant.

    (b) Remove excess into service drum.

(c) Water failure.

    (c) Investigate as to why water failure switch did not function.

(d) Pump failure.

    (d) Check over pump and motor.

(e) Stoppage in water line.

    (e) Check water strainer and valve settings.

Compressor Short-Cycles on Low Pressure Cutout

(a) Coils covered with excess frost.

    (a) Defrost coils.

(b) Clogged screens in liquid lines.

    (b) Clean screens.

(c) Improper setting of back-pressure valves.

    (c) Regulate properly.

(d) Also see causes listed for Low Suction Pressure.

Compressor Does Not Start

(a) Overload has tripped.

    (a) Push reset.

(b) High-pressure cutout open.

    (b) Push reset.

(c) Low charge of refrigerant.

    (c) Repair leak. Recharge system.

(d) Low-pressure cutout open.

    (d) See under Low Suction Pressure.

(e) Electric valve closed.

    (e) Check thermostat which may be in the "off" position.

(f) Water-failure switch open.

    (f) Check water supply.

Compressor Runs Continuously

(a) Shortage of refrigerant.

    (a) Charge additional refrigerant from service drum.

411

COMPRESSOR TROUBLE CHART (Con't)

Symptom and Possible Cause	Possible Remedy
(b) Compressor valve leakage.	(b) Check for leaks.
(c) Cylinders will overheat.	(c) Repair or replace valves.

SUMMARY

The principles of commercial refrigeration are based on circulation of refrigerants through the various components comprising the refrigeration plant. To understand how the system operates, it is necessary to study the essential parts of the system.

REVIEW QUESTIONS

1. Name three principal methods used for producing mechanical refrigeration.

2. What is the effect of pressure on refrigerants?

3. Name the difference between a direct and indirect expansion system.

4. What are the essential parts of a refrigeration plant?

5. Name the components of the high- and low-pressure side of a refrigeration system.

6. What is the function of the expansion valve and where is it located in the system?

7. What precautions should be observed prior to installation of condensers, fittings, and accessories?

8. What is the function of driers in the refrigeration system?

9. What is the function of heat exchangers?

10. What are the adopted standards in balancing a refrigeration system?

11. What precautions should be observed in the operation of a refrigeration system?

CHAPTER 18

Brine Systems

Refrigeration is distributed by several methods. In direct expansion systems, the volatile refrigerant is allowed to expand and evaporate in a pipe or coil placed in the space to be cooled where the refrigerant absorbs its latent heat of evaporation from the material to be cooled. This method of refrigeration is used in small cold-storage rooms, in constant-temperature rooms, in freezer rooms, and in other locations where possible losses due to the leakage of refrigerant would be low.

INDIRECT REFRIGERATION SYSTEM

In the indirect system, some refrigeration medium, such as brine, is cooled down by direct expansion of the refrigerant, and is then pumped through the material or space to be cooled, where it absorbs its sensible heat. Brine systems are used to advantage in large installations where the danger due to leakage of the large amount of refrigerant present is important, and in locations with fluctuating temperatures. For small refrigerated rooms or spaces, where it is desired to operate the refrigerating machine only part of each day, the brine coils are supplemented by holdover or congealing tanks.

HOLDOVER TANKS

A holdover tank is a steel tank containing strong brine in which are immersed direct expansion coils. During the period of operation of the refrigeration machine, the brine is cooled down and is capable of absorbing heat during the shut-down period, the amount depending upon the quantity of brine, its specific heat, and the temperature head.

CONGEALING TANKS

Congealing tanks serve the same general purpose as holdover tanks, but operate on a different principle. They contain, instead of a strong brine, a comparatively weak brine solution which freezes or congeals to a slushly mass of crystals during the period of operation. This mass of congealed brine, in addition to its sensible heat, is capable of absorbing heat equivalent to its latent heat of fusion. For the same refrigerating effect, congealing tanks can be made much smaller than holdover tanks for the reason that advantage is taken of the latent heat of fusion of the brine. Thus, there is a considerable saving in construction and maintenance cost for this type of system.

BRINE

In addition to acting as a simple carrying medium for refrigeration, a satisfactory brine should possess certain other properties desirable from an operating standpoint. Briefly, it should have a low enough freezing point to remain liquid at the lowest temperature encountered at any point in the refrigeration system. It should be as noncorrosive as possible, and should not be subject to precipitation when contaminated with the refrigerant through accidental leakage. The specific heat of the brine should be as high as possible in order to avoid the necessity of handling excessively large volumes.

Of the two materials commonly used for the making of brine, sodium chloride and calcium chloride, the latter is much to be preferred because it fulfills all the foregoing requirements in that it is less corrosive on equipment and is capable of furnishing brines having a much lower freezing point than sodium chloride. For practical purposes, the lowest freezing point possible with sodium (common salt) brines is $0°$ F, while the calcium brines, temperatures may safely run as low as $-58°$ F.

Calcium Chloride

The content of actual calcium chloride ($CaCl_2$) in commercial solid form is 73% to 75%, and in flake form it is 77% to 80%, the balance being

415

principally water of crystallization with a little salt (NaCl). Magnesium chloride is absent, thus precluding the danger of corrosion from mixed chlorides. About 0.1% of lime [Ca(OH)$_2$] is contained in the product; the alkalinity which this contributes to the brine is desirable for protection of iron and steel apparatus against corrosion.

Calcium chloride dissolves readily in water; in fact it has such an attraction for water that when exposed to air under ordinary conditions it absorbs atmospheric moisture until the calcium chloride is dissolved. Its solutions are relatively stable and, when made up to suitable strength, remain liquid at temperatures far below the lowest temperatures required in commercial refrigeration practice. The properties and freezing points of calcium chloride brines of various strengths are given in Table 1. The indicated freezing point of any brine is the temperature at which crystals begin to form in it; it does not mean that the brine solidifies completely at that temperature.

In the refrigeration industry, the specific gravity hydrometer is generally used to determine the strength of brine. In Table 1 corresponding salimeter and Baume hydrometer readings in degrees are given, in addition to specific gravities.

Initial Charges

The solid form was formerly the only type of calcium chloride on the market. This was delivered in steel drums and, in order to put it in convenient size for handling, it was necessary to break up the contents of the drum by pounding it about its circumference. After this, the end of the drum was cut off with an axe and the calcium chloride was dumped into the brine tank, where it dissolved in the water. A more convenient form of calcium chloride is now available. This is flake calcium chloride which, as its name indicates, is a loose flake material that does away with all the labor involved in preparing the solid type for solution. By the use of flakes, it is only necessary to pour the chloride from the drum or the convenient and easily handled 100-pound moisture-proof bag into the brine tank which has previously been partially filled with water. If the calcium chloride is placed directly in front of one of the brine agitators, it will soon go into solution.

Practically any water may be used to make up the brine, but for new installations, excessively hard water should be avoided if possible. In making up brine for new installations, it will be noticed that when calcium chloride is dissolved in water, considerable heat is developed.

416

Table 1. Properties of Calcium Chloride Brine

Density of Brine at 60° F.		Per Cent Anhydrous Calcium Chloride	Freezing Point ° F.	Weight of Brine at 60° F.		Pounds Dow 73-75% Solid Calcium Chloride		Pounds Dow 77-80% Flake Calcium Chloride	
Specific Gravity	Baume Scale			Pounds Gallon	Pounds Cu. Ft.	Per Gallon of Solution at 60° F.	Per Cu. Ft. of Solution at 60° F.	Per Gallon of Solution at 60° F.	Per Cu. Ft. of Solution at 60° F.
1.00	1.0	0.0	32.0	8.34	62.4	.00	0.0	.00	0.0
1.02	2.8	2.3	30.2	8.50	63.7	.26	2.0	.25	1.8
1.04	5.6	4.7	28.0	8.67	64.9	.55	4.1	.51	3.8
1.06	8.2	7.0	25.9	8.84	66.2	.84	6.3	.78	5.9
1.08	10.7	9.2	23.4	9.00	67.4	1.13	8.4	1.06	7.9
1.10	13.2	11.4	20.3	9.17	68.7	1.42	10.6	1.33	9.9
1.12	15.5	13.5	16.5	9.34	69.9	1.70	12.8	1.59	11.9
1.14	17.8	15.6	12.0	9.50	71.2	2.00	15.0	1.87	14.0
1.16	20.0	17.6	7.0	9.67	72.4	2.30	17.2	2.15	16.1
1.18	22.1	19.5	1.4	9.83	73.7	2.60	19.4	2.43	18.2
1.20	24.2	21.5	—5.8	10.01	74.9	2.90	21.7	2.72	20.3
1.22	26.2	23.3	—13.2	10.17	76.2	3.20	24.0	3.00	22.5
1.24	28.1	25.1	—21.3	10.33	77.4	3.51	26.2	3.29	24.6
1.26	29.9	26.9	—30.8	10.50	78.7	3.83	28.6	3.58	26.8
1.28	31.7	28.7	—44.1	10.66	79.9	4.14	31.0	3.88	29.0
1.29	32.6	29.6	—59.8	10.75	80.5	4.31	32.2	4.03	30.2
1.30	33.5	30.5	—41.8	10.84	81.2	4.47	33.4	4.19	31.3
1.32	35.2	32.2	—16.6	11.01	82.37	4.80	35.9	4.50	33.6
1.34	36.8	34.0	4.1	11.18	83.62	5.14	38.4	4.81	36.0
1.36	38.4	35.7	21.6	11.34	84.86	5.48	41.0	5.14	38.4
1.38	39.9	37.4	37.0	11.51	86.11	5.83	43.6	5.46	40.8
1.40	41.4	39.1	50.4	11.68	87.36	6.18	46.2	5.79	43.3
1.42	42.9	40.8	60.6	11.84	88.61	6.36	47.6	5.96	44.6

Warm alkaline brines readily attack zinc or galvanized coatings, so the best procedure is to dissolve the calcium chloride and cool the brine in a tank which contains no zinc-coated metal. In case no extra tank is available, the ice cans should be removed from the freezing tank before making up the brine.

The quantity of calcium chloride needed to charge a refrigeration system at a given strength is determined by the volume of brine needed. The capacity of a system is the sum of the volume of the brine tank plus the volume of the brine coils, coolers, auxiliary tanks, etc., less the brine displacement of ice cans, expansion coils, etc. Having determined the net capacity of the system in cubic feet, the amount of calcium chloride needed may be calculated by consulting Table 1 which shows the strength of brines and their corresponding freezing points. The brine should be of such a strength that it will not freeze at the lowest temperature in the system. The proper strength of calcium chloride brine should have a freezing point of from 10 to 15° F below the minimum brine temperature prevailing at the point of ammonia expansion.

To prepare the brine, find in the table the pounds of calcium chloride per cubic foot brine corresponding to that freezing point, and multiply by the net capacity of the system in cubic feet. The result is the pounds of calcium chloride required.

Foaming of Brine

The presence of foam on the surface of a refrigeration brine is often troublesome. It is due to air being given off by the brine or to a recent formation of a fine colloidal precipitate. Weak brines are capable of holding more air in solution than are strong brines; therefore, it is advisable to keep the brine up to strength at all times. When a weak brine is strengthened by the addition of calcium chloride, the air is forced out of solution in the form of tiny bubbles. The presence of insoluble corrosion particles tends to stabilize the foam or froth thus formed.

When calcium chloride has not recently been added to a brine, and froth or foam appears, it is an indication that corrosion is taking place or that an excess of air is being introduced into the brine. Steps should be taken at once to ascertain the cause and to prevent its continuance. The use of an electric light rather than an open flame for investigating the foaming condition of a brine is recommended on account of the possibility of hydrogen gas being present. The hydrogen is a by-product of electrolytic corrosion.

BRINE SYSTEM CORROSION

All metals used in construction of ice plants are subject to corrosion. Corrosion is most commonly an oxidation of the metal surface caused by the oxygen in the air, by the oxygen contained in absorbed air found in the water, or by the oxygen contained in water solutions. In the case of calcium chloride brine, the dissolved oxygen should be kept to a minimum. This may be accomplished by eliminating, as much as possible, the contact of the brine with air. Agitation and rapid circulation at such areas of contact should be avoided. Care should be taken that the brine lines discharge under the surface of the liquid at all times, since any dropping or splashing of the brine in the air tends to increase the amount of oxygen dissolved by the liquid. Naturally, brines used in open systems such as ice tanks are likely to develop corrosive tendencies more easily than those of closed systems. Nevertheless, care in design and operation of all brine refrigerating systems, so as to minimize the amount and time of contact with air, will reduce the attendant oxygen absorption and will help an appreciably in reducing corrosion.

Other factors which affect the absorption of oxygen in refrigerating brines are the density and temperature of the solution. Weaker brines dissolve more oxygen; therefore, brines always should be kept at full operating strength. While higher temperature brines, in themselves, dissolve slightly less oxygen than colder ones, the reaction between the oxygen and the hydrogen film is speeded up a great deal at warmer temperatures. Hence, in off seasons of the year, or under other conditions when refrigerating capacity is to be reduced, it is desirable to keep the temperature of brine about 30° F in any inactive units.

The second major cause of corrosion taking place in refrigerating brines is due to atoms of hydrogen combining on the surface of the metal to form molecules of the gas which escape as bubbles. Such action is largely a function of the acidity or alkalinity of the brine.

Corrosion Retarders

There are numerous rust and corrosion preventive compounds on the market. Many have sodium silicate or ''water glass'' as a base. These are sometimes useful in clarifying sodium chloride brines, but there seems to be no authorative data available which shows a lessening of corrosive effects resulting from their use. Sodium silicate, when added to calcium chloride brines, causes a precipitate of calcium silicate to form, which only adds to whatever sediment is already present.

The corrosion committee of the A.S.R.E. (American Society of Refrigerating Engineers) in their latest report recommended the use of from 100 to 150 pounds of sodium chromate per 1000 cubic feet of calcium brine. The commercial grade of sodium dichromate is usually employed for this purpose, but in order to change the dichromate to the normal chromate, it is necessary to add 27 pounds of caustic soda for each 100 pounds of dichromate of soda added. Both materials should be added in the form of strong solutions in order that they may be quickly mixed into the brine. The degree of protection afforded by this treatment is stated to be from 80% to 90%. In other words, the corrosion is calculated to be from 10% to 20% of what unavoidably occurs in an untreated brine system.

pH Values

The letters *pH* serve as a symbol denoting the negative logarithm of the concentration of the hydrogen ion in grains per liter. Because of the fact that this scale is based on a logarithmic equation, a variation of one unit indicates a tenfold variation in the actual acidity or alkalinity of the solution. A value of 7 on the *pH* scale is the neutral point, higher values indicate alkalinity and lower values acidity. Thus, for example, a solution having a *pH* value of 9 is ten times as alkaline as one having a *pH* value of 8, or on the acid side, *pH5* indicates ten times the acidity shown by *pH6*.

One method of determining the *pH* value of a solution is by using a *universal indicator* solution. A few drops of the solution added to a test tube half full of brine will show a color indicating the *pH* value of the sample. The best reading will be obtained by gently shaking the test tube just enough to spread the color about ½ inch from the top of the sample. If the mixture is shaken too much, the color will be diffused and it will be difficult to accurately identify the shade. Table 2 lists the *pH* values indicated by the main divisions of the various colors.

Table 2. pH Color Table

Condition	Color	pH Value
Acid	Red	4.0 or lower
"	Orange	5.5
"	Yellow	6.5
Neutral	Greenish yellow	7.0
Alkaline	Green	8.0
"	Blue	9.5
"	Violet	10 or higher

If a simpler test is desired, a phenolphthalein indicator solution may be used. This solution, containing 5 grains of phenolphthalein crystals per liter of denatured alcohol, may be obtained from any druggist. A few drops added to a test tube partially filled with brine will be colorless if the brine is neutral or will turn white if the brine is acid. If the brine is alkaline, the addition of phenolphthalein indicator will produce a red color, pink if slightly alkaline, and deeper red if more highly alkaline (above *pH 10.0*, no further color change will be noted).

A properly treated brine should be sufficiently alkaline to produce a slight pink color when the phenolphthalein test is made. This test can often be used to advantage in checking the results obtained with a universal indicator solution, as due to the green color of chrome-treated brine, values between *pH 7.0* and *pH 9.0* are often not readily distinguishable.

Acidity

Sodium dichromate, when dissolved in water or brine to retard corrosion, makes the solution acid. Before treating with dichromate, the acidity or alkalinity of the brine should be determined by one of the methods described. If the brine is acid or neutral (*pH 7.0* or lower), sufficient caustic soda should be added to increase the *pH* value of the brine to *pH 8.0*, or to produce an alkaline reaction if the phenolphthalein test is used. If the untreated brine has the proper *pH* value, the acidifying effect of the dichromate may be neutralized by adding 27 pounds of 76% commercial flake caustic soda for each 100 pounds of sodium dichromate used. Caustic soda must be thoroughly dissolved in warm water before it is added to the brine. An alternate method of neutralizing acid brine is to suspend a bag of slaked lime in the freezing tank until enough has dissolved in the brine to bring the *pH* value to the desired point.

When treating a brine with sodium dichromate, do not add the entire amount at one time; use approximately one-fourth of the desired amount and then test for alkalinity, adding caustic soda if needed before proceeding with the dichromate treatment.

Alkalinity

In order to most effectively retard corrosion, it is desirable to maintain refrigerating brine in a slightly alkaline condition (*pH 7.5* to *8.0*). Alkaline brines are generally less corrosive than neutral or acid brines, although with high alkalinity the activity may increase. Calcium chloride brines are usually highly alkaline and, in some cases, the normal dosage of sodium dichromate may not reduce this alkalinity to the most desirable point. Such brine can be brought to a more nearly neutral condition by

421

adding an excess of sodium dichromate, or by the addition of chromic, acetic, or hydrochloric acid. The first course is preferable, as there is less chance of damage to the equipment due to carelessness or accident, although acid may be used if desired. The direct use of acid has the advantages of producing a relatively large *ph* depression per pound of material added, and is less expensive.

The only common cause for a brine becoming alkaline in service is the leakage of ammonia where it can be absorbed by the brine. An increase in alkalinity does not prove the presence of ammonia, however, and for this reason a direct test should be made periodically. Naturally, if ammonia is found to be present, the leak should be located and promptly repaired. If there is any considerable amount of ammonia in the brine, arrange a temporary aerating system to eliminate as much of it as possible by heat, before using chemicals. The alkalinity may then be reduced to the recommended *pH* value by the addition of sodium dichromate or either of the acids mentioned.

Testing for Ammonia

In sodium chloride (table salt) brine or in water, traces of ammonia may be detected by the use of Nessler's solution in a strong concentrate caustic soda, which produces a brown precipitate or brownish-yellow (reddish-brown) coloration, depending on the amount of ammonia present. The reaction is produced when a few drops of the solution are added to a small sample (1 or 2 ounces) of the brine. This test cannot be made satisfactorily with calcium chloride brine. When the test is made on sodium chloride brine, no precipitate will be formed if ammonia is not present. A reddish-brown precipitate will be formed if ammonia is present. If the test is made on a calcium chloride brine, a white precipitate of calcium hydroxide will be formed if no ammonia is present. If ammonia is present, this precipitate will be marked by a reddish-brown color and will appear very similar to that produced by a salt brine. The reactions are quite definite and the distinciodistinction very easy to make.

The procedure depends upon whether or not the original sodium brine has been treated with dichromate to keep down corrosion. Begin by pouring a sample of the brine into one of two test tubes, until the tube is about half full. If the brine has been treated as recommended and contains dichromate, the orange or yellow color caused by the dichromate must first be removed. This is done by adding a reagent of barium salt, a drop at a time, until a heavy orange precipitate forms. When this solid sinks to the bottom of the tube, the liquid above it will be colorless. Do not add a large excess of the reagent, but only enough to precipitate all the dichromate and obtain a clear brine after the solids settle. Carefully pour off some of the top portion of the treated brine into the second test tube, being careful not to disturb the sediment, which should remain.

PREPARING NEW BRINE

When new brine is to be prepared for use in a freezing tank, great care must be exercised to avoid excessive damage to the galvanizing on the ice cans. This is particularly true when new cans are involved; older cans will usually have a slight coating of the products of corrosion that will protect them to some extent. In any case, the sodium dichromate should be added to the brine with the salt or calcium chloride before the brine comes in contact with the ice cans. Both salt and calcium chloride brines are very active electrically, and their acitivity increases as the temperature is increased. A brine that would not seriously damage the galvanizing at a temperature of 15° may be active enough to strip much of the galvanizing from new cans when brought to a temperature of 60° or 70°. The brine should be mixed outside of the freezing tank, cooled by the addition of ice or by other means to a temperature of 40° F or lower, and then pumped into the freezing tank.

An alternate method is to put the cans in position in the freezing tank, fill the tank with cold water to a height somewhat below the normal brine level, and apply refrigeration until the water temperature is below 40°. Be careful not to freeze up the coils or cooler. With the agitators running, slowly add the salt, caustic soda, and sodium dichromate. When calcium chloride is used, it must be added very slowly and the temperature of the solution checked frequently as this salt produces relatively large quantities of heat as it goes into solution.

SUMMARY

Because the brine system is of considerable importance in indirect refrigerating applications, much consideration has been devoted to it. In present day installations, calcium chloride is used in preference to sodium chloride. Thus, the properties of calcium chloride have been fully tabulated.

The problem of corrosion and treatment methods have been given in considerable detail. The determination of acidity or alkalinity in a brine solution by means of pH measurements is also included. The preparation of new brine and ice cans conclude the chapter.

REVIEW QUESTIONS

1. Name the various methods of refrigeration distribution.

2. What are the chemical constituents of brine?

3. Name the chemical formula of calcium chloride.

4. What is the purpose of congealing tanks in the brine system?

5. What is the proper strength of calcium chloride brine for a freezing point of from 10° to 15° F?

6. How can the presence of foam on the surface of a refrigeration brine be avoided?

7. What methods are used to lessen brine system corrosion?

8. Define pH values in determining the actual acidity or alkalinity of a brine solution.

9. What methods are used in testing for ammonia in the brine system?

10. Explain the methods used when preparing new brine to avoid excessive damage to the galvanizing of ice cans.

CHAPTER 19

Ice-Making Systems

The manufacturer of artificial ice will always occupy a large sector of the refrigeration industry. The ice-making industry is presently making ice by several different processes, dependent on whether distilled-water or raw-water ice is desired. Production of artificial ice takes two different forms, depending on requirements, as:

1. Can ice.
2. Plate ice.

CAN ICE

In the can system, the water is placed in galvanized iron cans or molds immersed in a brine tank which is kept cool by ammonia expansion coils. In this system, unless means are introduced to prevent it, air and other impurities have a tendency to collect in a core in the center of the can. Making ice from distilled water will eliminate this trouble, but due to its high cost, it is not widely used. At the present time, the tendency is to use raw water and to agitate it in order to eliminate the air and impurities. This agitation is usually accomplished by special agitation equipment, by a refrigerant air jet, or by special air piping.

PLATE ICE

In the plate system (which is still used but is not being installed in new plants to any large extent), hollow pans through which cold brine or ammonia circulates are immersed in a tank of water until ice eight to twelve inches thick is formed. The plate is arranged so as to allow the liquid ammonia to feed into it and the gas to return to the compressor in the usual manner.

AIR PIPING

Air piping, when used, must be carefully installed according to the manufacturer's specifications. The air to the blower is usually drawn from under the framework and returned directly from the blower to the laterals without any further conditioning. The only exception is in cold weather when it sometimes becomes necessary to take the air from a warm room or to heat the air slightly. The temperature of the air in the laterals must be kept from 46° to 50° F in order to prevent freezing of the moisture in them or freezing of ice on the drop pipes. Keep the air pressure as low as possible; usually from 1-¼ pounds to 1-½ pounds is sufficient. In exceptional cases, 2 pounds may be necessary.

One outstanding feature of the air tube is that the spring on the bracket is made to enter the lifting hole in the can, thus centering the tube. In addition, this spring clamps the bracket and keeps the tube in the center of the can, even though the can might tip lengthwise in the tank. The tube can swing crosswise in the can, but the can itself will not tip crosswise. The swinging action prevents freezing-in of the tube until the block is entirely closed. When placing the tube in the can, squeeze the spring down and start the end of the spring in the hole in the can, then slip the bracket over the edge of the can and the tube will be accurately placed without further adjustment.

Be careful, when putting the fitting into the drop pipe, to leave it sitting at an angle which insures a free swinging of the tube. The secret of good ice is in keeping the tube centered in the can. When pumping cores, be sure to wash the inside of the ice block and the dropline thoroughly so as to remove any deposits which may have started to collect. Have the core spotlessly clean before refilling with fresh water. Pull the ice regularly, a definite number of blocks per hour, on the hour. Work from one end of the tank to the other, drawing out every other row.

ICE CANS

Commercial sizes of ice cans vary with the weight of ice cakes required. The sizes shown in Table 1 are in accordance with standards adopted by the Ice Machine Builder Association of the United States.

Table 1.   Standard Welded and Riveted Ice Cans

Rated Capacity (lbs. of ice)	Inside Dimensions (inches)			Depth of Immersion (inches)	Displacement (cubic feet)
	Top	Bottom	Height		
50	5x12	4½x11½	31	28	0.9
50	6x10	5½x9½	31	28	0.9
50	8x8	7½x7½	31	28	1.0
60	5x14	4½x13½	31	28	1.1
100	8x16	7¼x15¼	31	28	2.0
200	11½x22½	10½x21½	31	28	4.0
200	14½x14½	13½x13½	35	31	3.5
300	11x22	10x21	48	46	6.0
300	11½x22½	10½x21½	44	40	5.7
400	11½x22½	10½x21½	57	52	7.4

Ice Removal

To remove the ice, the cans are lifted out of the brine and sprayed with, or dipped in, warm water. This loosens the ice so that when the can is inclined on its side, the cake slides out. The cans are usually tapered to facilitate removal of the ice.

Filling

The cans are usually filled by means of a filler device which is so constructed as to automatically shut off the water supply when the can is filled to the proper height. The filler is inserted in the can and the water is turned on. As the can fills, a floating ball rises until the can is filled to the right depth, when the ball automatically closes a valve. An automatic ice-can filling system of this type is shown in Fig. 1. This can filler reduces tank labor by filling a considerable number of cans at one time through the operation of a single lever. The life of the ice cans is increased by this method of handling as compared to the usual careless and injurious procedure followed when filling one can at a time.

A pressure-type can filler is shown in Fig. 2. In this system, the combination valves immediately under the pressure tanks will open when the operating lever is pulled forward to fill the ice cans. Observe that the water supply valve which fills the pressure tanks through the header at the bottom is closed. Also make sure that both the outlet and inlet valves are fully closed when the operating lever is in the intermediate position halfway between the emptying and refilling positions. After the ice cans are filled, the lever is normally pushed all the way by the operator. This allows the combination valves to open and opens the valve on the water supply line. The linkage to the water supply valve must be set with sufficient looseness or free play so that the bottom supply valve does not begin to open until the combination valves above have closed completely. Adjust the linkage so that all the combination valves have equal openings. (A check valve at the base of each of the combination valves makes all the water in the pressure tank above it run into the ice can which that particular tank is filling). The valves are opened by the lever and links, but are closed solely by the pressure of the springs inside the valve bodies. Keep the stem packing nuts just tight enough to prevent leakage and loose enough to permit the stems to slide freely.

Fig. 1. An automatic can filler.

INFERIOR ICE

Cloudy or milky ice is usually caused by the presence of air. It may be due to insufficient reboiling, the overworking of the reboiler or, more likely than either, to an insufficient supply of steam to the distilled water condenser, in which case the rapid condensation of the steam causes a vacuum, with the result that air is drawn in and mixed with the water.

Fig. 2. Operating levers and valves on a Frick pressure-type can filler.

The cause of a white core in the ice cake is usually caused by carbonate of lime and magnesia in the water. In many cases it comes from overworking the boiler, carrying too much water and also from not *"blowing off"* often enough.

FREEZING TANKS

Freezing tanks are made of such material as wood, steel, or concrete. Wooden tanks have a relatively short life and are subject to leaks. For this reason, freezing tanks made of steel coated with waterproof paint are preferable. Tanks made of reinforced concrete are also recommended as being superior to those made of wood.

The freezing tank (Fig. 3) contains direct-expansion freezing coils equally distributed throughout the tank and submerged in brine. The tank is provided with a suitable hardwood frame for supporting the ice cans and a propeller or agitator for keeping the brine in motion. The brine in the tank acts as a medium of contact only, the ammonia evaporating in the freezing coils extracts the heat from the brine, which again absorbs the heat from the water in the cans, thereby freezing it.

The tank itself should not be any larger than is necessary to hold the cans, the coils, and the agitator. About two inches should be left between the molds, and three inches between the pipes and the molds. Insulation of the freezing tank is accomplished by using twelve to eighteen inches of good insulating material on each of the sides and not less than twelve inches under the bottom.

The brine temperature should be maintained at 10° to 20° F and the back pressure in the ammonia coils from 20 to 28 pounds per square inch, which is equivalent to a temperature of 5° to 15° F in the coils.

Fig. 3. Arrangement of the accumulator and connections in a loose-can tank using flat coils between each row of cans.

CAN DUMPS

When harvesting a large number of cans in one lift, the can dump takes the form of a dump board, as shown in Fig. 4. Grids holding four to six cans are frequently handled with one dump. Such dumps are constructed of heavy steel channels lined with wood timbers to protect the can from wear. A dip tank is always used in connection with dumps handling more than one can at a time. The arrangement of cans is shown in Fig. 5.

Single-can dumps, Fig. 6, are furnished with or without sprinkler pipes for thawing the ice loose. When equipped with sprinkler pipes, the dump is usually so arranged that the water supply to the sprinkler pipes is automatically turned on when the can is in the dumping position and turned off as soon as the dump returns to its normal position. The weights are so distributed that the dump returns to or near its normal position by gravity after the ice has left the cans. When harvesting several cans at one time, the dump can be provided with hydraulic power which acts through

430

a piston to tilt and return the dump. Suitable pans are usually installed to catch the drip from the dump and protect the floor.

Fig. 4. Diagram showing the arrangement of the dip tank, dump board, crane and can filler for a group-lift plant.

PLATE ICE SYSTEMS

Plate ice is made by one of several methods, such as:

1. Direct-expansion plate system.
2. Direct-expansion plate system, using "still" brine.

431

3. The brine-cell plate system.
4. The brine-coil plate system.
5. The block system.

Fig. 5. Method of securing the ice cans in a can grid for a group lift.

Fig. 6. A single-can dumper with iron stand and dump box. The control valve of the thawing pipe in this case is connected to one of the trunnions of the dump box.

Direct-Expansion Plate System

The *direct-expansion plate system* is the simplest in construction and consists of direct-expansion zig-zag coils with ⅛'' plates of iron bolted or riveted in place. The thawing off of the face of the ice is accomplished by turning the hot ammonia gas from the machine directly into the tank coils.

Direct-Expansion Plate System (with "still" brine)

The *direct-expansion plate system with "still" brine,* known as the *Smith plate,* has coils immersed in a brine solution contained in a water- and brine-tight cell. Thawing off is accomplished by turning hot gas into the coils.

Brine-Cell Plate System

The *brine-cell plate system* consists of a tightly caulked and riveted cell or tank about four inches thick, provided with proper bulkheads or distributing pipes to give an even distribution of brine throughout the plate. The thawing off of the face of the ice is accomplished by circulating warm brine through the plate.

Brine-Coil Plate System

The *brine-coil plate system* is similar to the direct-expansion plate, except that brine is circulated through the coils instead of ammonia. Thawing off is accomplished by means of warm brine circulated through the coils.

Block System

In the *block system,* the ice is formed directly on the coils through which either ammonia or brine is circulated. After tempering, the ice is cut off in blocks the full depth of the plate by means of steam cutters which are guided through the ice close to the coils.

Harvesting

The method used in harvesting plate ice is similar in all of the previously enumerated systems, except that for harvesting block ice. One method consists in using hollow lifting rods and thawing them out with steam. Another method uses solid rods which are cut out when cutting up the ice. Still another method uses chains which are slipped around the cake when it floats up in the tank. The common size for plate ice is 16' × 8', 11'' thick, although other sizes are sometimes made.

The arrangement of plate-ice freezing tanks does not differ materially from that of can ice, except that they are divided into compartments of the required size and number. Each compartment is furnished with one or more plates on which the ice is frozen. The compartments are made of sufficient size to hold a supply of ice for one day so that each compartment may be emptied, cleaned, and refilled every twenty-four hours, less if demand requires.

Freezing Methods

The process of freezing ice in plate plants is principally as follows: Several vertical hollow iron walls are built in a large tank. The tank is filled with pure well water so that the iron walls are entirely submerged. The hollow iron walls are placed parallel to each other at a distance of from two to three feet. The freezing fluid, consisting either of cold brine or ammonia, is passed through the hollow walls with the result that the water will freeze on the outside of the walls. The water is kept in agitation by means of either a propeller, pump, or by compressed air, so that the water is kept continually on the move carrying the air with it to prevent it from being frozen in the ice. After the ice is frozen on the walls to the required thickness, the freezing fluid is shut off from the walls and a warm fluid passed through instead until the ice is loosened and taken out of the tank.

The plate walls differ in construction with the type of freezing fluid used. Thus, if cold brine is used, the brine has to be cooled in a separate refrigerator from which it is pumped through the walls and back to the refrigerator. This is the same method used for cooling rooms by means of brine pipes. The plate walls in such a case are generally constructed of iron. On account of the expansion and contraction caused by warm or cold brine being passed alternately through the hollow walls, it is very difficult to keep them tight.

Where ammonia gas is used as the freezing fluid, the walls are built up of expansion pipes, which are connected at each end by return bends to make one continuous zigzag coil. To obtain an even surface, the coils are covered with thin iron plates, on the outside of which the ice is frozen. To loosen the ice when thick enough, the cold ammonia gas is shut off and hot gas is passed through the pipes. Agitation is carried on by means of air jets located midway between the plates, sometimes in the center, sometimes three or four feet from one end, and sometimes at both ends of the plates.

Cutting

Cutting of the plates into cakes is accomplished in various ways, such as by steam cutters, power saws, and hand plows. In the block system, however, where the ice is cut off the plate in the tank, the cakes are removed by means of a light crane and hoist, and are then divided into the required sizes with an axe or bar.

AUTOMATIC ICE MAKERS

Automatic ice makers produce fragmented ice for processing shipping and for protection of a variety of food products. This ice is manufactured in various sizes and in quantities of up to 46 tons every 24 hours. As noted in Fig. 7, the ice maker consists basically of a *freezing section*, a *harvesting section*, and a *control panel*.

The freezing section includes a supporting frame, freezing tubes with liquid and gas headers, accumulator and float control valve, drip pan, oil and liquid traps, water distribution system, automatically operated suction, liquid and hot-gas valves, the necessary solenoid and stop valves, and a galvanized sheet-steel tube enclosure.

The harvesting section consists of a supporting base, ice-breaker assembly and motor, ice sizer, water drain pan with float valve and recirculating pump and motor, baffle plates, and water grid.

The control panel includes a timer, terminal board, relay, restarter mechanism, ice selector switch, and pressure gauges.

Operational Features

The accumulator float control valve feeds liquid refrigerant to the stainless-steel freezing tubes. Water is directed against the tops of the cold tubes. The water flowing over the tubes is frozen into solid, clear shells of ice. The cool excess water is collected in a pan at the base of the unit for recirculation. A float valve maintains the correct water level in the pan. When the ice reaches a predetermined thickness (after approximately 11 minutes), discharge gas is admitted to the tubes to force out the cold ammonia. The warm gas causes the ice to loosen (in less than 1 minute) and slide into the ice breaker. Ice thickness may be varied from ⅛ to ½ inch on standard units by regulating the duration of the freezing time.

435

Fig. 7. A typical shell ice maker (steel enclosure removed).

DRY ICE

Dry ice is merely solid carbon dioxide, which has a temperature of approximately 110° below zero Fahrenheit at normal atmospheric pressure. Carbon dioxide is produced either by heating certain limestones or by treating the stone with acid. Another method of production that is widely used commercially is a fermentation process, and still another is

436

by burning coke. Carbon dioxide is also developed in the petroleum industry as a byproduct of hydrogen production, and is also found in blast-furnace gases. Other large potential sources are the fields of natural gas or oil that produces carbon dioxide as a byproduct.

Because this refrigerant is so excessively cold, it is necessary to develop means to control its refrigerating effect. The most common methods employed for its control are:

1. By insulating it from the objects or space to be cooled.
2. By the use and control of a secondary or intermediate refrigerant.

In recent years this form of refrigerant has found a constantly increasing field of usage. In transportation of goods that must be held at low temperatures, more pay load can be carried because of the lighter weight of dry ice compared to ice-and-salt mixtures. Deterioration of equipment due to melting mixtures is avoided. The principal disadvantage is its rather high cost. This is being reduced, however, by development of more economical methods of manufacture. Another problem has been temperature control, but solutions of this difficulty are being found and several methods for proper control and economical use have been developed. A recently devised plan for using dry ice efficiently for cooling small compartments, truck bodies, and refrigerator cars consists essentially of a dry-ice compartment or bunker surrounded by a jacket. Bunker and jacket are heat insulated from the space to be cooled in the car, truck body, or other container, but constructed so that heat can easily pass from the fluid in the jacket to the dry ice in the bunker. From this jacket, cooling coils extend into the upper part of the space to be cooled. Temperature within the cooling coils, and therefore in the refrigerated space, is controlled at will by a thermostatic valve which regulates the flow of a secondary refrigerant in the cooling coils. This secondary refrigerant is a liquid with a high coefficient of thermal expansion which circulates by gravity flow down through the bunker jacket and up through the cooling coils. The large variation in density causes fairly rapid circulation.

A modification of this system is one in which dry ice is placed in an insulated metal container. This container is surrounded by a coil through which a liquid having a low freezing point circulates. The liquid circuit has a valve in it controlled by a thermostat in the refrigerated compartment. When refrigeration is needed, the thermostat opens the valve and allows the liquid to circulate. The liquid in the coil surrounding the dry-ice chamber gives up heat to the ice, causing it to vaporize. The vapor, in escaping, actuates a diaphragm-type of pump that circulates the liquid. When no refrigeration is required, the thermostat closes the valve, the resistance stops the pump, the liquid stops circulating, and the evaporation of the ice is very much retarded. This system is applicable to compartments of almost any size and quite a wide variety of temperatures can be maintained.

ICE-MAKING CALCULATIONS

Capacity

Ice-making capacity is usually equal to about 50% to 70% of the refrigeration capacity as expressed in tons of refrigeration per day. Of necessity, however, such operating conditions as initial temperature of the water supply, room temperature, and effectiveness of insulation will influence this ratio. If heat leakage losses are known, the ice-making capacity can be closely estimated. The following example illustrates the method:

Example: Assume that the water supply has a temperature of 60° F, and the temperature in the brine tank is 12° F. Calculate the ice-making capacity, assuming a 20% heat-leakage loss.

Solution: If it is remembered that the specific heat of water is 1, and that of ice 0.5, our calculation will be as follows: To lower the temperature of the water to the freezing point, the Btu per pound = $60 - 32 = 28$ Btu. To freeze one pound of water without change in temperature = 144 Btu. To reduce the temperature of one pound of ice to 12° $(32 - 12)$ $0.5 = 10$ Btu. The foregoing add up to 182 Btu per pound of ice made. To this must be added the heat leakage losses of 20%—that is, 1.2×182 or 218.4 Btu. Total refrigeration effect to manufacture one ton of ice is accordingly $2000 \times 218.4 = 436,800$ Btu. Since one ton of refrigeration = 288,000 Btu, it follows that the ice making capacity is

$$\frac{288,000}{436,800} = 0.659 \text{ or } 65.9\% \text{ of refrigerating capacity.}$$

Heat Losses

Heat losses in ice making usually varies from less than 10% in the best installations up to a possible 50% under poor conditions. When the ice production and the refrigeration rating of the machine is known, the heat losses may be calculated as follows:

Example: Assume an ice plant is producing 10 tons of ice daily from a refrigerating compressor rated at 15 tons refrigerating capacity, other conditions being equal to that given in the previous example, what will be the heat losses?

Solution: The total heat required to be removed per pound of ice, exclusive of heat losses is 182 Btu. Heat per ton of ice = 182 × 2000 or 364,000 Btu. A 15-ton plant has the capacity to withdraw 288,000 × 15, or 4,320,000 Btu. The capacity to produce ice exclusive of heat losses is consequently, $\dfrac{4,320,000}{364,000}$ = 11.9 tons of ice, approximately.

Tons of ice due to heat losses = 11.9 − 10 = 1.9.

The per cent heat loss in our example is therefore = $\dfrac{1.9}{11.9}$ = 0.16, or 16% approximately.

Freezing Time

The time required for freezing ice depends upon the temperature of the brine and the thickness of the cake. With brine at about 15° F, it will take approximately 50 hours to freeze 11-½ inches of ice. The longer time a given thickness is allowed to freeze, the better the quality of the ice will be. The time for freezing may be calculated mathematically if it is remembered that, for different thicknesses, the time required is proportional to the square of the thickness. In other words, eight-inch ice will require four times as long to freeze as will four-inch ice, and eleven-inch ice will require approximately seven and a half times as long to freeze as four-inch, etc. An empirical formula has been used to determine the foregoing time for ice in cans, as follows:

$$\text{Freezing time in hours} = \frac{At^2}{32\text{-}T}$$

where,

t = width of ice at the top (narrow way)
T = temperature of brine in F°
A = a constant usually taken as 7

Example: If it is assumed that the brine temperature is 14° F, and a standard 300-lb. can with inside dimensions of 11″ × 22″ at the top is used, calculate the hours of freezing time required.

Solution: A substitution of values in the freezing formula gives

$$\frac{7 \times 11^2}{32 - 14} = 47 \text{ hours approximately.}$$

ICE STORAGE

Manufactured ice is stored very much in the same manner as natural ice, insulation being the most important factor. Strips of lath, sawdust, hay, stray and, in the South, rice chaff, are used for packing it. The space allowed per ton of ice should be about fifty cubic feet. Manufactured ice is usually shipped in cars, where it is packed and insulated the same as when put in storage. Ice storage warehouses are usually equipped with a cooled anteroom, thus obviating the necessity of frequent opening of the main room door. A certain amount of ice can be kept in the anteroom as may be found necessary.

SUMMARY

Detailed methods of ice-making have been fully presented since the manufacture of artificial ice will always be necessary to fill a constant need in the various fields of food preservation. Modern methods of ice making, such as plant layout and component features, are dealt with in the first part of the text, whereas automatic ice-making units are presented in the latter part.

Dry ice and its importance in the protection of perishable foods during transportation, together with control methods, have been given. In the last part of the chapter, various calculations will provide information on the determination of plant size and time requirements when certain other factors are know.

REVIEW QUESTIONS

1. Name two different forms for the production of artificial ice.

2. What is the function of air piping in an ice making plant?

3. What methods are used for filling ice cans?

4. How can cloudy or milky ice be avoided?

5. What type of materials are used in freezing tanks?

6. Name the several methods used for the production of plate ice.

7. Name the method used in harvesting plate ice.

8. What are the freezing methods used in plate ice?

9. What are automatic ice makers?

10. What are the operational features of automatic ice makers?

11. What does dry ice consist of?

12. How may heat losses be calculated?

13. What is the method of storing manufactured ice?

CHAPTER 20

Supermarket and Grocery Refrigeration

Modern supermarket and grocery refrigeration generally consist of various prefabricated units designed to furnish the necessary cooling effect for a great variety of perishable foods, each unit being individually temperature controlled to prevent spoilage. Depending upon size and other requirements, the various refrigerated units usually consist of the following:

1. Display cases.
2. Walk-in coolers.
3. Storage freezers.

The placement of the various units depends upon such factors as available floor space, nonperishable food placement, etc., but, in general, the refrigerated units should be placed in locations to provide the maximum facility for display and customer convenience.

DISPLAY CASE ARRANGEMENT

Display cases, as the name implies, are used for display of perishable products. In the display of meat, for example, it is generally said by merchants that display cases are made to *sell* meats, not *keep* them.

If the customer has access to the product, it must be wrapped and priced. In wrapping, there are absorbent papers placed under the meat to absorb blood, which is bound to leak from the cut meats. The wrapping keeps the meats from the air and thus tends to stop discoloring. Also, the wrapping will show the weight, price per pound, and the product price. Customers like to pick out the piece of meat that they want without interference from the butcher. With this in mind, most meat counters are not enclosed. The refrigerant temperature is lower than the room temperature, and cold air does not tend to rise upward, but downward, so the meat cases may be open for the customer to pick out the piece of meat that he or she wishes to buy. A typical example of such a meat display counter is illustrated in Figure 1.

Courtesy Steffen's Market Windsor, Colorado

Fig. 1. Typical supermarket meat display counter.

The meats thus displayed are presented to the customer, and what is not sold one day will be in very presentable condition for the next day's sales, due to being thoroughly wrapped in transparent covering. The meat case must be well illuminated, with lighting having the proper color-temperature to bring out the right color of the meats. At night, many markets have covers which are drawn over the tops of the meat cases to also keep the meat at proper temperatures, but with modern designs, this is not too much of a problem.

The sides in the back are well insulated, as are the ends of the display case. The front is made of thermo-glass, so that the meat is clearly visible to the buying public.

A great variety of open and closed meat cases are presently available, the most recent development being that of the vertical shelf units, for

443

cold-cuts, bacons, hams, etc. Regardless of the case design however, the basic refrigeration requirements remain the same.

Because of the increase in self-service stores and supermarkets, open display cases, similar to the one in Fig. 2, are generally used in preference to the closed type, although closed cases are still used in many locations, particularly where higher than average incomes allow for greater consumption of the more valuable meat. Objective studies have indicated that, in general, the refrigeration performance is reasonably equal in both the open and closed types. The installation and adjustments made by service personnel, however, are not usually comparable, hence there is a possibility of a great variance in performance in any given make.

Courtesy Steffen's Market

Figure 2. Open, multishelf cold meats display case.

OPEN DISPLAY CASES

Open display cases keep the product cool by circulating chilled air from a coil. Double-duty cases feature a lower storage which furnishes a place to store packaged cuts prior to display. Reach-in or reach-through storage refrigerators may also be used to supply needed storage space.

Multishelf cases have proven very popular because they offer so much more display area in relation to floor space. Pre-packaged items, such as sandwich meats, are often equipped with holes or hooks so that they can be hung from metal pins. There is a dual gain in this vertical display because light intensities are diminished on the hanging package surface and this reduces fading. Light does fade processed meats such as ham, bacon, etc., but does not affect fresh meat by itself.

All open cases are affected by outside air turbulence caused by heating and cooling ducts, open doors, windows, etc. These drafts should be eliminated by directing ducts away from all open cases and by keeping outside doors closed.

The open, shelved display cases are also used for frozen food. Illustration of this type of display case is shown in Fig. 3. Dairy products and fresh produce cases are also of the open type display, where the customer has access to the products for picking out the produces to their liking. Fig. 4 illustrates a dairy products display, and Fig. 5 illustrates a fresh produce display case all with convenient access to the customer.

Courtesy Steffen's Market

Figure 3. Open, shelved frozen food case

DISPLAY CASE LIGHTING

One of the major sources of heat in a display case is that of radiant heat. Some of it comes from warm ceilings over the cases, but most of it comes from one thing the store manager could not do without, namely lighting. Brightly lit stores contribute greatly to the sales of meat products. Thus, foot-candle readings of 75 to 100 are commonly used in supermarkets and other food stores. In order to highlight the meat display, more lights are employed. Therefore, it is not unusual to have 100 to 150 foot-candles of light at the display level. The quantity of light does not affect fresh meats. The undesirable effect comes from the infrared (heat) rays in the light.

445

Courtesy Steffen's Market

Figure 4. Open faced dairy products case.

Courtesy Steffen's Market

Fig. 5. Easy-access produce case.

These rays penetrate the cold blanket of air, warming the packages in the same manner as winter-time sun bathers stay warm, although the air outside their shelters may be below freezing.

One method that quickly and dramatically illustrates the effect of radiant heat is the wrapping of a single cut of meat in aluminum foil. Within 24 hours it will be found that, because the aluminum rejects the radiant heat by reflection, the temperature of the wrapped package will very closely approach the temperature of the air within the display case. The temperature of adjacent conventionally-wrapped packages can rise substantially depending on the infrared radiators that are present.

Fluorescent lamps distribute the light more evenly than incandescant lamps. Deluxe Cool White, or Deluxe Warm White fluorescent lamps, will give truer reds and flesh colors, with less infrared than incandescent lamps. The lighting should be 100 foot-candles or better.

RADIANT HEAT EFFECT

Radiant heat transfer can be a major problem in food storage. This is particularly true in supermarkets where the self-service open meat cases are built with the maximum air flow permissible to hold ambient air mixing losses to a minimum. This air flow will remove the heat from the product at a fairly rapid rate, but if the rate of radiant heat input exceeds the rate of heat withdrawn by the case air, then the product temperature will rise. This can be demonstrated by wrapping a piece of meat in aluminum foil and placing it in the center of any open meat case. In a few hours, this particular cut indicates a temperature considerably lower than its neighbor in conventional wrapping material—a temperature fairly close to the temperature of the surrounding case air. However, no one can sell meat in aluminum foil.

If attempts are made to run the case colder, or if night covers are used, the product will freeze. This is so because the air temperature in open meat cases is normally kept as close to the freezing temperature of red meat as possible. The product adjacent to the source of cold air will obviously be somewhat colder than that at the return air opening, as the case air warms up by absorbing heat from the product as it passes across the case. Thus, lowering the entering air temperature will lower the return air temperature and the product temperature at that point, but it will also threaten the product at the entering side with freezing. Similarly, night covers which will alleviate the input of radiant heat at night will bring the meat temperature much closer to the case air temperature so that a cabinet running at optimum temperature levels during the day will freeze the meat at night.

It seems the only way to combat the problem is to eliminate or lessen the intensity of the heat source, or to find a way to change the heat-absorbing rate of the meat. This can only be done by the architects, store planners, and all the others involved in the creation of a food market.

Lighting is also an important transmitter of radiant heat. Thus, for example, the surface temperature of electric bulbs can run quite high (over 2000° F for incandescent lamps) and, although small in area, can contribute heavily to the load. This problem can be lessened by using diffusers and lenses that will cut down the intensity of the light and also provide a screen between the meat and high-temperature light source. The effect of light on the color of meat is actually a separate and distinct problem. In many cases, the product temperature may be excellent, but the meat may be changing color due to overlighting.

There are many other sources of radiant heat. However, for this discussion they are not as important as the ceiling and lights. There are many new heat sources to be watched. Among these are the now popular barbecue and bakery departments. They are potentially radiators that can add radiant heat to perishable products. Also, solar heat through windows in some cases has become a most important factor in some instances. It must be concluded that meat, a high profit maker and one of the most perishable items in the food stores, deserves the utmost care in the control of its environment which can only be controlled by careful preplanning and the use of good construction techniques.

CLOSED MEAT DISPLAY CASES

Closed meat display cases differ somewhat in operation from that of the open type. For example, a butcher who has become accustomed to a three-day meat life in an open display case with a minimum of product attention, now finds it necessary to constantly groom the display for maximum appeal. The unwrapped product is much more subject to contamination from a variety of sources. The butcher may also be accustomed to open meat-case temperatures of 28° to 34° F. Attempting to maintain these temperatures in a closed case results in coils icing up and in a very poor meat surface. Drying also occurs.

Coil Arrangement

The remaining problem is the application of refrigeration. Since the cold air will not rise higher than the coil from which it originates, it is necessary to have the coils as high as possible in the case. Visibility enters into this problem because it is necessary to let the customer see as much of the interior of the case as possible. Thus, the top containing the coil must be as narrow as possible. This reduces the volume of the top coil so that refrigeration must come from some other source. This involves introducing another coil lower down, either under the shelf, against the front, or imbedded in the floor or walls.

This problem has been solved differently for different shaped cases. The principal thing to be observed is that the circulation from the top coil is not directed immediately on the stock (which would tend to dry out the display on the top shelf) but is dispersed into the case, keeping up the relative humidity and preserving the original bright color of the stock displayed. The lower coils are in all cases arranged to cool the stock partly by circulation, but mostly by radiation which contributes to the same conditions just mentioned.

MULTIPLE CONNECTION

Two or more cases are often connected to one refrigeration machine, as shown in Fig. 6. Great care must be exercised in this type of installation to satisfy all of the conditions which may have to be fulfilled. It cannot be too strongly emphasized that unless the fixtures chosen are suitable for multiple connection, one fixture may operate too warm and another too cold, or the refrigeration machine may run all the time, resulting in spoilage losses and excessive power bills.

Courtesy Steffen's Market

Figure 6. Fresh fruits and vegetables open faced refrigerated display.

449

There are some installations where the temperature requirements are such that individual machines should be used for each fixture. There are, however, many installations where conditions are suitable for multiple connections. In general, the requirements for multiple operation are:

1. All multiple-connected fixtures must be in good repair.
2. Service on all fixtures must be uniform.
3. The coils must be balanced to operate at the same back pressure. In other words, all fixtures must give desired temperatures while operating at the same back pressure during the same running time.
4. The fixtures should not be too far away from the refrigeration machine. For best operation, the machine should be located as close to one fixture as another.
5. Three fixtures can be multiple connected, but the chances of successful operation are decreased as the number of fixtures is increased.

If conditions on a job will fulfill all these requirements, the fixtures may be satisfactorily multiple connected.

DISPLAY CASE MAINTENANCE

Regardless of the type, make or age of a meat case, one practice which will add to the display life of all meat is that of a regular weekly cleaning of the case interior. This is standard practice for many chain stores, and is recommended by most case manufacturers. The product should be removed, the machine shut down, and every screen, tray, and interior surface scrubbed down with a mild soap (detergent) or bicarbonate of soda solution. *Never use ammonia—the fumes will discolor meat rapidly.* Water should be used sparingly, and none should get on electrical parts. Use of a hose should be limited to flushing the drain periodically.

INSTALLATION

When installing refrigeration equipment in supermarkets and grocery stores, it is of the utmost importance that the liquid and suction lines be of ample size in order to avoid pressure drops. Refrigerant piping systems should be designed to accomplish the following:

1. Insure proper refrigerant feed to the evaporators.

2. Prevent excessive amounts of lubricating oil from being trapped in any part of the system.
3. Protect the compressor from loss of lubricating oil at all times.
4. Prevent liquid refrigerant from entering the compressor.
5. Maintain a clean and dry system.

Pressure drop in liquid lines is not as critical as it is in suction and discharge lines. The important factors to remember are that the pressure drop should not be so great as to cause gas formation in the line or to result in insufficient liquid pressure at the liquid feed device.

Suction Lines

In laying out the suction lines, the following should be adhered to:

1. The suction line should be sized for a practical pressure drop at full load.
2. It should be designed to return oil from the evaporator to the compressor under minimum load conditions.
3. It should be designed to prevent liquid from draining into the compressor during shutdown.
4. It should be designed so that oil from an active evaporator does not drain into an idle evaporator.

Most refrigerating piping systems contain a suction riser, either by reason of the relative location of the evaporator and compressor or because the riser is introduced for the purpose of minimizing the possibility of liquid draining from the evaporator into the compressor during compressor off cycles. Oil circulating in the system can be returned up gas risers only by entrainment with the returning gas.

Dehydrator

A permanent liquid-line dehydrator should be fitted into the line at the time of the installation. In the event that the dehydrating agent becomes saturated, it should be replaced. Dehydrators will eliminate moisture which may cause sticking expansion valves and internal corrosion of the various parts.

Oil Separators

An oil separator will greatly improve the operation of the system, but is not to be considered a service necessity.

Controls

In most installations of this type, display cases and walk-in coolers do not operate at the same temperature. Where the chill room is also connected to the same condensing unit, it will be necessary to add additional controls. When using a forced-air evaporator, experience has shown that an evaporator regulating valve and a snap-action valve are necessary. The evaporator regulating valve will maintain the operating pressure in the evaporator at or above a given point, while the snap-action valve will permit the evaporator to cycle and thus obtain defrosting. Check valves must be used in the suction line leading to each low-temperature evaporator. This may be either one large check valve installed in the main low-temperature suction line outside the low-temperature fixture, or an individual check valve after each bank of freezing plates.

Air Circulation

The results obtained with mechanical refrigeration depend largely on air circulation within the fixture. The importance of providing proper air circulation in supermarket display cases and cooling rooms cannot be too strongly emphasized. Air circulation is not only necessary for the proper distribution of refrigeration, but it also assists in removing gases and odors which are produced by the stored food products.

A fixture contains a cooling unit or heat absorber which, in the case of mechanical refrigeration, is the evaporator. As the heat is absorbed from the air in contact with the evaporator, it is cooled, becomes heavier, and falls. In order that refrigeration may take place, the cold air that falls must be circulated around the stored food. It is in this manner that the cold air absorbs the heat in the food product and the fixture. It expands, rises, and again comes into contact with the evaporator. This cycle is repeated over and over, the speed of the air flow depending on the relative temperature of the food and fixture.

Baffles

To assist and direct the circulating air, certain guides, termed "baffles," and cooling unit decks are constructed and installed within the refrigerator. They normally divide the fixture into two parts—one is the evaporator compartment, and the other is the food compartment. Circulation depends primarily on the existence of a temperature difference between the evaporator compartment and the food compartment. To assure the maintenance of this temperature difference, the best practice is to insulate the evaporator deck with 1″ to 1-½″ of corkboard (or its equivalent) for refrigerant temperatures below 20° F. Unless an evaporator deck is properly insulated, the temperature on both sides will be very nearly equal. Circulation under these circumstances will be very sluggish and often inadequate to maintain the desired temperatures in the food compartment. Foods may become slimy and the evaporator deck may drip water into the food compartment due to condensation on the bottom of the deck.

MECHANICAL CENTERS

A recent simplification in supermarket and grocery-store refrigeration consists of a complete refrigeration machinery package to serve an entire food store. The main advantage claimed for such prefabricated condensing units are that they cost less and, with proper planning, removes the machinery room from the selling area, thus offering additional display facilities.

Mechanical centers of this type (Fig. 7) are manufactured to suit various requirements. Horsepower sizes for both zero and normal-temperature applications are available, and although rooms have been built to accommodate units down to ½ horsepower, 30-horsepower remote condensing units have also been designed. Each unit is completely prewired, including circuit protection and prepiping to a specified terminal area. The machinery unit, being factory preassembled, may be placed in an appropriate remote location, such as on the roof or on a slab adjacent to the building or food store.

COMBINED SYSTEMS

Avoiding the use of compressors primarily designed for air conditioning eliminates the hazard of using unproven compressors for the critical low-temperature requirements of a food store. Combined system technique avoids electrical damper controls and shuns gadgets. It uses electrical defrost with no moving parts and proven reliability. Fig. 8 shows a typical prewired condensing unit.

Of significant importance to food operators is the availability of a food-protection alarm system. This protecting device is an electronic

Courtesy Sherer-Gillett Company

Fig. 7. A prewired mechanical center consisting of three
condensing units of 10, 3, and 1 HP, respectively.

sensor imitating the temperature reaction of food, and can precisely
determine the time a warning should be initiated. It is specifically designed to warn under precise conditions, but avoids nuisance calls. The
combined system along with a mechanical center employs principles
familiar to the mechanic serving the food store. This will avoid the need
for special maintenance training. Thus, the legal requirements for employment of a stationary engineer will, in most cases, be eliminated.

DISPLAY CASE DEFROSTING

The accepted defrosting method used in commercial display is almost
universally by the direct electric method. Basically, this system consists
of a resistance-type electric heater and an electrically operated time
switch to initiate the defrost. Several methods of termination are used, the
particular one chosen depending on the design of the refrigeration system
and, to some extent, on the customer's preference.

Courtesy Sherer-Gillett Company

Fig. 8. A 10-HP prewired condensing unit.

Straight Time Method

This system is controlled entirely by a clock. The length of defrost is determined by past experience with similar systems or by adjustment after installation. This system, while reliable, has the disadvantage of making no allowance for changing conditions of service load, humidity, or other variables. To insure proper operation at all times, the length of defrost must be set for the most severe conditions, which may result in excessively long defrost periods much of the time. The advantages of the straight time defrost are simplicity and low cost, although cost would not seem too serious a consideration in relation to the cost of the total system. The electrical diagram of this type of system is shown in Fig. 9.

Pressure Method

This method (Fig. 10) depends on the rise of the low-side refrigerant pressure as the coil temperature rises during defrost. (These are popularly termed TP controls). At a predetermined pressure (most often around 45 psig for low-temperature systems and 40 psig for medium-temperature systems), the defrost is terminated by a pressure-actuated switch. Pressure-terminated controls are usually provided with a ''fail-safe'' time

455

termination in case the pressure switch fails to operate. The pressure-terminated control has the advantage of adjusting automatically to the differing conditions of service load and humidity, and results in defrosts only of such duration as are necessary to clear the coil of frost. Its application should be limited to systems where the temperature/pressure relationship of the coil is accurately sensed by the control. This is because of the basic characteristic that the pressure in a system will usually respond to the area of coldest temperature. Hence, cold suction lines or cold machines (if they are colder than the coil or coils) will, in effect, be the controlling point. Faulty operation will often result when used on systems where part of the system is at a lower temperature than the termination temperature of the coil during defrost.

Fig. 9. A straight-time defrost control diagram.

Temperature Method

Termination on temperature rise depends on a thermostat in the refrigerator to be defrosted, usually located on the coil. Termination thermostats are mostly of the nonadjustable type and are often set for a temperature of around 50° F. Temperature-terminated controls have all the advantages of the pressure-terminated types and the additional advantage that they are not dependent on accurate reflection of pressures due to temperature at remote parts of the system. The cost of these controls is somewhat higher, but the additional advantages will, in most instances, be worth the extra cost. A diagram of a typical temperature-terminated control is shown in Fig. 11.

Fig. 10. Time/pressure defrost-
control diagram.

Fig. 11. A time/temperature defrost
control.

The foregoing exemplifies the versatility of the electric method of
defrosting and one of the numerous reasons for its acceptance and
popularity. No matter what type of control is used, the length of defrost is
dependent on the voltage at the heater. The amount of heat supplied by an
electric heater varies directly as the square of the voltage. Assume a
typical defrost heater delivers 2400 watts at 220 volts. This same heater
with the supply voltage reduced to 208 volts will deliver only 2140 watts;
this does not represent a serious reduction, but it is worthy of note.

The current draw of the heater is also related to the voltage. Citing the same example as before, at 220 volts the heater will draw 10.9 amperes; at 208 volts, 10.3 amperes; at 240 volts, 11.9 amperes. These differences may or may not be important in choosing the size of the supply wire. Supply voltage obviously has some effect on the length of the defrost period. As voltage is reduced, some increases in length of defrost results. Generally, the increase is not significant and is more than compensated for by the usual safety factors in the settings.

DRAINAGE REQUIREMENTS

An important item in refrigeration for display cases is the drain trap and associated piping. Drain water from the freezer coils should be conducted out of the refrigerator without the possibility of allowing warm air to enter. To accomplish this, a trap should be installed in the drain conductor. A drip space should be provided between the end of the drain pipe and the sewer to prevent sewer gas from backing up into the refrigerated space. A typical installation is shown in Fig. 12. Notice that all drain pipe and fittings must have a diameter of at least one inch. Each section of pipe which is connected directly to any walk-in cooler must have an adequate slope, with ¼ inch per foot being considered the minimum. The drain trap

VERTICAL · GIVE DRAIN 1/4" PER FOOT MINIMUM FALL

PROVIDE TRAP

12" MIN.

DO NOT CONNECT DIRECTLY TO SOIL PIPE. DRIP INTO FLOOR DRAIN OR FLOOR SINK

Fig. 12. A typical drain installation diagram. All drain pipe and fittings must have at least one-inch inside diameter. Every line connected directly to a display case must have a minimum slope of ¼ inch per foot. Drain lines should never be connected directly to the soil pipe, but should be allowed to drip into the floor drain or an open sink. On low-temperature equipment, a trap must be installed at least 12 inches from the fixture.

must *not* be located closer than 12 inches to the outlet from the cooler case.

WALK-IN COOLERS

Walk-in coolers, as the name implies, consist of a refrigerating unit provided with a suitable walk-in door to permit storing and preservation of large amounts of perishable foods. Refrigerating units of this type may be purchased from various manufacturers, complete with all the necessary accessories. Formerly, walk-in coolers were made of oak, maple, or other hardwood exteriors, and with a spruce lining. The insulation usually consisted of 3 to 4 inches of cork placed between the inside and outside surfaces. At the present time, however, they are made of light steel-clad *Fiberglas* insulated panels. Units of this type (Fig. 13) can be erected easily and put into operation simply by connecting wiring to the required electrical source. Large refrigerators of this type are usually equipped with an air-conditioning and humidifying unit.

Courtesy Sherer-Gillett Company

Fig. 13. Cutaway view of a Sherer-Gillett metal walk-in cooler.

Floor Arrangement

Sectional prefabricated walk-in coolers are furnished with or without an insulated floor. In the floorless units, it is estimated that about 1.3 Btu of heat per hour per square foot can be expected to come in from the ground. Thus, floor losses are considered negligible. If the unit is installed on a concrete floor over a space, insulation is necessary to prevent sweating beneath the cooler. This insulation can be a sprayed-on type (where equipment is available to apply it), or sheets of styrene can be cut and fastened to the area with mastic.

A level entry is becoming more important as the use of hand and electric trucks increases. The advantage of recessed floors and the convenience of a level entry afforded by a floorless cooler can also be had by recessing a sectional insulated floor. The supplier should be informed when the order is placed so that the door can be properly fitted.

An insulated concrete floor (on the ground) can be constructed for a low-temperature (freezer) cooler, but full consideration of all the facts must be taken. The only perfect insulation is a 100% vacuum, and this is not fully obtainable even in a *Thermos* bottle or in the laboratory. Regardless of the thickness of the insulation in the floor under a low-temperature cooler, a freeze zone will gradually extend into the ground to the extent that the heat of the earth permits. If the subsoil is or becomes wet, the inevitable icing and expansion will take place, and the floor is liable to heave (especially in clay subsoil). If the freezer is located along an outside wall, the danger of heaving and subsequent damage to the footings and walls is an added possibility in areas of severe winters.

The only sure way to prevent a freeze zone from reaching moisture is to provide a source of heat to act as a barrier. One method, shown in Fig. 14, is to lay a pipe grid below 4 inches of styrene foam insulation, and circulate a nonfreezing liquid (permanent antifreeze, brine, etc.) through it. The heat is furnished from the unit condenser through a heat exchanger. Another practical method is that of laying 4-inch drain tile in rows and connecting the rows to plenum chambers so that air can be circulated under the floor. The air can make an open circuit through the freezing room above. Small freezers generally do not have a freeze zone deep enough to cause trouble. The necessity of a heat barrier depends chiefly on the nature of the subsoil and the size of the freezer.

A tamped dry sand or gravel fill will make a good base on which to build an insulated floor. Some soil conditions may warrant a concrete floor below the insulation.

460

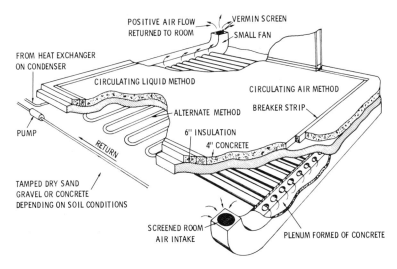

POSITIVE AIR FLOW
RETURNED TO ROOM

VERMIN SCREEN

SMALL FAN

FROM HEAT EXCHANGER
ON CONDENSER

CIRCULATING LIQUID METHOD

CIRCULATING AIR METHOD

ALTERNATE METHOD

BREAKER STRIP

PUMP

6" INSULATION

4" CONCRETE

RETURN

TAMPED DRY SAND
GRAVEL OR CONCRETE
DEPENDING ON SOIL CONDITIONS

SCREENED ROOM
AIR INTAKE

PLENUM FORMED OF CONCRETE

Fig. 14. An insulated concrete-floor arrangement for a walk-in cooler, showing plenum chambers for air circulation.

Walls

The simplest form of sectional walk-in coolers—four plain walls with a top and a door—will fill some customer requirements. However, there is a great variety of standard sections up to 10-½ feet high to make up just about any possible floor plan. Rolling cold conveyors can be fitted through special wall sections to facilitate the processing and packaging of meats and other perishables. Glass door display fronts (Fig. 15) bring the customer close to the refrigerated items, and the newest open-display multishelf cases, attached to the cooler, place tempting food items at the customer's finger tips. Thus, the walk-in cooler can become an effective selling tool as well as a holding area.

A freezer and cooler may share a common wall, saving the price of one wall. On the freezer side of the wall, temperatures may be 0° to—20° F, while on the cooler side they may be 40° at 90% relative humidity. The wall will be cooler than 40° F, and since the dewpoint for air at this condition is 37° F, the wall surface will sweat or frost. There will be less sweating if a thicker insulation is used in the freezer walls, but additional insulation will not eliminate the condition entirely. It is better to place a cooler against an erected freezer rather than vice versa. If there is a door

461

through the common wall, sweating or frosting will be more noticeable. This condensed moisture usually is no great problem, but it is good to be aware of it. If separate coolers and/or freezers are to be adjacent, it is good to maintain a space between when installing to allow air circulation. This will prevent moisture collecting and damaging the sections. Clear spans are possible up to 20 inches in coolers, and up to 16 inches in freezers. Widths greater than these are attained with I-beams. The tops are self supporting, but extra weight, such as oversize coils or overhead meat tracks, must be fastened to supports above the top by means of rods. In addition to meat tracks, accessories such as meat rails, shelving, and floor and platter racks are also available from the manufacturer.

Courtesy Sherer-Gillett Company

Fig. 15. Exterior view of a sectional walk-in cooler permitting additional units to be installed according to requirements.

Wholesale Storage

Inasmuch as the wholesale storage of meat is generally in larger pieces—quarters or halves, or even whole animals—and the time of storage is longer, both the temperature and the relative humidity should be lower. Wholesale storage temperatures are generally kept right on the edge of freezing. Because the rooms are larger, forced convection is readily applicable since the circulation can be more readily dispersed, cooling the room without blowing directly on the stock. Common practice, however, is still largely in favor of steel pipes or cold plates, the frost being brushed off instead of melted as a means of defrosting. The general practice in wholesale storage is to separate the refrigeration department into a number of rooms, each designed for a different and specific purpose, such as chilling, storage of cured meats, storage of carcasses, etc.

Frozen Meat Storage

In the present stage of development, this is a special problem. Bacteria are not killed at low temperatures. Their growth, however, is greatly reduced. Loss of color is also reduced as the temperature is lowered. In general, it may be said that the temperatures required are the lowest economically obtainable, generally around 10° below zero, but $-20°$ F is better. It is necessary to provide a vestibule for entering such low-temperature storage rooms to prevent the inrush of warm air, both from an economy standpoint and from deposit of moisture on the stock and coils. Generally, it may be said that steel-pipe coils, or plate coils composing the shelves on which the stock is piled, is good practice. Small storage amounts are generally best in chests rather than refrigerators with swinging doors, because the cold air tends to lie in a chest when the lid is open, but spills out when a vertical door is swung. The application resembles ice-cream hardening rooms, and the same general rules will apply.

MEAT PREPARATION

Chilling Fresh-Killed Meats

This problem presents difficulties, particularly in the retail market where a certain amount of killing is done and no particular provision is made for separating the warm fresh-killed meat from the storage stock. In the larger markets, a separate chill room is always provided, and it is best to do this in every case if economically possible. The reason is that, as long as the animal heat is present, the fresh-killed carcass steams like one's breath when introduced into a refrigerated atmosphere. This steam circulates and condenses on every cold thing in its path, thus wetting the surfaces of all the stored meat that happens to be in the same room.

If the amount of fresh-killed meat is small as compared to the capacity of the refrigeration and the size of the room in which it is introduced, this moisture will be absorbed by the air circulation carried to the coils and down the drain. On the average, it may be said that if the carcass is allowed to hang outside for a few hours, a single beef or a couple of hogs can be hung in a $6' \times 8'$ market cooler, if heavy-duty coils have been installed. If four or five animals are to be killed at one time, however, a separate chilling room should be provided and it should be cleared of storage stock when a new batch of carcasses is put in.

Forced circulation coils are especially adaptable to meat-chilling rooms, as they have the ability to take the heat out of the meat rapidly without excessive dehydration because of the moisture given off by the warm carcasses. When the machine equipment for coolers in which fresh-killed meat will be chilled is selected, the hot meat load should be added to the normal heat load of the cooler.

Packaged Meat

By the time packaged meat is put on display, it has usually come a long way. Before being packaged it has been cooled, warmed, and transferred two or three times, then cut and maybe chopped or ground. By this time, the forces of loss have certainly gained some headway. The amount of heat gain depends to a great extent on the treatment that the product has received prior to this step. It may be stated that any harm that has occurred to the product prior to being put on display cannot be overcome simply because it is put in a pretty package.

Each properly wrapped cut will maintain its own humidity and is protected from further bacterial and mold spore contamination. Therefore, when the packaged meat is placed in the open display case, it is subject primarily to temperature. One authority states that the ideal meat temperature for minimum bloom life is from 32° to 34° F. From 34° to 38° F is a semidanger zone, speeding up oxygenation and hastening fading, but not to the point where troubles are created if the store turnover is sufficiently high. If the meat temperature rises above 38° F, the fading becomes so rapid that the wrapped portions can become discolored and unsaleable in well under eight hours. These temperatures also may have been reached at some time before the cut was displayed. This is the reason much stress is placed on the quick transfer of packages, the use of cool cutting rooms and/or package conveyors, and minimizing other contacts which warm the meat.

SUMMARY

In this chapter an attempt has been made to furnish detailed information about modern methods of food preservation as presently practiced, as well as display-case arrangements, placing of condensing units, and related items. Since the radiant heating effect is a major problem in the field of food preservation in storage, the methods used for air circulation, insulation, and other factors to prevent heat penetration have been discussed in detail.

Because of space consideration, mechanical centers housing complete equipment units (usually in remote locations) have been presented. These self-contained units provide a late development in the field of food preservation. Additional features, such as defrosting methods, controls, drainage requirements, walk-in coolers and floor arrangements, meat preparation, etc., conclude the chapter.

REVIEW QUESTIONS

1. Name three methods for the preservation of foods in supermarkets and grocery stores.

2. What is the display case arrangement?

3. What are the advantages in using open display cases in supermarkets?

4. State the number of foot candles of light used in display cases for meat.

5. What are the sources of radiant heat and their effect on food-storage display cases?

6. What is the proper coil arrangement in meat display cases?

7. What precautions should be observed in the multiple connection of condensing units?

8. How may pressure drops be avoided in refrigeration lines?

9. Name methods of controls used in supermarket refrigeration.

10. What is the advantage of mechanical centers?

11. What are the accepted defrosting methods used in commercial display cases?

12. What provision is made for drainage for display cases?

13. What does a walk-in cooler consist of?

465

14. What type of insulation should be provided for walk-in coolers?

15. State the temperature requirements for frozen-meat storage.

16. Give the methods of meat refrigeration prior to storage.

CHAPTER 21

Locker Plants

A technique of food preservation that has become very popular during the last decade is that of storing food in special freezing lockers. One general method of doing this in common use at the present time is by means of community lockers. The community locker or locker plants are usually equipped very much in the same manner as large refrigerating plants, but they are equipped with special facilities, such as locker rooms, quick-freeze rooms, cutting rooms, aging rooms, chill rooms, customer rooms, etc.

LOCKER PLANT CONSTRUCTION

There are many types of locker plants in operation, and additional plants are designed to suit the requirements of the particular location and trade under consideration. If an attempt is made to classify these types of refrigeration plants, it will be found that most of them can be grouped as follows:

1. Plants equipped exclusively for locker rental with all the related facilities.
2. Plants combining locker rental with a retail establishment, such as a grocery or butcher store.
3. Plants combining locker rental service with certain manufacturing processes, such as slaughtering, curing of meats, and freezing of fruits and vegetables.
4. Locker storage rooms or buildings attached to refrigeration plants, such as ice plants and creameries.

The primary consideration when planning a locker plant is to determine whether a plant of any type will prove economically successful in the location under consideration. This requires a thorough study of local factors relating not only to the prices and proximity to production of food products, but also to consumer habits, including the sale of meats, fresh foods, and commercially frozen food.

DESIGN CONSIDERATIONS

Every locker storage plant presents individual engineering problems, but each should be designed with economy of space and location of facilities previously determined as a primary consideration. In addition, the probability of expansion must be kept in mind to make the smallest possible additional investment when required. Since refrigeration space is the most expensive, it is necessary that the part of the plant that is lined with insulation and equipped with coils be constructed with the smallest amount of space possible. In this instance, however, there are certain minimum requirements that must be considered. Thus, if a certain minimum requirement for insulation or floor is exceeded, it may become impossible to operate economically.

The floor plan of a properly designed and equipped locker storage plant is shown in Fig. 1, and should include the following features:

1. A *chill room* with an air circulation of ample capacity.
2. A well equipped *processing room* (not refrigerated) in which to clean, cut, wrap, and label products to be frozen and stored.
3. A suitable *quick-freeze room* held at a temperature of 5° to 15° below zero.
4. A *locker room* with a closely controlled temperature.
5. Refrigerating machinery of adequate capacity that is automatically controlled.

In addition to these, such additional facilities as a customer room, aging room, pickling room, vegetable storage room, and other auxiliary refrigerated space should be considered.

Chill Room

The function of the chill room is to remove the animal heat from the meat. The temperature is held at 30° to 38° F. Frequently, a chill room and an aging room are combined in one. The main advantage in having a

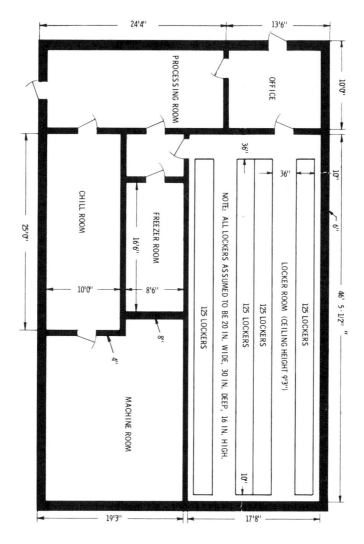

Fig. 1. Layout of rooms in a typical locker plant.

separate aging room is the ability of maintaining an even temperature. Thus, when warm products are placed in the chill room, the temperature of the room will tend to increase a certain amount, depending, of course, on the amount of the heat load and frequency of operation.

469

The size of the chill room depends on the size of the plant. It is considered good practice to allow ½ sq. ft. per locker. Thus, a 300-locker plant should have about 150 sq. ft. of floor space. The height of the ceiling should be sufficient to accommodate meat rails or overhead tracks. However, if no meat tracks are used, 10 ft. will be a good average height.

Processing Room

This room (sometimes termed the cutting room) is where the cutting, cleaning, wrapping, and labeling of the products take place. This room is insulated but is not refrigerated. The usual equipment for the room consists of regular butcher tools, such as cutting blocks, grinding machines, saws, knives, scales, etc. If the processing room is part of a large plant, a smoke house for the curing of hams and bacon, and a pickling room for the processing of corned beef are usually included.

Quick-Freeze Room

The quick-freeze room, as the name implies, is where the meats, fruits, and vegetables are quickly frozen before being placed in the locker compartment. The quick-freeze room is held at a temperature of 5° to 15° F below zero. In this room, the floor space allowance should be approximately 1/5 sq. ft. per locker. Thus, a 300-locker plant should have about 60 sq. ft. of floor space. The height of the ceiling usually averages 9 to 10 ft. The time required for quick freezing ranges from 3 to 6 hours.

Locker Room

The locker room is the heart of the locker-plant system. The size of this room depends on the dimensions of the individual lockers and the amount of space allowed between the locker rows. Locker sizes have now become more or less standard, although each manufacturer has his own individual design intended to make his equipment the most attractive to the user and the plant operator. Metal is used almost exclusively in locker construction. Most lockers today are either 20″ × 30″ × 15″ or 20″ × 30″ × 17″. Some are of the rolling-drawer type and others are of the straight-compartment type with locking doors. Lockers are usually supplied with either padlocks or with locks built into the door proper in order that the customer may be reasonably sure that his storage space is safe. The plant operator always has a master key for all lockers, which enables him to place frozen foods in lockers after processing.

The aisles between the locker rows should be from 32 to 38 inches wide. There should be an allowance of approximately 2 feet above the lockers for the cooling coils or units. It has been found convenient not to make the lockers more than six tiers high. The inside finish of the locker room should be of odor-free material. This is important, since in long storage, even a slight trace of odor will taint the food. Such finishes as hard cement plaster, mastic, or an odor-free wood, such as spruce, is most commonly used.

INSULATION REQUIREMENTS

The arrangement of any plant should be such that it reduces to a minimum the insulation needed and yet has the correct amount of insulation in each room so as to insure the most economical operation possible. Refrigeration doors should be located so as to permit as little heat as possible to enter the refrigerated rooms from the outside. The quick-freeze room should be planned with as little outside area as possible, and with the door opening either into the locker room or anteroom. This design will reduce the service load on this section and prevent excessive condensation on floor and walls of adjacent high-temperature rooms.

One of the primary considerations to be dealt with when designing a cold-storage or refrigeration plant will be the type of insulation material. There are several materials that are in common use to prevent heat leakage through walls, floors, and ceilings. Cork has long been used for low-temperature insulation. It has a high resistance to heat flow. This property is largely due to the fact that cork is a cellular material, and each of the cells contains air. Air is one of the best known insulators, and any material that contains small air cells is a good insulator. The more thoroughly the air is confined, the better are the insulating characteristics. Many so-called dead-air spaces actually allow circulation of the air to the extent that their value is reduced.

Insulating materials can be divided into two general classes—those that have both insulating value and structural strength, and those that have insulating value only. Certain insulating materials, such as cork and rock cork, are available in board or block form and have sufficient structural strength to be used for self-supporting partition walls. Other materials, such as shavings, sawdust, flakewood, redwood-bark fiber, and quilt materials must be held in place by an additional structure. When erecting any kind of insulation, the recommendations of the manufacturers should be carefully followed. Fig. 2 shows method of insulation used for masonry walls, while Fig. 3 shows how a concrete ceiling can be insulated.

The most important consideration in any cold-storage insulation is moisture. Every possible effort should be made to build cold-storage walls airtight; if they are not airtight, they will not be moisture proof. Moisture tends to travel through a wall from the warm to the cold side. It enters the wall on the warm side as a vapor, but as it penetrates the wall, its temperature drops, and when the dew point is reached, condensation begins. If the temperature is below freezing, the moisture freezes and not only reduces the resistance of the wall to heat flow, but causes rapid deterioration.

Fig. 2. Method of cold-storage insulation in masonry-wall construction.

Fig. 3. Method of cold-storage insulation as applied to concrete ceilings.

EQUIPMENT AND CONTROLS

A locker refrigerating plant, in order to function effectively, must incorporate certain temperature controls and other devices in addition to the condensing unit, piping valves, and evaporator coils. Control equipment has been developed to a high state of effectiveness and there are several companies that manufacture such equipment exclusively. The arrangement of equipment in a typical locker plant is shown in Fig. 4.

Evaporators

Although there are many forms of evaporators in use, the plate type has found a constantly increasing use in locker-plant installations. A vacuum plate-type evaporator installation is shown in Fig. 5. Evaporators of this type are constructed especially for low-temperature applications and are made of steel tubing which has been flattened to give a greater area in contact with the plate, and these tubes are encased in a welded steel shell. After fabrication, a vacuum is drawn on the plate which insures a positive contact between the tubing and the plate. All joints are welded and the outside surfaces of the plate are sprayed with aluminum paint or molten zinc to provide a lasting and durable finish. The use of zinc-finished plates is recommended where they are above freezing (32° F) temperatures. The standard size locker-room plates are 12 inches wide with four passes of tubing inside, and are usually available in lengths of 6, 7, 9, and 12 feet.

473

Fig. 4. Arrangement of equipment in a typical locker plant.

Fig. 5. Details of evaporator installation in a typical locker room.

Standard size freezer-room plates are 22 inches wide with eight passes of tubing, and are available in 4, 5, 6, and 9-foot lengths. The refrigerant tubing on the outside of all plates to which the refrigerant lines are connected is of ⅝-inch copper.

Locker-room plates are usually packed separately by the manufacturer and must be installed in banks by the installer. The plates in the banks may be connected together in series or parallel as desired. With a series hookup, the number of plates per bank must be limited to the amount of tubing which can be successfully handled by one expansion valve. Experience shows that four 12″ × 12′ plates, or equivalent in tubing, is the limit per bank in a series hook-up.

With a parallel hook-up, it is recommended that not more than six 12″ × 12′ plates be connected to one manifold, although there have been instances where eight plates operated successfully. Always use one expansion valve for each bank of plates. The two methods of assembling plates are shown in Fig. 6. After the plates are assembled they should be suspended from the ceiling and provision made to raise or lower each corner of the bank to permit leveling, thus insuring uniform frosting of the plates.

PLATES CONNECTED IN PARALLEL

PLATES CONNECTED IN SERIES

Fig. 6. Diagram showing the method of connecting evaporators.

Freezer plates are always connected in series and installed as shelves in the freezer section. (See Fig. 7.) A suitable rack should be made locally to support the plates. This rack may be of galvanized iron pipe or angle iron. The direction of expansion through the freezer plates should be from the bottom upwards, using a suitable expansion valve. No more than four 6' × 22" plates should be used on one expansion valve. The spacing of the plates will depend on the type of products which are being frozen. For example, if packages the size of 2-½ gallon ice-cream containers are to be frozen, then the plates must be spaced accordingly to accommodate them. For small packages of meat, the plates can be spaced much closer. The common practice is to install the plates with varied spacings to accommodate any size package.

Liquid and Suction Lines

These lines must be of ample size to avoid pressure drop. The suction line leaving each bank of locker-room plates must be brought up above the top of the plates in order to form a trap. The expansion-valve bulb should be located at a point above the suction manifold (or on a series hook-up, above the plate outlet) to insure flooding the tubes in the top plate. (See Fig. 6.)

It is advisable to install a drier coil on each bank of plates in order to allow the expansion valve to be properly adjusted without excessive frost back. This will also permit each expansion valve to operate independently of the others without spill-over, which would affect all the other expansion valves. In some instances, suction lines may be long enough within the room to accomplish this without a drier coil. The suction line from the freezer plates should also be installed so as to permit each expansion valve to function independently. The connection into the main suction line should be made so as to prevent oil logging due to faulty oil return. (See Fig. 7.)

Hand Valves

Hand valves are not a necessity, but will aid in servicing the equipment. Common practice is to install a hand shut-off valve ahead of each expansion valve, and another one at the end of each bank of plates. (See Fig. 7.)

Accumulators

Accumulators should be used on every installation. A combination accumulator and heat interchanger should be installed outside the low-temperature fixture, but as close to the fixture as possible. Its purpose is to permit complete flooding of all plates with refrigerant without permitting the liquid refrigerant to enter the crankcase. It will also level out the usual intermittent oil return which is common on low-temperature evaporators.

476

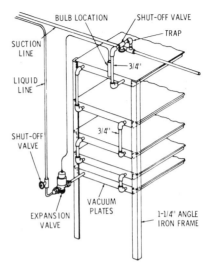

Fig. 7. Method of connecting freezer plates.

Dehydrator

A permanent liquid-line dehydrator should be installed at the time of the original installation. In the event the dehydrating agent becomes saturated, it should be replaced. Dehydrators will eliminate moisture which may cause sticking expansion valves and internal corrosion of various parts.

Oil Separators

An oil separator will greatly improve the operation of a locker system, but it is not to be considered a service necessity.

Controls

In most locker installations, the freezer plates and the locker-room plates operate at the same refrigerant temperature. This will usually freeze products in the desired time unless there are some unusual freezing demands placed on the freezing-room plates. However, there are times, especially during cold weather, when the condensing unit does not operate a sufficient length of time to give the desired freezing.

In either instance, it will be necessary to control the operation of the condensing unit by a cold-control knob installed on the low-pressure switch. This will enable the user to change the switch settings to obtain the desired freezing time. The locker room in this instance must be

477

controlled by a room thermostat operating a liquid-line solenoid valve (Fig. 8). Receiver capacities must be sufficient to hold the amount of refrigerant pumped out when the liquid-line solenoid closes.

Fig. 8. Schematic arrangement of the multiple installation of solenoid valves in liquid lines. In the illustration, S represents a solenoid valve; T thermostat; ERV evaporator regulating valve. When using solenoid valves on dc circuits of 64 volts or higher, it is necessary to use a surge preventer to properly protect the solenoid windings against the reactive surges set up when a dc circuit is opened.

Where the chill room is also connected to the same condensing unit, it will be necessary to add additional controls. When using a forced-air evaporator, experience has shown that an evaporator regulating valve and a snap-action valve are necessary. The evaporator regulating valve will maintain the operating pressure in the evaporator at or above a given point, while the snap-action valve will allow the evaporator to cycle and thus obtain defrosting.

Check valves must be used in the suction line leading to each low-temperature evaporator. This may be either one large check valve installed in the main low-temperature suction line outside the low-temperature fixture, or an individual check valve after each bank of plates. The second recommendation is somewhat less desirable than the first, as it allows refrigerant to condense in the length of suction line in the locker room. Unless a check valve is used, the relatively warm gas from the chill-room evaporator will condense in the colder evaporators and keep the low-pressure switch from operating.

478

SUMMARY

Locker plants provide yet another method of food preservation. These are usually equipped in the same manner as are large refrigerating plants, except that they are provided with locker rooms and special customer rooms in addition to the customary cutting, chill, and quick-freeze rooms. Construction and design methods, together with complete locker-plant layouts, are fully described, including insulation requirements, controls, and such accessories as valves, accumulators, and oil separators as part of the condensing unit operation.

REVIEW QUESTIONS

1. Describe the construction of community locker plants.

2. What is the floor plan arrangement in locker plants?

3. What is the function of the chilled processing and quick-freeze rooms in locker plants?

4. State the approximate dimensions of individual lockers.

5. Describe the insulation requirements in locker plants.

6. What is the control arrangement in locker plants?

7. Describe the arrangement of liquid and suction lines.

8. Name accessories used in locker-plant operation.

CHAPTER 22

Special Refrigeration Applications

WATER COOLERS

Water coolers are manufactured in many sizes and models to suit various requirements. Water cooling equipment is divided into two classes:

1. Bottle coolers.
2. Pressure or tap coolers.

Bottle Coolers

The bottle type of water cooler differs from the pressure type mainly in that the former does not require any plumbing connection since the bottled water (usually distilled) is delivered to the customer's premises as required, whereas the pressure type depends on tap water connected to the cooler by means of a water supply line.

Bottle coolers (Fig. 1) lend themselves to office use and particularly in temporary locations, since no plumbing is required. The only requirement is an electric source for the condensing unit and the availability of distilled water which is usually delivered in five-gallon glass bottles.

The Condensing Unit—The condensing unit normally used in bottle coolers is of the hermetic or sealed rotary type, with a 110-volt, 60-cycle, alternating current motor of approximately 1/12 horsepower. Most condensing units of this type are equipped with a natural-draft finned-type condenser which is so arranged and baffled that efficient circulation is assured.

Fig. 1. A bottle water cooler with condensing unit and associated connections.

Cooling Chamber—With reference to Fig. 2, the evaporator consists of two concentric shells soldered together at the top and bottom. The liquid refrigerant enters the space between the shells at the bottom and the suction line is connected at the top of the outside shell. A restrictor-type refrigerant control is located in the insulation of the cooler and the thermostatic switch-bulb well is soldered to the evaporator. The inner shell forms the water storage tank of ½-gallon capacity, and all parts of this cooling chamber are hot tinned to prevent corrosion.

Controls—The operation of the condensing unit is controlled by a thermostatic switch and cold control for temperature adjustment by the user. Motor protective relays are normally incorporated in the condensing unit.

481

Fig. 2. Phantom view of a typical bottle water cooler.

Pressure Water Coolers

The pressure-type water cooler differs from the bottle cooler mainly in that the water is supplied from the building water lines as fast as it is used, whereas in the bottle cooler the water must be replenished from time to time. As shown in Fig. 3, the individual pressure cooler is a self-contained, complete refrigerating system and water-cooling unit. It is equipped with all the necessary fittings for dispensing drinking water as well as catching and disposing of the waste water. This type of water cooler is usually equipped with a compact fractional-horsepower com-

482

pressor (from ⅛ to ⅓ horsepower), and has capacities ranging from 3 to 19 gallons per hour.

Fig. 3. A pressure water cooler showing the water and refrigerant connections.

Precooler—A precooler is a device in which the incoming water receives partial cooling from the waste water, thus decreasing the quantity of heat to be removed by the compressor. The amount of heat removed by the precooler depends on the volume of waste water as compared to the water drawn from the cooler. The amount of heat which the precooler will remove also depends on the length of the tubing comprising the precooler. In various designs of precoolers and water coolers, it may be said the precooler increases the capacity of the unit from 25% to 75%.

Remote Bubbler—There are times when an individual water cooler is so located that a proportion of those drinking from it are in another room. This means that they have a long way to go for a drink or that they do not get the amount of water they should have. To avoid this, individual water coolers have been designed for remote bubbler connection and are supplied by various manufacturers.

The water cooler should be located at the point where the greatest demand for water exists and the remote fountain (bubbler) run to the area of lesser demand. This run should never exceed a total of 15 feet. The pipe running from the water-cooler unit to the remote bubbler should be brass and no larger than ⅜ in. and preferably ¼ in. The pipe must be insulated for ice water with cork or its equivalent.

Plumbing Connections—If the water inlet valve is brought through the floor, a shut-off valve should be installed under the flooring. After installing the shut-off valve, bring the water inlet line through the floor and up through the hole in the base of the cooler where a reducing connector should be installed for connection to the copper tubing in the cooling unit.

To facilitate the installation of the drain circuit, a union should be installed in the pipe so that the union is accessible through the panel opening in the rear of the cabinet. The pipe should then be continued through the hole in the cooler base, through the floor to a trap, and then to the main drain.

If water connections cannot be brought through the floor, but must run through the wall, it may be necessary to bring the piping through the cabinet grilling in the rear of the cooler rather than through the cabinet shell. This will permit removal of the panel without disconnecting the water lines.

To install the drain line when the piping is to be brought out the back of the cooler, first install the connection by means of a nipple leading into the drip pan. A street ell should then be screwed into the elbow and, by proper adjustment of the elbow, the street ell can be brought out through the service panel opening.

Cycle of Operation (typical bottle water cooler)

On a bottle-type model, such as that illustrated in Fig. 1, drinking water is supplied from an inverted bottle which rests on the top of the water cooler. The neck of this bottle extends down into a watertight extension above the cooling chamber. When water is drawn from the chamber, the water level falls below the outlet of the bottle, allowing air to enter the bottle and water to flow out until the water level covers the bottle opening. Water from the bottle passes down to the cooling chamber through a baffle which directs the incoming water along the refrigerated sides of the cooling chamber. Cooled water is drawn from the bottom of the cooling chamber and passes through a tube to the faucet at the front of the cooler. Waste water from the faucet is caught in a receptacle which has a drain tube running from the chamber outlet to a connection at the rear of the water cooler, where it is sealed with a special safety plug.

Cycle of Operation (typical pressure water cooler)

On pressure models, city water enters the cooler through the inlet connection at the rear of the cooler. Here, the water passes through the precooling jacket surrounding the drain tube, where it is precooled by the waste water in the drain. The precooled water from this jacket enters the cooling chamber at the top. A baffle inside the cooling chamber directs the incoming water down the sides of the cooling chamber close to the refrigerated walls. The outlet water tube runs close to the bottom of the cooling chamber where the coolest water lodges. The cooled water passes through this tube to the self-closing valve pressure regulator and bubbler on some models, and to the faucet on other models.

Air which finds its way into the cooling chamber is released through a small hole in the outlet tube near the top of the chamber. The glass filler connection is connected in the outlet line between the cooling chamber and the self-closing valve. Water for a remote bubbler passes through a tube leading from the bottom of the water-cooling chamber to a connection at the rear of the water cooler. This connection is sealed with a special safety plug when a remote bubbler is not used. From the bubbler or faucet, waste water passes through the catch basin strainer to a short length of drain tubing leading directly to a drain connection at the rear of the cooler.

ICE-CREAM REFRIGERATION

Refrigeration employed in ice-cream dispensing, as used in stores, consists of special prefabricated freezer cabinets of steel-shell construction and usually finished in a white painted finish. Sometimes the freezer is mounted on a stand only, but more often, a unit is made up of a storage cabinet and a hardening cabinet along with the freezer. The insulation

(Fig. 4) usually consists of approximately 4 inches of corkboard on the sides, ends, and bottom supported by a steel framework. This insulation in a well-designed cabinet is encased in a steel shell and sealed against infiltration of moisture by an asphalt compound. The lids are commonly of the hinged type, having a soft rubber lip on the outer edge which forms a tight seal against the soft rubber insulting collar at the top of each compartment.

Fig. 4. Cross-sectional view of a typical ice-cream cabinet.

The freezer capacity is usually 1, 2-½, or 5 gallons per batch and is usually built integral with a hardening and storage space for 40, 60, or 80 gallons, depending on individual requirements. In the combination plant setup, the hardening space is provided as an integral part of the unit. Often, it is necessary to supply one or two extra hardening cabinets for additional hardening and storage space. The load can be figured on these individual units exactly as in the large plants and summed up to give the total hardening load.

Condensing Units

Depending on the particular requirements, the condensing unit may be obtained as a part of the cabinet or mounted in a remote location. As a general rule, counter freezers are applicable for use with *Freon*-12 refrigerants. There are, however, some of the old ammonia-type brine counter freezers still in use. Although condensing unit sizes may vary, Table 1 shows the approximate compressor size requirements for various ice-cream cabinet freezers.

Table 1. Freezer Requirements

Freezer	Hardening Space	Gal/Hr.	Daily Capacity	Approx. Comp. Size
2½ Gal.	None	12½	Continuous	1 HP
2½ Gal.	40 Gal.	12½	40 Gal.	1 HP
2½ Gal.	60 Gal.	12½	60 Gal.	1½ HP
5 Gal.	None	20	Continuous	1½ HP
5 Gal.	40 Gal.	20	40 Gal.	1½ HP
5 Gal.	60 Gal.	25	60 Gal.	2 HP

*NOTE Compressors are to run at −20° suction and 100° condensing pressure on hardening cabinets.

It must be understood that Table 1 has been prepared by one manufacturer for his own equipment and is offered only as an example. Different freezer and cabinet constructions may result in slightly different machine recommendations. A typical condenser unit used in ice-cream cabinets is shown in Fig. 5.

Courtesy Tecumseh Products Company

Fig. 5. Exterior view of an air-cooled condensing unit suitable for remote operation of an ice cream cabinet.

Control Methods

Controls employed consist of the usual automatic expansion valve to control the amount of refrigerant entering the evaporator. This valve is

487

normally insulated with a rubber and cork composition cover and installed in the control compartment. In addition, most units provide a combination accumulator and heat interchanger for improved expansion-valve operation and for higher overall efficiency.

Cold-control adjustment is made by a conventional thermostatic switch which, in most cases, contains the "on" and "off" feature and motor-protector element.

Multiplexing Ice-Cream Cabinets

Multiplexing remote-control ice-cream cabinets has been found satisfactory where the user knows which cabinet will be subjected to the heaviest service or require the coldest temperature. Where, however, the operating conditions cannot be determined by any degree of certainty, a method has been employed which is very flexible in operation. This method, shown in Fig. 6, employs a temperature-regulating valve on all cabinets with condensing unit operations controlled by the low-pressure switch. The automatic expansion valves are still used to control the refrigerant to each cabinet. It is suggested that manifolds be used, as this permits the servicing of any one of the cabinets without materially affecting the operation of the other cabinets connected in the system.

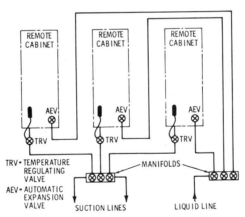

Fig. 6. The correct method of making a multiple installation of two or more ice-cream cabinets on one compressor.

BOTTLED AND CANNED BEVERAGE COOLERS

The day of the chest-type beverage cooler is past, and the present-day trend is to upright automatic dispensers, which are coin operated. These dispensers are of the dry type, with a fan circulating the cold air. They are self-contained, operating on 120 volt, 60-hertz AC power. The compressor and motor in self-contained bottle and can coolers are usually of the hermetic type with the condenser, fan, fan motor, shut-off valves, check valve, and relay all located in the machine compartment and assembled on a unit base.

There are some reach-in units, where beer and liquors are cooled in liquor stores. The refrigeration equipment on these is very similar to the grocery type units, and servicing is the same.

MILK COOLERS

There are three general types of milk coolers used to cool milk on the farm, such as:

1. The aerator type.
2. The vat type.
3. The tube cooler.

Aerator Type

Aeration is the cooling of milk by allowing it to flow by gravity over a surface cooler through which either cold water or brine is pumped to reduce the temperature. Principally, it consists of a series of tubes located one above the other so that the milk can be distributed over the top tube and allowed to drip from one tube to the other and be cooled by surface contact. Usually, this aerator is in two general sections—the top section which uses water for a cooling medium, and the bottom section usually supplied with refrigeration. In most localities, water between 70° and 80° F may be obtained from city mains or from a cooling tower, and is used to cool the pasteurized milk from an initial temperature of 145° F down to some point between 75° and 80° F, depending entirely on the temperature and quantity of water available. Milk then enters the bottom section of the aerator which is supplied with refrigeration, either directly in the direct expansion, or indirectly as in the brine or sweet-water type of aerator.

Vat Type

The *vat type* of pasteurizer and cooler combined, known generally as the vat type of pasteurizer, consists of a large porcelain or glass-lined vat containing a helical coil which is submerged and rotating during the processing. During pasteurizing, either hot water or steam is circulated through this coil to supply the heat necessary to pasteurize the milk. Immediately after the pasteurizing is accomplished, tap water is circulated through the coil to bring the milk down to a point between 75° and 85° F, after which chilled brine or fresh water is circulated to furnish further reduction down to 40° or 45° F. This type of equipment has an advantage in that the milk is handled only once, making it unnecessary to put it through a separate pasteurizer, precooler, and aerator.

The chief disadvantage with this type of equipment lies in its high initial cost. Some slight difficulty is also sometimes experienced in cleaning it. The calculation of the refrigeration loads for the two set-ups is practically identical.

Tube Cooler

The third type of cooling equipment consists of a *concentric tube cooler* wherein the milk passes through a center tube, while water for precooling and final chilling is passed through the angular space around the center tube within the outside tube. Here, again, the refrigerator load calculations are identical to those previously described.

Condensing Unit

The condensing unit (Fig. 7) is normally of the hermetic preassembled type, the size of the motor depending on the requirements. The condensing unit of this type of cooler is usually mounted on top of the cooling cabinet. In some instances, however, it may be desirable to install a milk cooler with the condensing unit at some location other than on top of the cabinet. In other instances, the milk-cooling tank may require two units to meet the heavy load conditions or to get the proper water circulation for a pull-down in the required time.

Agitator Unit

The agitator, as the name implies, provides the necessary water circulation by impeller action. The agitator assembly is designed for placement in the cooler tank through a hole cut in the top of the stationary lid. Since the agitator is independent of the condensing unit, it is driven by a small AC motor, usually about 1/10 HP, operating on 110-volt, 60-cycle, single-phase current, and is normally equipped with a wound-coil (magnetic) reset-type of overload protector.

Courtesy Tecumseh Products Company

Fig. 7. Exterior view of a 2-HP air-cooled hermetic-type condensing unit for bulk-milk cooler units.

SERVICING AND REPAIRS

It is very important that every serviceman be able to properly diagnose any difficulty in special refrigeration units. As a general rule, service complaints will fall under three general headings:

1. **Unit Does Not Run**—The failure of a unit to run when plugged into an electrical outlet may be due to one of several reasons which must be determined by the serviceman before changing the unit.

   a. The source of power should be checked with a test cord.

   b. Check for broken wires in the lead-in cord.

   c. Check the "off" and "on" switch to make sure the switch is in the "on" position.

   d. The proper type of power must be used.

   e. Check for a defective thermostatic switch which is not making contact at this point in the circuit.

   f. Check the motor protector relay.

   g. Test the capacitor to determine if it is functioning properly. If all the foregoing points have been checked very carefully, and still the unit refuses to run, then the unit must be changed.

2. **Unit Runs but Does Not Refrigerate**—This condition may occur just after installation if the cabinets have been stored in an extremely cold place or exposed to low temperatures during delivery

491

and do not have time to warm up after being installed in the customer's premises. It has happened, in some instances, that the cabinet was installed in a location where low outside temperatures were encountered. If this is not the cause of the unit failure, it will be necessary to change the unit.

3. **Unit Does Not Refrigerate Properly**—This condition may be due to one of several causes which must be diagnosed very carefully by the serviceman before assuming the unit to be defective. Any one of the following may be at fault:

   a.   Unit not properly charged.

   b.   Temporary low line voltage.

   c.   Defective switch.

   d.   Dirty condenser.

   e.   Condenser fan motor does not operate.

Leaks in the System

In the event that any welded part of the condensing unit develops a leak, it will be necessary to change the unit. The leak will usually be discovered through the presence of oil around the point at which the leak develops. It must not be assumed, however, that the presence of oil on any part of the unit is an indication of a leak. An actual refrigerant leak may be discovered with a detector. A shortage of refrigerant in the system may result in partial frosting of the evaporator, which should be checked with the cold control set in the coldest position. If the entire evaporator does not cool when the cold control is set in this position, the unit may be short of refrigerant.

Electrical Tests

Testing a motor in a sealed condensing unit to determine why it does not operate becomes a very simple process if the correct procedure is followed. Each test made should be one of a series of eliminations to determine what part of the system is defective. By checking other parts of the wiring system before checking the unit itself, a great deal of time can be saved since, in most cases, the trouble will be in the wiring or controls rather than in the unit.

In order to make a complete electrical test on electrical outlets and on the unit itself, it is advisable to make a test cord in the manner shown in Fig. 9. By connecting the black and white terminal clips together and placing a light bulb in the socket, the cord may be used to check the wall outlet into which the unit is connected. By connecting the white and red terminal clips, this same test may be made by depressing the push button. This will serve as a test to make certain the push button is in working order.

Fig. 8. A standard test cord.

This test cord should have a capacitor installed in the red lead to the pushbutton, if the compressor is a capacitor start type unit.

When these tests have been completed, and it is known that current is being supplied to the unit, the next step is to check the three wires on the base of the compressor unit. Pull the plug from the wall receptacle and carefully examine the nuts which hold the wires in position. Then try each wire to be sure it is held firmly in place, as a loose wire may keep the unit from operating. Test the thermostatic switch to determine whether or not contact is being made at that point. Turn the cold-control knob several times. If this fails to start the unit, then short across the thermostatic switch terminals on the switch. To do this, it will be necessary to remove the switch cover from the top of the switch. If the unit starts, it is an indication that the thermostatic switch is not operating properly and must be repaired or replaced. After the thermostatic switch has been checked, and if the trouble is not located, it will be necessary to determine whether the trouble is in the motor, motor protector relay, or capacitor.

Capacitor Test

The capacitor must be checked before testing the unit itself. This is done in the following manner:

1.   Disconnect the capacitor wires from the motor protector relay.

493

2. Connect these two wires to the black and white terminals of the test cord.
3. Put a 150-watt light bulb in the receptacle on the test cord and plug it into an outlet.

If the 150-watt bulb does not light, it is an indication that the capacitor has an *open circuit* and must be replaced. However, if the bulb does light, it is not an indication that the capacitor is perfect. This must be checked further by shorting across the two terminals of the capacitor with a screwdriver having an insulated handle. If the brilliancy of the light changes—that is, if the light bulb burns brighter when the terminals of the capacitor are shorted—it indicates that the capacitor is in proper operating condition. A decided sparking of the terminal will also be noticed when the terminals are shorted. If the brilliancy of the bulb does not change, it is an indication that the capacitor has an internal *short circuit* and must be replaced.

Replacing the Motor Protective Relay

If the motor protective relay is found defective during the preceding test, replace it. The motor protector relay is usually accessible for replacement from the rear of the cabinet. Remove the nuts holding the relay in position and remove the lead clamp. Notice the position of the electrical leads before replacing them. Place the electrical leads on the replacement relay and check them for correct connections against the wiring diagram.

MILK PASTEURIZATION

Although the farm dairy is not primarily interested in pasteurization of the milk, it will be well to bring you up to date on pasteurization methods. The newest and most accepted method of pasteurization is what is called the High Temperature-Short Time Method. This involves temperatures of 155° F to 185° F. Ordinary milk for consumption is usually raised to a temperature of 161° F for a minimum of 15 seconds, and then quickly brought down to a temperature of 40° to 45° F. This is often accomplished by running the hot milk slowly over a series of horizontal refrigerated tubes, thus allowing the removal of the higher temperature, until the desired final temperature is reached.

494

Simple Load Calculations

Assume that the processing plant handles 800 gallons of milk daily. To cool the milk from a pasteurizing temperature of 161° F to 40° F, the following assumptions are made:

1. Specific gravity of milk is 1.03
2. Weight of milk is 8.6 pounds per gallon
3. Weight of water is 8.34 pounds per gallon
4. Specific heat of milk is 0.95.

The equation for heat load is:

$$H = WS (t_1 - t_2),$$

where,

H = Btu required
W = Weight of fluid being cooled,
S = Specific heat Btu per pound cooled,
t_1 = Higher Temperature,
t_2 = Lower Temperature.

Substituting values in equation (1) we obtain:

$$H = 800 \times 8.6 \times 0.95 (161 - 40) = 790,856 \text{ Btu.}$$

If it is assumed that ample water is available, it is entirely feasible to cool milk with water to 80° F. The heat to be removed by water consequently is:

$$H = \frac{790,856 \times 81}{121} = 529,415 \text{ Btu.}$$

If the water used for cooling is allowed to raise 10° F, the amount of water necessary is:

$$\frac{529,415}{10 \times 8.34} = 6348 \text{ gallons}$$

495

If a four hour period is allowed for the cooling, the amount of water necessary per hour is 6348/4 or 1587 gallons. If 25% of water is allowed for losses, the gpm requirements of water finally is:

$$\frac{1587 \times 1.25}{60} = 33 \text{ gallons per minute (approximately.)}$$

To cool milk from 80 to 40° F, the amount of heat to be removed is:

$$H = 800 \times 8.6 \times 0.95 \, (80 - 40) = 261,440 \text{ Btu.}$$

The above figures of course will not apply to the home dairyman, as he is not interested in the pasteurization of the milk. What he is interested in is the size of condensing units and tank capacities to take care of two milkings a day. The manufacturer of the tank and condensing unit will arrive at the sizes of the equipment, given the gallons per day that are to be processed.

SUMMARY

Special refrigeration applications covered include water coolers, ice-cream cabinets, bottled-beverage coolers, and milk coolers. In each of the foregoing, condensing-unit equipment and control features have received particular attention.

Insulation requirements, plumbing methods, and multiple condensing-unit operation are fully described, in addition to a service and repair section. This latter part of the chapter provides instructions for various testing methods to use to correct operation difficulties.

REVIEW QUESTIONS

1. Name the various classes of water coolers.

2. Describe the difference between bottle and pressure water coolers.

3. What is the arrangement of the plumbing connections in pressure water coolers?

4. Describe the cycle of operation in bottle and pressure water coolers.

5. State the arrangement of the refrigerating components in ice-cream dispensing cabinets.

6. What are the control methods used in ice-cream refrigeration?

7. What is meant by multiplexing ice-cream cabinets?

8. Describe bottled beverage cooler cabinets.

9. What are the three general types of milk coolers used on the farm?

10. Describe the servicing and repair methods used for special refrigeration units.

CHAPTER 23

Cold Storage Practice

The primary reason for cold storage of perishable foods is its preservation from decay and spoilage which will impair its usage. Refrigeration has not only saved incalculable quantities of meat, fish, fruits, eggs, milk, and cream from spoilage, but has played an important part in changing the diet of the world. No longer are the inhabitants of one hemisphere, country, or locality dependent on local foodstuffs, but may draw on the entire world as a source. This has made possible great developments in agriculture and livestock raising in countries far distant from potential markets and has permitted the utilization of the great productivity of the tropics in supplying fruits and foods for other zones.

Nearly all of the industries involved in the preparation of food and drinks make large use of artificial ice or mechanical refrigeration. The dairy industry, for example, finds that for precooling milk and cream and in the manufacture of butter and ice cream refrigeration is indispensable.

QUICK FREEZING

There are several reasons for the popularity of the quick freezing method for preserving fruits and vegetables. Perhaps the most important is the time and labor saved as compared to the usual home-canning method. Other reasons are the preservation of flavor and color so as to approach very nearly the fresh product. Some of the farm products which may be frozen and stored are berries, peaches, apricots, plums, cherries, citrus and tomato juices, asparagus, peas, lima beans, snap beans, corn, cauliflower, broccoli, spinach, Brussels sprouts, and squash. When handling farm products on a large scale, it is important that the planting and supervision of the crops in the field is timed so that a properly ripened and matured product will flow in an even manner to the freezing plant and thus avoid large peak freezing loads, which the plant may not be equipped to handle.

In general, it may be said that the greater the speed of the freezing, the better the products will be in texture and taste upon thawing. At present, locker-plant freezing is sometimes of one variety and sometimes of another. It is proper, however, to quick freeze everything going into a locker. The usual apparatus for quick freezing in a locker plant is a series of cold plates or shelves through which air may or may not be blown.

The temperature of the room in which meats, fruits, or vegetables are quickly frozen before they are placed in the locker, varies from 5° to 15° F below zero. Most items are individually wrapped in small packages to enable quick penetration of the cold. The time required for quick freezing ranges from 3 to 6 hours.

Frozen foods has become a very large industry, so much so that one may purchase most any item in the frozen foods dept. of a super market. Prepared dinners, fish, vegetables, juices, ice cream, and most any other item that one should wish for can be purchased frozen. Frozen foods however, should not be thawed and then re-frozen.

FOOD PREPARATION FOR QUICK FREEZING

Blanching

The method of preparing most vegetables for quick freezing is known as *blanching*. This procedure consists in partly cooking and softening the vegetables for freezing, aids in cleansing, and produces a nearly natural color. The main purpose, however, of the blanch is to kill bacteria and mold. Bacteria and mold once killed do not multiply at the freezer or storage temperatures, but they should be killed by blanching before freezing.

Blanching is done by placing the vegetables in boiling water the length of time specified for the different varieties. After blanching, they should immediately be placed in cold water to remove the heat as rapidly as possible. It is also highly important that products for freezing be placed in the freezer as soon after blanching and packaging as possible. Fruit jars, jelly glasses, and waxed cups are generally satisfactory for the packaging of frozen products. Waxed containers are the most satisfactory for all-around purposes. Glass containers should have the lids loose during the freezing period to prevent breakage. In filling any container, care should be used not to fill to the top. An air space should be allowed for expansion during freezing.

Sugar and Salt Packing

Sugar can be used in two ways—dry or in solution. When used dry, the sugar should be fairly well distributed over the product. This method serves partly to control oxidation when applied to strawberries, red raspberries, and loganberries, and is particularly effective with fruits such as apricots and peaches.

Sugar solutions are made hot and then cooled before being applied to the fruit. Densities are determined on the basis of weight. To make a 50% density solution, for instance, use 5 lbs. of sugar and 5 lbs. of water, while to make a 60% solution, use 6 lbs. of sugar and 4 lbs. of water. The solutions must be cooled before pouring over the fruit. Salt solutions are also made on a weight basis.

Preparation of Fruits

Fruits are packed either well drained or in syrup. The use of syrup is found advantageous in most cases because it protects the fruit from the air.

Blackberries—Wash and sort the berries. For the best results, pack the fruit with sugar 3-to-1, or with a 50% syrup solution. Paraffined cups are recommended.

Cranberries—Select the riper and more colored berries, sort, and wash. Barely cover the fruit with a 50% syrup. Cranberries may be frozen, however, without sugar or syrup. Paraffined cups are recommended.

Loganberries—The loganberry is well adapted to freezing, with ripe, firm berries being best. Wash the fruit carefully and pack in paraffined cups, glass jars, etc. Add sugar at the rate of 2-to-1, taking care to cover the fruit well. If syrup is used, a 50% to 60% solution is best.

Black Raspberries—Wash, sort, and pack in containers, covering with sugar 3-to-1. If syrup is used, a 40% to 50% solution is satisfactory.

Red Raspberries—wash, sort, and pack. Sugar 3-to-1, or use a 50% syrup solution.

Strawberries—The fruit should be picked when well colored and ripe, but not soft. Cap and wash, and then pack in containers with a 3-to-1 sugar ratio, or a 60% syrup solution. Wax cups may be used.

Youngberries—The youngberry is a mild flavored large berry that is very well adapted to freezing. After washing and sorting, pack 3-to-1 with sugar, or use a 30% to 40% syrup solution.

Black Cherries—Use only well ripened fruit that has been washed and sorted. Stems may be left on or removed. The syrup pack is recommended for best results, using a 40% to 50% solution.

Sour Cherries—Use only bright red, tree-ripened fruit with a slightly acid taste. Wash, stem, pit, and sort carefully, then pack the fruit in waxed cups, using sugar 5-to-1, or a cold syrup of 60%.

Apricots—Apricots are in the best condition for freezing when firm and ripe, showing good color and maturity. Soft fruit is to be avoided, because freezing contributes to a loss of firmness. Keep the fruit cool and handle it quickly. Avoid bruising. Wash the fruit carefully, then halve and pit. Peeling is not necessary—in fact the skins help to hold the halves more firmly together. If peeling is desired, however, the fruit can be dipped in boiling water and subjected to steam or lye, as practiced in commercial canning. In case lye is used, the fruit should be rinsed in a weak citric-acid bath. A syrup pack is recommended for this fruit due to its rapid oxidation. A 40% to 50% solution is satisfactory.

Figs—Wash and sort carefully, removing the stem up to the base of the fig. Pack the fruit without peeling in paraffined cups. Use the dry pack or syrup with a density of 35%. In most cases, the syrup pack will be found better.

Grapes—Many of the popular varieties of grapes can be frozen. Tokay, Concord, Muscat, and others are suitable. Maturity is essential to obtain full flavor. Wash, sort, and stem carefully, placing the fruit in glass jars or wax containers. Cover the grapes with a syrup of 40% density.

Peaches—When selecting peaches, choose only those of high quality with predominant flavor. The fruit should be firm and ripe. Handle it quickly to prevent oxidation. Sliced peaches are very much better than halves, but require more care. A 50% syrup pack is recommended. They should be placed in the quick freezer as soon as possible.

Preparation of Vegetables

Vegetables are packed either well drained or in brine. The use of brine is recommended in most cases.

Asparagus—Frozen green asparagus, when properly handled, is satisfactory. Careful sorting is essential to obtain good, succulent, and tender stalks. Prepare and pack quickly to avoid shrinkage or shriveling. Blanch for two or three minutes in boiling water and then chill quickly in cold water. Pack and seal in airtight containers without further treatment. If a brine pack is desired, containers that do not seal airtight may be used. A 2% brine is preferred.

Beans—Green or waxed beans are very satisfactory to quick freeze. Wash and blanch for 2 or 3 minutes. Dip in cold water, drain and chill quickly, then pack in glass jars or waxed containers.

Table 1. Condensed List of Storage Temperatures

Products	How Packed	Temp., °F.
Almonds	In shells	34°—38°
Apples	Barrels	30°—32°
Apricots	Baskets	40°—45°
Beans	Green, bushel baskets	36°—40°
Berries	Short Periods, 3 or 4 days	36°—40°
Butter	Boxes or crates	38°—40°
Cabbage	Bulk	32°—36°
Cabbage	Boxes or crates	32°—33°
Calves	Fresh, held in chill	30°—36°
Calves	Fresh, short time	36°—38°
Candy	Bulk or boxed	60°—65°
Canteloupes	Box or crate, 3 weeks	34°—36°
Carrots	Bushel crates	34°—40°
Cherries	Fresh, in boxes or crates	36°—40°
Cherries	Frozen	0°— 5°
Cheese	Cream, Limburger, etc.	38°—40°
Condensed Milk	Bulk	36°—40°
Corn	Green, bushel crates	36°—40°
Corn	Dried, sacks	40°—45°
Cream	Fresh, 40 qt. cans	32°—36°
Cream	Condensed	36°—40°
Currants	In boxes	36°—40°
Dates	Cured, boxes	40°—45°
Eggs	Crates, 30 dozen	28°—30°
Figs	Dried	40°—45°
Fish	Fresh, short periods	25°—30°
Fish	Frozen, regular storing	0°—10°
Fish	Dried, in boxes or barrels	35°—40°
Flour	Any kind, from cereals	35°—40°
Fruit	Usual boxes or baskets	35°—40°

Table 1. Condensed List of Storage Temperatures

Products	How Packed	Temp., °F.
Fruits & Vegetables	Dried	40°—45°
Furs	Coats, for summer season	25°—30°
Furs	Rugs of animals, or stuffed	25°—30°
Game	Frozen	0°—10°
Game	In cooler, short period	25°—30°
Grapes	Large baskets	34°—40°
Grapefruit	Boxes	35°—40°
Hams, etc.	Cured	35°—40°
Hominy	Boxes	35°—40°
Lard	Barrels or cans	36°—40°
Lemons, Limes	Boxes	35°—40°
Lettuce	Bushel crates	35°—37°
Maple sugar	Boxes or Syrup	40°—45°
Meat	Frozen	0°— 5°
Meat	In chill	30°—32°
Meat	In cooler	36°—38°
Meat	Cured or smoked	40°—45°
Milk	Fresh, 40 qt. cans or bottles	38°—40°
Milk	Condensed	36°—40°
Molasses	Barrels	40°—45°
Mutton	Fresh, short period	30°—32°
Nuts	In shells	35°—40°
Oils, lard, cottonseed, etc.	Barrels or cans	35°—40°
Oleomargarine	Boxes	25°—30°
Olive oil	Barrels	35°—40°
Onions	Sacks	32°—36°
Oranges	Boxes	35°—40°
Oysters	In shells, sack	36°—42°
Oysters	In tubs or cans	30°—34°
Peaches, Pears	Bushel baskets	35°—40°
Peas	Green, bushel crates	36°—40°
Pineapples, Plums	Boxes	40°—45°
Poultry	In freezer	0°—10°
Poultry	Cooler, short period	36°—38°
Potatoes	Sacks	34°—38°
Rice	Sacks	40°—45°
Skins	Dried	25°—30°
Skins	Uncured	30°—35°
Strawberries	Qt. boxes, crated	36°—40°
Vegetables	In barrels or crates	36°—40°
Watermelons	Short period	34°—36°
Wines	Barrels	40°—45°

Lima Beans—Lima beans should be harvested while still young and tender, shelled, blanched in boiling water for 2 or 3 minutes, then cooled in water. Drain well and pack in glass or waxed containers.

Broccoli—Broccoli has been frozen with unusual success, the product never losing its characteristic fresh green color. Use only the tender stalks with compact heads, sorting carefully, and cutting back the stems to the part that is tender. Blanch in boiling water 3 or 4 minutes and cool in fresh rinse water. Drain and pack in glass or waxed cups or boxes.

Cauliflower—Cauliflower can be frozen after being carefully trimmed. Remove all the green leaves, cut the larger flowerets apart, then soak for a short time in a weak brine solution. Blanch in boiling water for 2 or 3 minutes, cool and drain promptly, and pack in glass or waxed containers. Freeze in either a 2% brine solution, or as a dry pack.

Sweet Corn—Obtain corn of proper maturity, in the full milk stage. Blanch in boiling water for not more than 6 minutes, then cool in fresh water. Wrap each ear in wax or parchment paper, and pack in tin cans or waxed boxes. For cut corn, blanch as outlined and cut from the cob. Use care not to cut or scrape the cob. Pack the cut corn in glass or in waxed cups or boxes, covering with a 2% brine solution.

Mushrooms—The small button-sized mushrooms are found best for freezing. Care must be taken in handling so as not to damage the caps lest discoloration appear. Sort for size, if necessary, and wash carefully. Blanch in boiling water for 2 to 4 minutes (depending on the size of the mushrooms) and cool rapidly in water. Pack in glass or waxed cups, preferably with a 2% brine solution.

Peas—Obtain peas that are absolutely fresh from the vine, otherwise they will be of a high starch content. Hull and wash. Blanch in boiling water for 1 minute, cooling quickly in plenty of cold water. The peas can be packed dry or in 2% brine. Pack in glass jars, parchment lined wax boxes, or in waxed cups.

Spinach—Although somewhat difficult to handle, spinach makes a very good frozen product. Care must be taken to see that the spinach is not too advanced in maturity. It should be well washed to remove all sand and grit. Blanch in boiling water for 2 to 2-½ minutes. Rinse well in cold water, drain, and pack without added liquid in glass jars, parchment lined waxed boxes, or in waxed cups.

Table 1 lists the storage temperatures for various foods, and is supplied through courtesy of *The Frick Company*, Waynesboro, Pa.

SUMMARY

Cold storage, its purpose for storage of perishable food, and the methods used have been fully described in this chapter. Quick freezing and the preparation of food, including "blanching," have been listed in detail. Thus, the preparation and packaging of common fruits and vegetables prior to freezing have been given considerable space and it is hoped that this important feature will be helpful for the individual home owner as well as locker-plant operators and others interested in modern cold-storage practice.

REVIEW QUESTIONS

1. Describe quick-freezing methods for fruits and vegetables.

2. What are the preparations necessary for quick freezing?

3. In general, how does the speed of the quick-freezing method affect the product?

4. What should be the temperature of fruits and vegetables prior to food-locker placement?

5. What is meant by the term "blanching" and what is the procedure?

6. Describe the reasons for adding salt and sugar solutions to various products.

7. What types of containers are used for storage of quick-frozen foods?

8. Why is brine used in the preparation of vegetables for quick freezing?

9. Enumerate the several vegetables that are successfully stored by means of the quick-freezing method.

CHAPTER 24

Fans and Blowers

Fans and blowers are used in refrigeration and air-conditioning plants to provide cooling and air flow. Depending on the particular application and construction principles, they are sometimes termed exhausters or propeller fans. Fans may be classified according to the construction as:

1. Centrifugal or radial flow.
2. Axial flow.

In the centrifugal fan, the air flows radially through the impeller within a cylinder or ring. Fan performance may be stated in various ways, the air volume per unit time, total pressure, static pressure, speed, and power input being the most important. The terms, as defined by the National Association of Fan Manufacturers, are as follows:

1. The *volume* handled by a fan is the number of cubic feet of air per minute expressed as fan outlet conditions.
2. The *total pressure* of a fan is the rise of pressure from fan inlet to fan outlet.
3. The *velocity pressure* of a fan is the pressure corresponding to the average velocity determination from the volume of air flow at the fan outlet area.
4. The *static pressure* of a fan is the total pressure diminished by the fan velocity pressure.

5. The *power output* of a fan is expressed in horsepower and is based on fan volume and the fan total pressure.
6. The *power input* to a fan is expressed in horsepower and is measured horsepower delivered to the fan shaft.
7. The *mechanical efficiency* of a fan is the ratio of power output to power input.
8. The *static efficiency* of a fan is the mechanical efficiency multiplied by the ratio of static pressure to the total pressure.
9. The *fan outlet area* is the inside area of the fan outlet.
10. The *fan inlet area* is the inside area of the inlet collar.

With respect to the foregoing, it should be noted that, while the total pressure truly represents the actual pressure developed by the fan, the static pressure may best represent the useful pressure for overcoming resistance.

PROPELLER FANS

A propeller fan, as noted in Fig. 1, consists of a propeller or disk-type wheel within a mounting ring or plate, and includes the mechanical supports for either a belt drive or direct connection. Propeller fans move the air in an axial direction by propulsive force, in an action similar to that of a ship's propeller. Fans of this type are usually termed exhaust or ventilating fans. When propeller fans are made with flat, or nearly flat blades, they are sometimes termed disk fans.

MOUNTING RING

MOTOR

Fig. 1. Mounting arrangement of a propeller fan.

CENTRIFUGAL FANS

A centrifugal fan (Fig. 2) consists of a fan rotor or wheel with a scroll type housing, and includes drive mechanism supports for either belt drive or direct connection. This type of fan may have an inlet on one or both sides. Those with one inlet are usually termed *exhausters,* while those with two inlets are called *blowers.*

Fig. 2. Mounting arrangement of a centrifugal fan.

FAN CHARACTERISTICS

Fans and blowers may operate with equal efficiency at any speed which their mechanical strength allows. The efficiency however, varies with the different orifice area, but not according to any fixed law for all designs. The performance curves of a 24-inch centrifugal fan is shown in Fig. 3. When designing or estimating the performance of any fan, such performance characteristics are the basis for selection of the motor and associated controls.

POWER REQUIREMENTS

When determining the horsepower requirements for a fan or blower, it is essential to remember that, for a certain discharge orifice, the power required varies directly as the cube of the speed. The horsepower requirements of a centrifugal fan generally decreases with a decrease in the area of the discharge orifice, if the speed remains constant. The horsepower requirements of a propeller fan increases as the area of the discharge orifice decreases, if the speed remains unchanged.

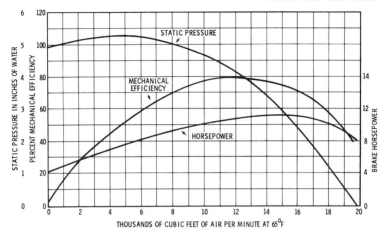

Fig. 3. Performance characteristics of a 24-inch centrifugal fan with backwardly inclined blades when running at 1600 rpm.

MOTOR TYPES

Fans used in refrigeration and air conditioning are usually driven by electric motors. Small size fans in the higher speed ranges are usually equipped with direct-connected motors, whereas larger size fans, and those operated at lower speed, are generally V-belt driven.

When selecting a motor for fan operation, it is advisable to select the next larger standard size than the fan requires. It should, however, be kept in mind that direct-connected fans do not require as great a safety factor as that of belted units. It is desirable to employ the belt drive when the required fan speed or horsepower requirement is in doubt, since a change in pulley size is relatively inexpensive if an error is made.

Directly-connected small fans for various applications are usually driven by a single-phase ac motor of the split-phase, capacitor, or shaded-pole type. The capacitor motor is more efficient electrically, and is used in districts where there are current limitations. Such motors, however, are usually arranged to operate at only one speed. In such cases, where necessary to vary the air volume or pressure of the fan or blower, the throttling of air by a damper installation is usually made. In large installations, such as when mechanical draft fans are required, there are various drive methods in use such as:

1.   By the use of a slip-ring motor to vary the speed.

509

2. By the use of a constant-speed, directly connected motor, which, by means of movable guide vanes in the fan inlet, serves to regulate the pressure and air volume.

FAN SELECTION

Most often, the service determines the type of fan to use. When operation occurs with little or no resistance, and particularly when no duct system is required, the propeller fan is commonly used because of its simplicity and economy in operation. When a duct system is involved, a centrifugal or axial type of fan is usually employed. In general, centrifugal and axial fans are comparable with respect to sound level, but the axial fans are somewhat lighter and require considerably less space. The following information is usually required for proper fan selection:

1. Capacity requirement in cubic feet per minute.
2. Static pressure or system resistance.
3. Type of application or service.
4. Mounting arrangement of system.
5. Sound level or use of space to be served.
6. Nature of load and available drive.

The various fan manufacturers generally supply tables or characteristics which usually show a wide range of operating particulars for each fan size. The usual tabulated data includes static pressure, outlet velocity, revolution per minute, brake horsepower, tip or peripheral speed, etc.

SUMMARY

Fans and blowers, their classification and characteristics, including terms as promulgated by the National Association of Fan Manufacturers, have been fully discussed in this chapter. In addition, power requirements and motor types suitable for fan and blower operation have been listed. Information on fan selection and detailed data to be supplied by the fan manufacturer, such as capacity requirement in cubic feet per minute, static pressure or system resistance, nature of load, etc., are other useful items included that are designed to assist the practical maintenance man or operating engineer.

REVIEW QUESTIONS

1. Describe the function of fans and blowers in a refrigeration system.

2. Give two classifications of fans as employed in a refrigeration system.

3. Describe terms as defined by the National Association of Fan Manufacturers.

4. What is a propeller fan and how does it operate?

5. How is the efficiency of a fan affected by the different orifice areas?

6. What are the horsepower requirements for a fan?

7. What drive methods are employed in fan operation?

8. Describe the various types of motors employed in fan operation.

9. What criteria is used in fan selection?

CHAPTER 25

Refrigeration Piping

The success of any refrigeration plant depends largely on the proper design of the refrigeration piping and a thorough understanding of the necessary accessories and their function in the system. In sizing refrigerant lines, it is necessary to consider the optimum sizes with respect to economics, friction losses, and oil return. It is desirable to have line sizes as small as possible from the cost standpoint. On the other hand, suction- and discharge-line pressure drops cause a loss of compressor capacity, and excessive liquid-line pressure drops may cause flashing of the liquid refrigerant with consequent faulty expansion-valve operation.

Refrigerant piping systems, to operate successfully, should satisfy the following:

1.   Assure the proper refrigerant feed to the evaporators.
2.   The refrigerant lines should be of sufficient size to prevent an excessive pressure drop.
3.   An excessive amount of lubricating oil should be prevented from being trapped in any part of the system.
4.   Liquid refrigerant should be prevented from entering the compressor at all times.

PRESSURE DROP CONSIDERATIONS

Pressure drop in liquid lines is not as critical as it is in the suction and discharge lines. The important thing to remember is that the pressure drop should not be so great as to cause gas formation in the liquid line and/or insufficient liquid pressure at the liquid feed device. A system should normally be designed so that the pressure drop due to friction in the liquid line is not greater than that corresponding to a one- to two-degree change in saturation temperature. Friction pressure drops in the liquid line include the drop in accessories, such as the solenoid valve, strainer-drier, hand valves, etc., as well as in the actual pipe and fittings from the receiver outlet to the refrigerant feed device at the evaporator.

Friction pressure drop in the suction line means a loss in system capacity because it forces the compressor to operate at a lower suction pressure to maintain the desired evaporating temperature in the coil. It is usually standard practice to size the suction line to have a pressure drop due to friction not any greater than the equivalent of a one- to two-degree change in saturation temperature.

LIQUID REFRIGERANT LINES

The liquid lines do not generally present any design problems. Refrigeration oil is sufficiently miscible with commonly used refrigerants in the liquid form to assure adequate mixture and positive oil return. The following factors should be considered when designating liquid lines:

1. The liquid lines, including the interconnected valves and accessories, must be of sufficient size to prevent excessive pressure drops.
2. When interconnecting condensing units with condenser receivers or evaporative condensers, the liquid lines from each unit should be brought into a common liquid line, as shown in Fig. 1.
3. Each unit should join the common liquid line as far below the receivers as possible, with a minimum of two feet preferred. The common liquid line should rise to the ceiling of the machine room. The added head of liquid is provided to prevent, as far as possible, hot gas blowing back from the receivers.
4. All liquid lines from the receivers to the common line should have equal pressure drops in order to provide, as nearly as possible, equal liquid flow and prevent the blowing of gas.
5. Remove all liquid-line filters from the condensing units and install them in parallel in the common liquid line at the ceiling level.
6. Hot gas blowing from the receivers can be condensed in reasonable quantities by liquid subcoolers, as specified for the regular

condensing units, having a minimum lift of 60 feet at 80 degrees condensing medium temperature.

7. Interconnect all the liquid receivers of the evaporative condensers above the liquid level to equalize the gas pressure.
8. The common and interconnecting liquid line should have an area equal to the sum of the areas of the individual lines.
9. Install a hand shut-off valve in the liquid line from each receiver.

Where a reduction in pipe size is necessary in order to provide sufficient gas velocity to entrain oil up the vertical risers at partial loads, greater pressure drops will be imposed at full load. These can usually be compensated for by oversizing the horizontal and downcomer lines to keep the total pressure drop within the desired limits.

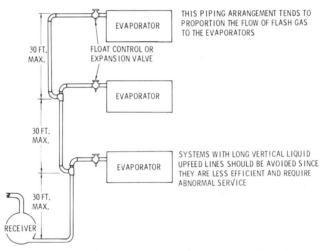

Courtesy General Electric Co.

Fig. 1. Piping diagram illustrating liquid riser connections when evaporators are located at different levels above the receiver.

INTERCONNECTION OF SUCTION LINES

When designing suction lines, the following important considerations should be observed:

1. The lines should be of sufficient capacity to prevent any considerable pressure drop at full load.

2. In multiple-unit installations, all suction lines should be brought to a common manifold at the compressor, as in Fig. 2.

3. The pressure drop between each compressor and main suction line should be the same in order to insure a proportionate amount of refrigerant gas to each compressor, as well as a proper return of oil to each compressor.

4. Equal pipe lengths, sizes, and spacings should be provided.

5. All manifolds should be level.

6. The inlet and outlet pipes should be staggered.

7. Never connect branch lines at a cross or tee.

8. A common manifold should have an area equal to the sum of the areas of the individual suction lines.

9. The suction lines should be designed so as to prevent liquid from draining into the compressor during shutdown of the refrigeration system.

Fig. 2. A double suction riser arrangement providing oil return at minimum load.

DISCHARGE LINES

The hot-gas loop accomplishes two functions, in that it prevents gas which may condense in the hot-gas line from draining back into the heads of the compressor during the off cycles, and it prevents oil which leaves one compressor from draining down into the head of an idle machine. It is important to reduce the pressure loss in hot-gas lines because losses in these lines increase the required compressor horsepower per ton of refrigeration and decrease the compressor capacity. The pressure drop is kept to a minimum by sizing the lines generously to avoid friction losses, but still making sure that refrigerant line velocities are sufficient to entrain and carry along oil at all load conditions. In addition, the following pointers should be observed:

1. The compressor hot-gas discharge lines should be connected as shown in Fig. 3.
2. The maximum length of the risers to the horizontal manifold should not exceed 6 feet.
3. The manifold size should be at least equal to the size of the common hot-gas line to the evaporative condenser.
4. If water-cooled condensers are interconnected, the hot-gas manifolds should be at least equal to the size of the discharge of the largest compressor.
5. If evaporative condensers are interconnected, a single gas line should be run to the evaporative condensers, and the same type of manifold as provided at the compressors should be installed.
6. Always stagger and install the piping at the condensers.
7. When the condensers are above the compressors, install a loop having a minimum depth of 3 feet in the hot-gas main line.
8. Install a hand shut-off valve in the hot-gas line at each compressor.

Fig. 3. A method of interconnecting hot-gas discharge lines.

Note—For tabulated data on pipe size for various refrigerants see the Appendix.

WATER VALVES

The water regulating valve is the control used with water-cooled condensers. When installing water valves, the following should be observed:

1. The condenser water for interconnected compressor condensers should be applied from a common water line.
2. Single automatic water valves, or multiple valves in parallel (Fig. 4), should be installed in the common water line.
3. Pressure control tubing from the water valves should be connected to a common line which, in turn, should be connected to one of the receivers or to the common liquid line.

Fig. 4. A method of interconnecting water valves.

MULTIPLE-UNIT INSTALLATION

Multiple compressors operating in parallel must be carefully piped to insure proper operation. The suction piping at parallel compressors should be designed so that all compressors run at the same suction pressure and so that oil is returned in equal proportions to the running compressors. All suction lines should be brought into a common suction header in order to return the oil to each crankcase as uniformly as possible.

The suction header should be run above the level of the compressor suction inlets so that oil can drain into the compressors by gravity. The header should not be below the compressor suction inlets because it can become an oil trap. Branch suction lines to the compressors should be taken off from the side of the header. Care should be taken to make sure that the return mains from the evaporators are not connected into the suction header so as to form crosses with the branch suction lines to the compressors. The suction header should be run full size along its entire length. The horizontal take-offs to the various compressors should be the same size as the suction header. No reduction should be made in the branch suction lines to the compressors until the vertical drop is reached. Fig. 5 shows the suction and hot-gas header arrangements for two compressors operating in parallel. Take-offs to each compressor from the common suction header should be horizontal and from the side to assure equal distribution of oil, and to prevent accumulating liquid refrigerant in an idle compressor in case of slop-over.

VERTICAL RISER OF SUCTION LINE
TO BE CENTRALLY LOCATED WITH EACH
SUCTION LINE EXACTLY THE SAME
LENGTH

CONNECT AS
CLOSE AS POSSIBLE

IF CONDENSER IS LOCATED
BELOW COMPRESSOR
DISCHARGE LINE, TRAPS
(AS SHOWN) ARE NOT
REQUIRED

VAPOR EQUALIZER
LINE

VALVE STEM TO BE LOCATED
IN A HORIZONTAL
POSITION

OIL EQUALIZER LINE LOCATED
AT FLOOR LEVEL. COMPRESSORS
MUST BE ON SAME LEVEL.

Fig. 5. Connections for the suction and hot gas headers in a multiple compressor installation.

Discharge Piping

Hot-gas piping at paralleled compressors should be arranged to prevent excessive vibration and to prevent oil and liquid refrigerant from draining back to the head of any of the compressors. The branch hot-gas lines from the compressors should be connected into a common header. This hot-gas header should be run at some level below that of the compressor discharge connections. For convenience, it is often at the floor. This is equivalent to the hot-gas loop for the single compressor.

The hot-gas loop accomplishes two functions:

1. It prevents gas, which may condense in the hot-gas line during the off cycles, from draining back onto the heads of the compressors, thus eliminating compressor damage from this potential hazard.
2. It prevents the oil, which leaves one compressor, from draining down onto the head of an idle machine.

The solid lines in Fig. 6 show the correct hot-gas piping from two paralleled compressors. Each broken line has a muffler for reducing hot-gas pulsations, and each line drops down into the hot-gas header at the floor before rising and going to the condenser. The piping shown by the broken lines in Fig. 6 is incorrect and can result in three kinds of trouble:

1. The use of a *bull-headed* tee at the junction of the two branches and the main will greatly increase the gas turbulence and pressure drop at this point. Hammering in the line may also result.
2. Any gas condensed in the discharge riser during an off cycle can run down and accumulate on the heads of the compressors, causing possible valve breakage on start up.
3. Oil leaving one operating compressor will be thrown into the branch line of the other compressor (which may be idle). This situation will cause the oil to drain down onto the head of the idle compressor.

Fig. 6. Correct and incorrect hot-gas connection when compressors are connected in parallel.

519

Interconnecting Piping

In addition to properly designing the suction and hot-gas piping at paralleled compressors, there are also certain other piping connections required between compressors and condensing units for proper parallel operation. These piping connections are shown in Fig. 7.

When two or more compressors are to be interconnected, they should be placed on foundations such that all oil-equalizer tappings are exactly

Fig. 7. Piping interconnection for multiple condensing units.

level. An interconnecting oil-equalization line should connect all of the crankcases in order to maintain uniform oil levels and adequate lubrication in all compressors. The oil equalizer may be run level with the tappings or, for convenient access to the compressors, it may be run at the floor. Under no conditions should the equalizer be run at a level higher than that of the tappings.

For the oil-equalizer line to work successfully, it is necessary to equalize the crankcase pressures. This is accomplished through the use of a gas equalizer line which is installed above the oil level. This line may be run to provide head room or run level with the tappings on the compressors. It should be piped so that oil or condensed vapor will not be trapped. Both lines should be the same size as the tappings on the largest compressor and should be valved so that any one machine can be taken out for repair without shutting down the entire system. Neither line should be run directly from one crankcase into another without forming some sort of U-bend or hairpin to absorb vibration.

Condensing Pressure Equalization

When multiple condensing units are interconnected as shown in Fig. 7, it is also necessary to equalize the pressure in the condensers to prevent hot gas from blowing through one of the condensers and into the liquid line. To do this, install a condensing-pressure equalizer line, as shown. If the piping is looped as shown, vibration should not be a problem. The bottoms of all condensers should be at the same level to prevent backing liquid into the lowest one.

The condensing pressure equalizer line between units should be the same size as the largest hot-gas line to which it connects. The liquid lines from the condensers should be piped as shown in Fig. 7 to insure equal level in all condenser receivers.

Multiple-Compressor Controls

The control of reciprocating compressors must be such that excessive accumulation of liquid refrigerant in the crankcase during off cycles is prevented. In addition, the following pointers should be observed:

1. Step control by a modulating thermostat in the conditioned space should be provided to start and stop each compressor according to the load variation. No solenoid valves should be used.

521

2. A step controller operated by a modulating low-pressure switch should be provided to start and stop each individual compressor according to variations in the suction pressure. Solenoid valves should be used to control the refrigeration to coils and conditioners.

3. When evaporative condensers are interconnected, each condenser should be started and stopped along with its compressor.

4. In case condensers and compressors cannot be properly balanced, all compressors and condensers should be operated as a unit.

PIPE CONNECTIONS

Steel pipe has been used with success on many installations, and particularly on large jobs where the cost of copper pipe or tubing prohibits their use. The inside of steel pipe or steel tubing should be free of scale and dirt. (Do not use pipe with a mill-oil coating. It must be internally clean, to the bare metal). Refrigerant acts as a cleansing agent and will quickly loosen all scale and dirt from the interior of pipe and fittings. This dirt not only clogs the strainers in the system, but may possibly damage the compressor.

As far as possible, all steel pipe lines should be welded to prevent leaks at the joints. Welded pipe connections should be carefully made and proper care exercised so that scale and welding particles may be removed from the lines before putting the plant into operation. Try to make the final welding joint near a flange connection so the line can be opened and cleaned after the welding has been completed. Be sure, however, to keep the welds at least a foot or more away from threaded joints, because the welding heat on the pipe is apt to ruin the threaded connections.

Wherever threaded joints are used, the threads should be cut full length and should be of standard taper and size, so that they will tighten gradually and not shoulder at the end of the threads. The threads should be cleaned with benzine, rubbed bright and dry, and all rust removed with steel wool. Make threaded joints of extra heavy pipe by welding a section of extra heavy weight to the standard weight, if necessary. On sizes over 2 inches, peen the inside of the pipe where flanges are made up. Make the flanges as tight as possible and do this peening carefully. If extra heavy pipe is used, trouble from leaks will not occur. Standard weight pipe, while suitable for pressures involved, very often loosens up due to vibration. Pipe intended for use should be kept dry, the ends being plugged at all times until the pipe is placed in the line.

Soldered Joints

The making of soldered joints is of particular importance. The surface of the tubing, and especially the inside of the fittings being joined, must be absolutely clean and free from oxidation, dirt, moisture, or grease. Steel wool is considered best for cleaning the fittings and pipe, and immediately after they have been cleaned, the joints should be made up.

One should be especially careful of the flux used in making a soldered joint. It has been found that *No-korode* and *Streamline* solder pastes give better service than fluxes. *Warning: Under no circumstances use any of the commercial solders with an acid core, since the fluxes in them are corrosive.* A thin, even coating of this flux should be distributed over all surfaces to be joined, the flux preferably being applied by means of a dry brush. When soldering paste is applied, the pipe should be warm (heated, if necessary, to approximately 80° in advance). The end of the pipe or tubing should be cut square in order that it may fit tightly against the ledge which usually forms a part of the fitting. The tubing should be perfectly round, with as little clearance as possible between the fitting and the pipe or tube wall. Poor fits will give trouble. Use a miter vise when cutting tubing to be soldered.

To solder, heat the entire surface evenly. Do not apply the solder until the entire surface around the joint is at an even temperature and just above the melting point of the solder. Do not have the surface too hot, and be sure there is enough solder to fill the entire space. Soldered joints require skill and care on the part of the mechanic. Fewer defects can be seen in a soldered joint than can be observed in a welded joint. The type of solder to be used will depend upon local ordinances and suggestions of the tubing manufacturers.

Flared Joints

Flared joints must be made with care. If the flaring tool is used without proper oiling, it will thin out the tubing to a point where there may be danger of breakage. Observe whether or not the tubing is being thinned at the flare. If the tubing is hard and tends to crack, then anneal the end before making the flare. Be sure to use the nonfreezing or vented nuts on all frosted flare connections. All tubing for flared joints must be soft copper.

Hanging Pipe Lines

The *Freons,* being heavy, will cause vibration in pipe lines unless care is taken to avoid sharp bends and unless proper hanging and bracing is done. Make the runs of pipe as short as possible. Make all hangers rigid, but avoid tying into floors or columns that will cause noise to be carried through the building. In such cases, pipes should be braced from the machine column, the foundations, or from one another; special supports can be devised to suit each condition. Make the hangers solid and firm. Wherever possible, set the condensers on concrete piers. Noise in condenser tubes can invariably be traced to vibration, either in pipe lines or in the whole condenser.

Where compressors are installed at a distance of 20 feet or more from the condenser, and particularly if the condenser is higher than the compressors, an antipulsation drum or vibration eliminator is necessary and should be installed in the compressor discharge line.

Strainers

Strainers should be placed in the line ahead of all expansion valves of the automatic type. Put the cleanout cap on the bottom. Suction traps are standard equipment with all compressors and are intended to remove the fine particles of dirt which may accumulate in the system. As a means of eliminating dirt, a so-called sump leg is quite effective. Dirt or scale will tend to separate whenever the velocity of the fluid decreases or whenever a change in direction takes place.

Pipe Line Repairs

If at any time a leak in a pipe line, in a weld, or in a soldered fitting is to be repaired, be sure that the line has been completely emptied of *Freon* before applying any heat to it. Open this line at a point near a shut-off valve. After the repair has been made and the joint has completely cooled, blow out the air and torch gases in the line by admitting a little *Freon* from the opposite end. Always remember that high temperatures obtained when welding or soldering will tend to break down any *Freon* that is present, releasing dangerous fumes. It is very necessary to have the line cool before admitting any of the refrigerant to it, and to get rid of any gases that may have been generated in the line during the repair procedure.

PIPING INSULATION

Insulation is required for refrigeration piping to prevent moisture condensation and to prevent heat gain from the surrounding air. The desirable properties of insulation are that it should have a low coefficient of heat transmission, it should be easy to apply, have a high degree of permanency, and should provide protection against air and moisture infiltration. Finally, it should have a reasonable installation cost.

The type and thickness of insulation used depends on the temperature difference between the surface of the pipe and the surrounding air, and also on the relative humidity of the air. It should be clearly understood that, although a system is designed to operate at a high suction temperature, it is quite difficult to prevent colder temperatures occurring from time to time. This may be due to a carrying over of some liquid from the evaporator or the operation of an evaporator pressure valve. Interchangers are preferable to insulation, in this case.

The safest pipe insulation available is molded cork or rock cork of the proper thickness. Hair-felt insulation may be used, but great care must be taken to have it properly sealed. For temperatures above 40° F, wool felt or a similar insulation may be used, but here again, success depends on the proper seal against air and moisture infiltration.

Liquid refrigerant lines carry much higher temperature refrigerant than suction lines, and if this temperature is above the temperature of the space through which they pass, no insulation is usually necessary. However, if there is danger of the liquid lines going below the surrounding air temperatures and causing condensation, they should be insulated when condensation will be objectionable. If they must unavoidably pass through highly-heated spaces, such as those adjacent to steam pipes, through boiler rooms, etc., then the liquid lines should also be insulated to assure a solid column of liquid to the expansion valve.

There are four types of insulation available commercially and in general use for refrigerator piping, namely:

1. Cork.
2. Rock cork.
3. Wool felt with waterproof jacket.
4. Hair felt with waterproof jacket.

Cork Insulation

Cork pipe covering is prepared by pressing dried and granulated cork in metal molds. The natural resins in the cork bind the entire mass into its new shape. In the case of cheaper cork, an artificial binder is used. The cork may be molded to fit pipe and fittings, or it may be made into flat boards of varying sizes and thicknesses. Cork has a low thermal conductivity. The natural binder in the material itself makes cork highly water resistant and its structure insures a low capillarity. It can be made practically impervious to water by surfacing with an odorless asphalt.

All fittings in the piping as well as the pipe itself should be thoroughly insulated to prevent heat gain, to protect the pipe insulation from moisture infiltration and deterioration, and to eliminate condensation problems. Molded cork covering made especially for this purpose is available for all the common types of fittings. Each covering should be the same in every respect as the pipe insulation, with the exception of the shape, and should be formed so that it joins to the pipe insulation with a break. Typical cork fitting covers are furnished in three standard thicknesses for ice water, brine, and special brine.

To secure maximum efficiency and long life from cork covering, it must be correctly applied and serviced, as well as properly selected. Hence, it is essential that the manufacturer's recommendations and instructions are followed in detail. The following general information is a summary of the data which are of general interest:

All pipe lines should be thoroughly cleaned, dried, and free from all leaks. It is also advisable to paint the piping with waterproof paint before applying the insulation, although this is not recommended by all manufacturers. All joints should be sealed with waterproof cement when applied. Fitting insulation should be applied in substantially the same manner, with the addition of a mixture of hot crude paraffin and granulated cork used to fill the space between the fittings, as shown in Fig. 8.

Rock Cork Insulation

This type of insulation is manufactured commercially by molding a mixture of rock wool and a waterproof binder into any shape or thickness desired. The rock wool is made from limestone melted at about 3000° F, and then blown into fibers by high-pressure steam. It is mixed with an asphaltum binder and molded into the various commercial forms. The heat conductivity is about the same as cork, and the installed price may be less. Because of its mineral composition, it is odorless, vermin proof, and free from decay. Like cork, it can be made completely waterproof by surfacing with an odorless asphalt. The pipe covering fabricated from rock wool and a binder is premolded in single layer sections 36 inches long to fit all standard pipe sizes, and is usually furnished with a factory-applied waterproof jacket.

When pipe lines are insulated with rock cork covering, the fittings are generally insulated with built-up rock wool impregnated with asphalt. This material is generally supplied in felted form, having a nominal thickness of about one inch, and a width of about eighteen inches. It can be readily adapted to any type of fitting and is efficient as an insulator when properly applied.

Fig. 8. Various fittings insulated with cork covering.

Before applying the formed rock-cork insulation (Fig. 9), it is first necessary to thoroughly clean and dry the piping and then paint it with a waterproof asphalt paint. The straight lengths of piping are next covered with the insulation which has the two longitudinal joints and one end joint of each section coated with a plastic cement. The sections are butted tightly together with the longitudinal joints at the top and bottom, and temporarily held in place by staples. That part of the exterior area of each section to be covered by the waterproof lap should be coated by the plastic cement and the lap pressed smoothly into it. The end joints should be sealed with a waterproof fabric embedded in a coat of the plastic cement.

Each section should then be secured permanently in place with three to six loops of copper-plated annealed steel wire.

Fig. 9. Rock-cork pipe insulation.

Wool-Felt Insulation

Wool felt is a relatively inexpensive type of pipe insulation, and is made up of successive layers of waterproof wool felt which are indented in the manufacturing process to form air spaces. The inner layer is a waterproof asphalt-saturated felt, while the outside layer is an integral waterproof jacket. This insulating material is satisfactory when it can be kept air-and-moisture tight. If air is allowed to penetrate, condensation will take place in the wool felt, and it will quickly deteriorate. Thus, it is advisable to use it only where temperatures above 40° F are encountered, and when it is perfectly sealed. Under all conditions, it should carry the manufacturer's guarantee for the duty which it is to perform.

After all the piping is thoroughly cleaned and dried, the sectional covering is usually applied directly to the pipe with the outer layer slipped back and turned so that all joints are staggered. The joints should be sealed with plastic cement, and the flap of the waterproof jacket should be sealed in place with the same material. Staples and copper-clad steel wire should be provided to permanently hold the insulation in place, and then the circular joints should be covered with at least two layers of waterproof tape to which plastic cement is applied.

The pipe fittings should be insulated with at least two layers of hair felt (Fig. 10) built up to the thickness of the pipe covering, but before the felt is placed around the fittings, the exposed ends of the pipe insulation should be coated with plastic cement. After the felt is in place, two layers of waterproof tape and plastic cement should be applied for protection from moisture infiltration.

Insulation of this type is designed for installation in buildings where it is normally protected against outside weather conditions. When outside pipes are to be insulated, one of the better types of pipe covering should be used. In all cases, the manufacturer's recommendations should be followed as to the application.

Fig. 10. Pipe fitting insulated with preshrunk wool felt.

Hair-Felt Insulation

Insulation of this type is usually made from pure cattle hair that has been especially prepared and cleaned. It is a very good insulator against heat, having a low thermal conductivity. Its installed cost is somewhat lower than cork, but it is more difficult to install and seal properly, and hence its use must be considered a hazard with the average type of workmanship. Prior to installation, the piping should be cleaned and dried and then prepared by applying a thickness of waterproof paper or tape wound spirally, over which the hair felt of approximately one-inch thickness is spirally wound for the desired length of pipe. It is then tightly bound with jute twine, wrapped with a sealing tape to make it entirely airtight, and finally painted with waterproof paint. If more than one thickness of hair felt is desired, it should be built up in layers with tar paper between. When it is necessary to make joints around fittings, the termination of the hair felt should be tapered down with sealing tape and the insulation applied to the fittings should overlap this taper, thus insuring a permanently tight fit.

The important point to remember is that this type of insulation must be carefully sealed against any air or moisture infiltration, and even then difficulty may occur after it has been installed. At any point where air infiltration (or breathing, as it is called) is permitted to occur, condensation will start and travel great distances along the pipe, even undermining the insulation that is properly sealed.

There are several other types of pipe insulation available, but not extensively used. These include various types of wrapped and felt insulation, but they seldom can be applied with success. Whatever insulation is used, it should be critically examined to see whether it will provide the protection and permanency required of it, otherwise it should never be considered. While all refrigerant piping, joints, and fittings should be covered, it is not advisable to do so until the system has been thoroughly leak tested and operated for a time.

SUMMARY

In this chapter, a detailed account of pipe selection and installation procedure has been treated. Pressure drop in the various parts of commercial refrigeration systems due to pipe friction, and the proper dimensioning to obtain the best operating results are items which are of importance when installation of equipment is made. These have been fully covered. Tabulated data on pipe sizes for various refrigerants are given in the Appendix.

Multiunit installation of compressors along with piping diagrams, provide information for the installer. Pipe-connection methods, treatment of pipe joints, soldering of joints, and support of pipes are additional features worthy of note. A part of this chapter deals with piping insulation, various types of insulation material, methods of application, etc.

By careful observation of the foregoing detailed description of refrigeration piping and methods of installation, the piping problem will be greatly simplified and will also result in proper system operation.

REVIEW QUESTIONS

1. What are the general precautions to employ in the selection of refrigeration piping?

2. Describe the causes of friction pressure drop in refrigeration lines.

3. What factors should be considered when installing liquid lines?

4. Describe the interconnection of suction lines in a typical refrigeration system.

5. Describe a typical water valve and its operation.

6. What precautions should be observed in a multiple-unit installation?

7. Describe the controls employed when installing multiple-unit compressors.

8. Why are steel pipes used in preference to copper pipes for refrigeration lines?

9. What precaution should be observed when hanging pipe lines?

10. What is the function of piping insulation?

11. Describe wool-felt insulation and its method of application.

CHAPTER 26

Commercial Absorption Systems

Commercial absorption refrigeration, is similar in operation to the domestic absorption refrigeration, covered in Chapter 5. Each manufacturer of this equipment will have a different design and one should consult the manufacturer of the unit for service information, etc.

The theory behind commercial absorption refrigeration can be secured from ASHRAE's Handbook of Fundamentals (1972) from which this material is reprinted by permission.

ABSORPTION REFRIGERATION CYCLES

Absorption refrigeration cycles are heat-operated cycles in which a secondary fluid, the absorbent, is employed to absorb the primary fluid, gaseous refrigerant, which has been vaporized in the evaporator. Such cycles are comparable in some ways to a combination of a heat-engine cycle and a mechanical refrigeration cycle. Simplified diagrams for the two heat-operated methods of obtaining refrigeration are shown in Fig. 1 and Fig. 2, respectively.

As is evident from the figures, a similarity exists between the five main components of the absorption method and five of the components of the combination method.

1. **Generator and Boiler.** *Heat from a high-temperature source is transferred into the generator of the absorption method and into the boiler of the combination method to obtain relatively high pressure vapor. Absorbent is also regenerated in the generator of the former method.*

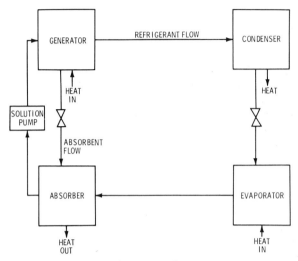

Fig. 1. Basic absorption refrigeration cycle.

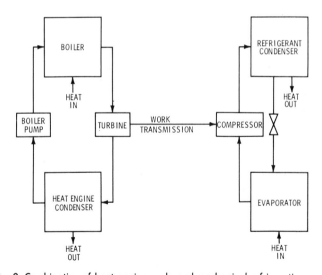

Fig. 2. Combination of heat engine cycle and mechanical refrigeration cycle.

2. **Condenser and Refrigerant Condenser.** *Heat is transferred out of the condenser of the absorption method and out of the refrigerant*

533

condenser of the combination method to an intermediate temperature sink in order to condense refrigerant at relatively high pressure.

3. **Evaporator and Evaporator.** *Heat from a low temperature source is transferred into the evaporator of both methods as a result of vaporization of refrigerant at relatively low pressure.*

4. **Absorber and Heat-Engine Condenser.** *Heat is transferred out of the absorber of the absorption method and out of the heat-engine condenser of the combination method to an intermediate temperature sink to enable the conversion of relatively low pressure vapor to a liquid state. Absorbent is added to the absorber of the former method to assist in this conversion.*

5. **Solution Pump and Feed Pump.** *A small amount of work is put into the solution pump of the absorption method and into the feed pump of the combination method to pressurize liquid.*

The turbine (or other expansion machine) and the compressor of the combination method do not have counterparts in the absorption method.

The relation between work and heat for an ideal heat engine operating in a Carnot cycle is given by the Second Law, as follows:

$$W = q_g \frac{T_h - T_s}{T_h} \tag{56}$$

where

$W = work, Btuh.$
$q_s = heat\ input\ rate,\ Btuh.$
$T_h = temperature\ of\ heat\ source,\ Rankine.$
$T_s = temperature\ of\ heat\ sink,\ Rankine.$

The relation between work required and refrigeration load for an ideal mechanical refrigeration machine operating as a reverse Carnot cycle is as follows:

$$-W = q_e \frac{T_s - T_l}{T_l} \tag{57}$$

where

$-W$ = *work input rate, Btuh.*

q_e = *refrigeration load, Btuh.*

T_l = *temperature of refrigeration load, Rankine.*

T_s = *temperature of heat sink (assume to be same as for heat engine), Rankine.*

The coefficient of performance for the combination of the two ideal cycles is:

$$(COP) = \frac{q_e}{q_g} = \frac{T_l(T_h - T_s)}{T_h(T_s - T_l)} \tag{58}$$

This equation also applies to the ideal absorption refrigeration process. It is plotted as a function of sink temperature with the temperature of the heat source as the parameter for the case of refrigeration load at 40 F in Fig. 3.

There are many inefficiencies in both methods of refrigeration. Although the overall thermodynamic efficiency tends to be higher for the absorption method, this method lacks the versatility of the combination method and, by itself, is less capable of effectively utilizing high temperature energy.

The way in which absorption refrigeration works, characteristics of fluids employed, and thermodynamic and other considerations will be discussed in the remainder of this part of the chapter.

BASIC ABSORPTION REFRIGERATION CYCLE

The distinctive parts of the basic absorption cycle are the absorber and generator. As has been described, there is a certain similarity between these components and those of the heat-engine condenser and boiler. However, there are differences in the internal processes and in the function of the components. The refrigerant condensation and evaporation steps are the same for the two cycles, however, and any difference in these components is related to the particular properties and purity of the refrigerant.

In the basic absorption cycle, low pressure refrigerant vapor is converted to a liquid phase (solution) while still at the low pressure. The

535

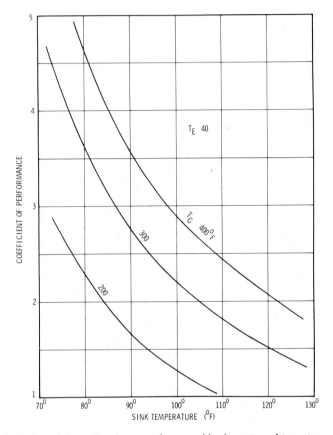

Fig. 3. Ideal coefficients of performance for reversible absorption refrigeration cycles.

conversion is made possible by the vapor being absorbed by a secondary fluid, the absorbent. The absorption proceeds because of the mixing tendency of miscible substances, and generally, because of an affinity between the absorbent molecules and the refrigerant molecules. Thermal energy which is released during the absorption process must be disposed of to a sink. This energy arises from the heat of condensation, sensible heats, and normally, heat of dilution.

The refrigerant-absorbent solution is pressurized in the solution pump. Since the volume is very much less than that of the refrigerant vapor, the work required by the solution pump is small in comparison with the work of compression required in a mechanical refrigeration cycle.

536

The refrigerant-absorbent solution is conveyed to the generator where the refrigerant and the absorbent are separated, i.e., regenerated, by a distillation process. A simple still is adequate for the separation when the pure absorbent material in nonvolatile, as in the water-lithium bromide system. However, fractional distillation equipment is required when the pure absorbent material is volatile, as in the ammonia-water system, because the refrigerant should be essentially free of absorbent. Otherwise, vaporization in the evaporator is hampered. The absorbent which is regenerated normally contains a substantial amount of refrigerant. If the absorbent material tends to become solid, as in the water-lithium bromide system, it is necessary to have enough refrigerant present to keep the pure absorbent material in a dissolved state at all times. Certain practical considerations, particularly the avoiding of excessively high temperatures in the generator, make it generally desirable to leave a moderate amount of refrigerant in the regenerated absorbent. The high temperature energy required for regeneration is approximately equal to the intermediate temperature energy released in absorption.

As shown in Fig. 1, the refrigerant and the absorbent have different circulating patterns. The refrigerant goes from the generator to the condenser, to the evaporator, then to the absorber, to the solution pump and back to the generator. The absorbent short-circuits from the generator to the absorber. The absorbent may be thought of as a carrier fluid which carries spent refrigerant from the low pressure side of the cycle to the high pressure side.

PRACTICAL ABSORPTION REFRIGERATION CYCLES

The principal source of inefficiency in the basic absorption refrigeration cycle is sensible heat effects. The conveying of hot absorbent from the generator into the absorber causes a considerable amount of thermal energy to be wasted. A liquid-to-liquid heat exchanger (generally called the liquid heat exchanger), which transfers energy from this stream to the refrigerant-absorbent solution being pumped back to the generator, saves a major portion of the energy for useful purposes.

The use of this heat exchanger is shown in the flow diagram for a water-lithium bromide cycle, Fig. 4.

The conveying of warm condensate from the condenser into the evaporator similarly causes a reduction in refrigerating effect. Although the loss here is not as great as described previously, it is sometimes desirable to transfer energy from the condensate to the refrigerant vapor

537

leaving the evaporator by means of a heat exchanger, commonly called a **precooler.**

Fig. 4. Lithium Bromide (LiBr) water absorption refrigeration cycle.

There is further need of heat exchange in an ammonia-water machine. Absorbent reaches such a high temperature in the generator that it is desirable to transfer some heat by means of a coil in the analyzer or stripping section. Also heat must be removed in a partial condenser which forms reflux for the system. Means of accomplishing these two purposes, as well as the other changes, are shown in the flow diagram for an ammonia-water machine, Fig. 5.

In some modifications of the ammonia-water cycle, part of the heat of absorption is utilized for preheating the pressurized refrigerant-absorbent solution. The component which makes use of this heat of absorption is known as an **absorber heat exchanger** *or a* **solution cooled absorber.**

538

The heat recovery modifications for the basic cycle which have been described do not bring the coefficient of performance over a threshold level of unity. In other words, the heat required to generate one pound of refrigerant is not less than the heat taken up when this one pound of refrigerant evaporates in the evaporator. A method of breaking through this threshold level is the use of a double-effect generator. It is possible with the water-lithium bromide pair to employ two generators: one, at a high temperature and pressure, which is heated by an external source of thermal energy; and a second, at a lower pressure and temperature, which is heated by heat of condensation of the vapor from the first generator. Condensate from both effects are conveyed to the evaporator. The coefficient of performance is nearly double that of an ordinary cycle.

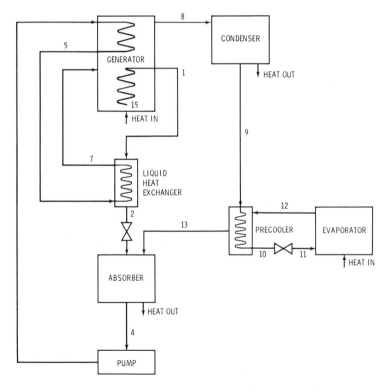

Fig. 5. Ammonia-water absorption refrigeration cycle.

CHARACTERISTICS OF THE REFRIGERANT-ABSORBENT PAIR

The two materials which make up the refrigerant-absorbent pair should meet nearly all of a number of requirements in order to be suitable for absorption refrigeration. Chief among these requirements are:

1. **Absence of Solid Phase.** *The refrigerant-absorbent pair should not form a solid phase over the range of composition and temperature to which it might be subjected. If solid forms it presumably would stop flow and cause a shutdown of the equipment.*

2. **Volatility Ratio.** *The refrigerant should be much more volatile than the absorbent in order that the two can be separated easily in the generator. Otherwise the cost of the generator and the heat requirement for separation become large.*

3. **Affinity.** *It is commonly considered desirable that the absorbent have a strong affinity for the refrigerant under conditions in which absorption takes place. This affinity causes a negative deviation from Raoult's Law and results in an activity coefficient of less than unity for the refrigerant. It reduces the amount of absorbent which has to be circulated and consequently the waste of thermal energy due to sensible heat effects. Also the size of the liquid heat exchanger which transfers heat from the absorbent to pressurized refrigerant-absorbent solution in practical cycles is reduced. Recently Jacob, Albright, and Tucker have made calculations which indicate that strong affinity does have disadvantages. With this affinity is associated a high heat of dilution, and consequently extra heat is required in the generator to separate the refrigerant from the absorbent.*

4. **Pressure.** *It is desirable that operating pressures, largely established by the physical properties of the refrigerant, be moderate. High pressures necessitate the use of heavy-walled equipment. Low pressures (vacuum) necessitate the use of large volume equipment and special means of reducing pressure drop in the flow of refrigerant vapor.*

5. **Stability.** *Almost absolute chemical stability is required because the fluids are subjected to rather severe conditions over many years of service. Undesirable results of instability could be the formation of gas, formation of a solid, or formation of a corrosive substance.*

6. **Corrosion.** *It is especially important that the fluids themselves or any substance resulting from instability do not corrode the materials used in constructing the equipment.*

540

7. **Safety.** *The fluids must be substantially nontoxic and non-flammable if they are to be in an occupied dwelling.*

8. **Viscosity.** *Low viscosity for the fluids is desirable to promote heat and mass transfer, and, to some extent, to reduce pumping problems.*

9. **Latent Heat.** *It is desirable for the latent heat of the refrigerant to be high so that the circulation rate of the absorbent can be kept at a minimum.*

No known refrigerant-absorbent pair meets all the requirements that have been listed. Two pairs, ammonia-water and water-lithium bromide, are considered to come the nearest, and these are the only two which have found extensive commercial use. As has been indicated, the volatility ratio for ammonia-water is smaller than desired. Also ammonia is in ANS B9 Safety Code Group 2, and thus its use inside a dwelling is restricted. The pressure encountered with this pair are somewhat high. Other requirements are quite well met.

The main problem with the water-lithium bromide pair is the possibility of solid formation. Since the refrigerant turns to ice at 32 F, the pair cannot be used for low temperature refrigeration. Lithium bromide crystallizes at moderate concentrations. When the absorber is air cooled, these concentrations tend to be reached, and thus the pair is usually limited to applications in which the absorber is water cooled. There is a possibility that the use of a combination of salts as the absorbent will reduce the crystallizing tendency enough to permit air cooling. Other disadvantages of the water-lithium bromide pair are those associated with low pressure and with the high viscosity of the lithium bromide solution. These latter disadvantages are largely overcome by proper design of equipment. The combination does have the advantages of high safety, high volatility ratio, high affinity, high stability, and high latent heat.

Of the other refrigerant-absorbent pairs which have been investigated, the following types are of some promise:

1. *Ammonia-salts.*
2. *Methylamine-salts.*
3. *Alcohols-salts.*
4. *Ammonia-organic solvents.*
5. *Sulfur dioxide-organic solvents.*
6. *Halogenated hydrocarbons-organic solvents.*

Several of these types appear to be suitable for a relatively simple cycle and may not give as much crystallization problem as the water-lithium bromide pair does. However, stability and corrosion information on most of them is very sketchy. Also the refrigerants, except for fluoro-refrigerants in the last type, are somewhat hazardous.

WATER-LITHIUM BROMIDE MACHINE

The cycle for a water-lithium bromide absorption refrigeration machine is shown in Fig. 4. This cycle includes a liquid heat exchanger for the absorbent and refrigerant-absorbent streams. Some of the limitations and details of the cycle are illustrated in **Example 1.**

Example 1: *Make a First Law analysis of a water-lithium bromide machine operating with the cycle shown in Fig. 4 and having the following requirements:*

1.	*Refrigeration load*	100 tons
2.	*Evaporator temperature*	40 F

(This low temperature is desirable for achieving good dehumidification as well as cooling in air conditioning.)

3. *Absorber outlet temperature*   90 F

(The absorber temperature should be kept at about 100 F or lower to reduce the danger of crystallization.)

4. *Condenser temperature*   110 F

(This temperature is not critical and may be set higher than the absorber temperature to achieve better use of the cooling water.)

5. *Generator temperature*   192 F

(This temperature is related to the condenser temperature in such a way that the absorbent will be in the feasible concentration range.)

Solution: *A number of assumptions are made. Among these is that the two phases are in equilibrium with each other at the points for which temperatures are given. Pressure drops, except at expansion devices, are negligible. The approach at the low temperature end of the liquid heat exchanger is taken to be 10 F deg.*

Table 1 is set up and a start is made at filling in the values. Low side and high side pressures are the water vapor pressures for the evaporator and condenser temperatures, respectively. Enthalpies for water and steam are found in the steam tables.

Table 1.   Conditions in Lithium Bromide Cycle of Example 1

Point	Temp, F	Pressure mm Hg	Weight Fraction LiBr	Flow lb/lb Re- frigerant	Enthalpy Btu/lb
1	192	66	0.61	11.2	−30
2	100	66	0.61	11.2	−70
4	90	6.3	0.56	12.2	−75
7	163	66	0.56	12.2	−38.3
8	192	66	0.0	1.0	1147
9	110	66	0.0	1.0	78
11	40	6.3	0.0	1.0	78
12	40	6.3	0.0	1.0	1079

Relative flow rates are determined from material balances as follows:

$$w_a/w_d = x_b/(x_a - x_b) \qquad (60)$$

where

w_a = *flow rate of absorbent, lb/hr.*

w_d = *flow rate of refrigerant, lb/hr.*

x_a = *concentration of LiBr in absorbent, lb/lb solution.*

x_b = *concentration of LiBr in refrigerant-absorbent solution, lb/lb solution.*

$w_a/w_d = 0.56/(0.61 - 0.56) = 11.2$

$$w_b/w_d = (w_a/w_d) + 1 \qquad (61)$$

where

w_b = *flow rate of refrigerant-absorbent solution, lb/hr*

$w_b/w_d = 11.2 + = 12.2$

The enthalpy of the refrigerant-absorbent solution leaving the liquid heat exchanger is calculated from an energy balance as follows:

$$\begin{aligned} h_7 &= h_4 + [(h_1 - h_2) \times w_a/w_b] \\ &= -75 + [(-30 - (-70)) \times 11.2/12.2] = -38.3 \qquad (62) \end{aligned}$$

The temperature corresponding to this enthalpy is found from Fig. 49 of Chapter 31, ASHRAE Handbook of Fundamentals.

The refrigerant flow rate is calculated from an energy balance as follows:

$$w_d = q_e/(h_{12} - h_{11})$$
$$= 1,200,000/(1079 - 78) = 1200 \text{ lb/hr} \qquad (63)$$

The absorbent and refrigerant-absorbent solution rates are next calculated as follows:

$$w_a = 11.2w_d = 11.2 \times 1200 = 13,400 \text{ lb/hr}$$
$$w_b = 12.2w_d = 12.2 \times 1200 = 14,600 \text{ lb/hr}$$

The net heat input to the generator is calculated from an energy balance as follows:

$$q_g = w_d h_8 + w_a h_1 - w_b h_7$$
$$= 1200 \times 1147 + 13,400(-30) - 14,600(-38.3)$$
$$= 1,540,000 \text{ Btuh} \qquad (64)$$

The coefficient of performance on a net basis is:

$$(\text{COP}) = q_e/q_g$$
$$= 1,200,000/1,540,000 = 0.780 \qquad (65)$$

Heat transfer rates for the other components are:

Liquid heat exchanger

$$q_l = w_a(h_1 - h_2)$$
$$= 13,400(-30 - (-70)) = 540,000 \text{ Btuh} \qquad (66)$$

Condenser

$$-q_c = w_d(h_8 - h_9)$$
$$= 1200(1147 - 78) = 1,280,000 \text{ Btuh} \qquad (67)$$

544

Absorber

$$-q_a = q_g + q_e + q_c$$
$$= 1,540,000 + 1,200,000 - 1,280,000$$
$$= 1,460,000 \text{ Btuh} \qquad (68)$$

In actual practice, somewhat lower concentrations of lithium bromide solutions than those of **Example 1** *are commonly used in small commercial units to insure against crystallization of the salt, particularly on shutdown. These more typically are 54 percent and 58.5 percent. Higher concentrations, normally about 60 percent and 64.5 percent, are used in large commercial units in order to operate at higher absorber temperatures and thus save on heat exchanger costs. Controls and a shutdown dilution cycle are employed with these large units to prevent crystallization.*

AMMONIA-WATER CYCLE

As has already been indicated, the ammonia-water cycle must be more complex than the water-lithium bromide cycle to provide acceptable performance. More heat recovery means are required, and the simple still must be replaced by a complete distillation system. This system will be referred to as the generator, since it is housed in one component. Because of the complexity of the cycle, the design for optimum performance, based on a given set of design parameters, requires extensive calculations.

Characteristics and calculating methods for an ammonia-water machine are demonstrated in Example 2.

Example 2: *Make a First Law analysis of an ammonia-water machine operating with the cycle on which the following test data are available:*

Refrigeration load	36,000 Btuh
Precooler liquid	80 F
Evaporator inlet temperature	38 F
Evaporator outlet temperature	50 F
Absorber outlet temperature	130 F
Condenser outlet temperature	123 F
Refrigerant-absorbent solution temperature leaving reflux condenser	168 F

Generator bottom temperature	364 F
Absorbent leaving generator temperature	270 F
Absorbent leaving heat exchanger temperature	170 F
Vapor leaving reflux condenser temperature	172 F
Condenser pressure	300 psia
A b s o r b e r pressure	70 psia

Solution: *It will be assumed that phases are in equilibrium with each other at the evaporator inlet, absorber outlet, bottom of generator, vapor outlet from reflux condenser, and condenser outlet. Pressure drops, except in expansion devices, are negligible.*

Table 2 is set up, and values are filled in this table as they are determined. Enthalpies are likewise determined in either of these sources for the points for which sufficient physical data are now known.

The enthalpy at Point 12 is the ordinate at which the 50 F, 70 psia, tie line crosses the abscissa representing the refrigerant composition (0.991).

The enthalpy at Point 13 is calculated from an energy balance around the precooler.

$$h_{13} = h_{12} + H_g - h_{10}$$
$$= 527 + 102 - 51 = 578$$

Table 2. Conditions in Cycle of Example 2

Point	Temp, F	Pressure psia	Weight Fraction NH_3	Flow lb/lb Refrigerant	Enthalpy Btu/lb
1	270	300	0.10	1.97	215
2	170	300	0.10	1.97	110
4	130	70	0.40	2.97	0
5	168	300	0.40	2.97	43
7	230	300	0.40	2.97	113
8	172	300	0.991	1.00	599
9	123	300	0.991	1.00	102
10	80	300	0.991	1.00	51
11	38	70	0.991	1.00	51
12	50	70	0.991	1.00	527
13	92	70	0.991	1.00	578
15	364	300	0.10	1.97	315

The temperature at Point 13 is determined by a tie-line procedure which is the reverse of that used to find the enthalpy at Point 12.

Relative flow rates are determined from material balances as follows:

$$w_a/w_d = (x_d - x_b)/(x_b - x_a) \tag{70}$$

where

x_d = concentration of ammonia in refrigerant, lb/lb.
x_b = concentration of ammonia in refrigerant-absorbent solution, lb/lb.
x_a = concentration of ammonia in absorbent, lb/.
$w_a/w_d = (0.991 - 0.400)(0.400 - 0.100) = 1.97$
$w_b/w_d = w_a/w_d + 1 = 1.97 + 1 = 2.97$

The enthalpy at Point 7 is calculated from an energy balance around the liquid heat exchanger.

$$\begin{aligned} h_7 &= h_5 + (h_1 - h_2)(w_a/w_b) \\ &= 43 + (215 - 110)(1.97/2.97) = 113 \end{aligned} \tag{71}$$

The refrigerant flow rate is calculated from an energy balance as follows:

$$\begin{aligned} w_d &= q_c/(h_{18} - h_{11}) \\ &= 36,000/(527 - 51) = 75.5 \text{ lb/hr} \end{aligned} \tag{72}$$

The absorbent and refrigerant-absorbent solution flow rates are next calculated as follows:

$$w_a = 1.97 w_d = 1.97 \times 75.5 = 149 \text{ lb/hr}$$
$$w_b = 2.97 w_d = 2.97 \times 75.5 = 224.5 \text{ lb/hr}$$

The net heat input to the generator may now be calculated from an energy balance around the generator reflux condenser and liquid heat exchanger as follows:

$$\begin{aligned} q_s &= w_a h_s + w_a h_2 - w_b h_4 \\ &= 75.5 \times 599 + 149 \times 110 - 224.5 \times (0) \\ &= 61,600 \text{ Btuh} \end{aligned} \tag{73}$$

547

The coefficient of performance on a net basis is

$$(COP) = q_c / q_g$$
$$= 36,000 / 61,600 = 0.584 \qquad (74)$$

Heat transfer rates for other parts of the cycle are:

Liquid Heat Exchanger

$$q_l = w_d(h_1 - h_2) = 149(215 - 110)$$
$$= 15,600 \text{ Btuh} \qquad (75)$$

Condenser

$$-q_c = w_d(h_s - h_g)$$
$$= 75.5(599 - 102)$$
$$= 37,500 \text{ Btuh} \qquad (76)$$

Reflux Condenser

$$q_r = w_b(h_5 - h_4)$$
$$= 224.5(43 - 0) = 9,700 \text{ Btuh} \qquad (77)$$

Absorber

$$-q_a = q_g + q_e + q_c$$
$$= 61,600 + 36,000 - 37,500$$
$$= 60,100 \text{ Btuh} \qquad (78)$$

Precooler

$$q_p = w_d(h_9 - h_{10})$$
$$= 75.5(102 - 51) = 3,850 \text{ Btuh} \qquad (79)$$

CHAPTER 27

Circulating Pumps

Circulating pumps are an important part of commercial and industrial refrigeration systems where brine is used, because of the expensive equipment involved to refrigerate or freeze, with the commonly used refrigerant. The refrigerant is contained in a comparatively small area and is used to cool the brine with the refrigerant coils, to a point where the brine will take over for the further cooling or freezing. A loss of the brine is economically inexpensive compared to the loss of the refrigerant. Pumps for this type of service may be classified according to service requirements and uses as:

1. Rotary.
2. Centrifugal.
3. Reciprocating.

ROTARY PUMPS

Rotary pumps are usually employed for pressure lubrication service on ammonia compressors. They deliver a continuous liquid flow like the centrifugal type, but operate at much slower speeds. Rotary pumps should be operated within the assigned speed limits to avoid rapid wear and depreciation.

Rotary pumps, sometimes called *gear pumps* or *positive-displacement pumps,* depend for their operation on the principle that a rotating plunger, impeller, gear, or cam traps the liquid in the suction side of the pump casing and forces it to the discharge side. In the simple gear pump, there are two cams or gears meshed together and revolving in opposite directions, as shown in Fig. 1. The case surrounding these cams forms a close fit on the ends of the teeth and also on the sides.

ADMISSION   DISCHARGE

Fig. 1. The working principle of a rotary gear pump.

In operation, each pair of meshing teeth separate, a space forms with vacuum, and atmospheric pressure forces in the liquid to fill these spaces. Any liquid which fills the space between any two adjacent teeth must follow along as they revolve and be forced out of the discharge opening, since the meshing of the teeth during rotation forms a seal separating the admission and discharge parts of the secondary chamber. Because of the number of teeth on the cams, this sequence of operation is practically continuous, some spaces carrying liquid to the discharge pipe and some spaces filling from the suction pipe. The action is positive. Each revolution transfers a quantity of liquid from suction to discharge because of the size and shape of the cam teeth. There are no valves in a rotary pump, and it is therefore better adapted to handling viscous liquid, such as lubricating oils, which would cause large frictional losses in passing through the restricted openings in valves of a reciprocating pump.

Another type of rotary pump is known as the *helical screw gear type*. This pump, shown in Fig. 2, consists of three rotors—that is, one power rotor and two idle rotors. The convex surfaces of the power rotor mesh with the concave surfaces on the sealing rotors in such a way that a nearly fluid-tight closure between the rotors is obtained. In these pumps, the standard direction of rotation is clockwise when the observer is standing at the driver and looking toward the coupling end of the pump.

550

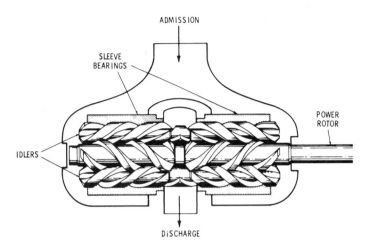

Fig. 2. The rotor arrangement of a helical-screw gear pump.

The direction of rotation varies according to type of rotary pump. For instance, in the double-helical type, the direction of rotation is as follows: The standard direction of rotation is counterclockwise when facing the shaft extension end. This is indicated by an arrow on the pump body. The rotation of internal roller-bearing pumps may be reversed by removing the outside bearing cover and stuffing box and transferring the small plug in the side-plate casting to the opposite side. These plugs (one in each side plate) should be on the discharge side to induce circulation through the bearings to the inlet, and to maintain inlet pressure on the stuffing box and ends of the drive shafts. Another pump, working on the internal-gear principle, makes note of the following directions:

1. In determining the desired rotation, the observer stands at the shaft end of the pump.
2. Note that the balancing groove in the shoe must be on the inlet side.
3. If a change in the direction of rotation is desired, it is necessary only to remove the cover, slip out both the top and bottom shoes, turn them end for end so that the grooves will be on the new inlet side, and re-assemble the pump.

Another model of the same make is listed as an automatic reversing pump. It has a uni-directional flow regardless of the direction of shaft rotation and without the use of check valves. Finally, for a rotary pump of the helical gear type, the instructions for determining the direction of rotation are as follows: To determine the direction of rotation, stand at the driving end facing the pump. If the shaft revolves from left to right, the rotation is clockwise. If the shaft revolves from right to left, the rotation is counterclockwise. The diagram in Fig. 3 is fundamental of the flow of liquid in this pump. Changing the direction of rotation of the pump drive shaft reverses the direction of the flow of liquid, causing the position of the inlet and discharge openings to be reversed. If the direction of rotation is not specified, the pump will be furnished for clockwise rotation.

The standard rotation of motors is in a counterclockwise direction. The direction of motor rotation is determined from a position at the end of the motor which couples on to the pump.

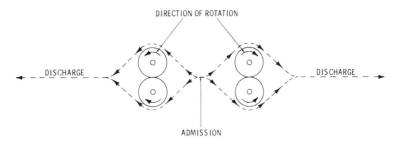

Fig. 3. Diagram of the liquid flow in a rotary gear pump.

CENTRIFUGAL PUMPS

A centrifugal pump (Fig. 4) is a pump in which the pressure is developed principally by the action of centrifugal force. Such pumps are commonly used where large quantities of water or brine are to be delivered at moderate pressures.

The rotating member of a centrifugal pump gives a rapid rotary motion to a mass of water contained in a surrounding case; the centrifugal force forces the water out of the case through the discharge outlet. The vacuum thus created makes available atmospheric pressure to force in more water through the center. The process continues as long as motion is given to the rotor and there is a supply of water to draw upon. From this, a centrifugal pump may be defined as one in which vanes or impellers, rotating inside a close-fitting casing, draw in liquid at the center and, by centrifugal force, throw it out through an opening at the periphery of the casing.

Fig. 4. The drive arrangement for a typical centrifugal pump, with the motor and pump mounted on a large sub-base.

Centrifugal pumps operate at high speeds and may therefore be connected directly to electric motors. Where noiseless operation is desired and little space is available, the centrifugal pump is usually the most logical choice.

After suction has been established by priming, the suction lift with a centrifugal pump is as high as with a reciprocating pump. It is absolutely necessary, however, to prime centrifugal pumps, which can be accomplished by submerging the unit, or by equipping the inlet pipe with a foot or check valve, so the suction pipe and casing may be filled with water or brine.

Single-stage centrifugal pumps are usually used for up to 100 feet total head. Above 100 feet, multiple-stage or turbine pumps are used. Multiple-stage or turbine pumps may also be used with advantage for heads less than 100 feet, particularly two-stage pumps, as they operate at about half the speed of the single-stage pump for the same capacity.

The design of impellers in centrifugal pumps varies greatly to meet the great variety of service conditions. The correct impeller for a given installation is of prime importance in order to secure economical and satisfactory operation. A high degree of efficiency can be obtained with open impellers, under certain conditions, by carefully proportioning the curvature of the blades and by reducing the side clearances to a minimum with accurate machining of the impeller edges and side plates. In general, however, more efficient results can be obtained by the use of enclosed or shrouded impellers. Fig. 5 shows both an open and an enclosed impeller.

553

PLATE ON ONE SIDE   BOTH SIDES SHROUDED

Fig. 5. Semi-open and enclosed impellers.

Most centrifugal pumps are fitted with ball bearings. A typical construction consists of a single-row deep-groove ball bearing of ample size to withstand all axial and radial loads. The bearing housings are of the cartridge type in order that the entire rotating element may be removed from the pump without disturbing the alignment or exposing the bearings to water or dirt. The bearing housings are positioned by means of dowel pins in the lower half of the casing and are securely clamped by covers split on the same horizontal plane as the pump casing. In this construction, the entire bearing may be removed from the shaft without damage by using the sleeve nut as a puller.

RECIPROCATING PUMPS

This type of pump may be gear driven by electric motor or other prime mover. They may be single-acting, single-double-acting, duplex-double-acting, triple-single-acting, and triple-double-acting. The first two types should not be geared direct to an electric motor.

Steam pumps should not be considered for continuous duty because they are extremely wasteful, and should be considered only where exhaust steam is required for heating purposes, and where they are required intermittently for emergency purposes. Reciprocating pumps are generally employed where the amount of liquid (water or brine) is moderate and where the delivery pressure is high. They should be connected to secure a full and uniform supply and where the suction lift seldom exceeds 22 feet. The less the suction lift, the better it will be for economy. To avoid air pockets, the suction line should have a uniform upward grade toward the pump of at least 6 inches in one hundred feet. Every precaution should be taken to prevent foreign matter from entering the pump; if necessary, a strainer should be provided. Long-radius elbows and bends should be used on both suction and discharge lines, with valves provided near the pump.

POWER REQUIREMENTS

The following suggestions will be found useful for the selection of the proper motive power.

For reciprocating pumps, the motor should have approximately 5% more power than that actually required. Manufacturers of these pumps give the best speed of the motor to be geared to them, and state the pump rpm in the event of a belt drive. For centrifugal and rotary pumps, the motor should have approximately 10% more power than actually required.

Shunt-wound DC motors are satisfactory for ordinary service, but a compound-wound motor should be furnished for single-stage pumps having a fluctuating load, and for pumps frequently stopped and started under load. A compound-wound motor is preferred in all cases where there is a high starting torque, and where a practically constant speed is desired. This is also one reason why compound motors are also preferable for centrifugal pumps.

For centrifugal pumps, either hand-started or automotic-control, two-phase or three-phase induction motors of the squirrel-cage or internal-resistance type are satisfactory. For single-phase current, any of the ordinary types of motors capable of starting the pump under load may be used.

PUMP MAINTENANCE

In general refrigeration practice, motor-driven centrifugal pumps are most generally used for handling water and brine in cold storage plants. Regardless of the type used, all will render satisfactory service if properly installed and maintained, assuming the pump has been properly selected and that it fits the actual operating requirements for any given situation. It is necessary, however, to make an occasional and thorough inspection of pumps of all types in order that defects or signs of trouble may be detected and corrected before either partial or complete breakdown of the equipment occurs.

As a guide for diagnosis of pump trouble, as well as for methods and a schedule of checks and tests, the following pointers will be of assistance:

No Liquid Delivered

1. Pump not primed.
2. Suction lift too high. Check with a vacuum gauge at the pump suction.
3. Rotating in wrong direction.

Not Enough Liquid Delivered

1. Air leaks in suction line or through stuffing box.
2. Speed too slow.
3. Suction lift too high.
4. Too much suction lift (for hot liquids).
5. Foot valve or end of suction pipe not immersed deeply enough.
6. Foot valve too small or obstructed.
7. Piping improperly installed, permitting air or gas to pocket in pump.
8. Mechanical defects:
    a. Pump damaged.
    b. Pump badly worn.
    c. Packing defective.

Intermittent Delivery

1. Leaky suction lines.
2. Suction lift too high.
3. Air or grease in liquid.
4. Pump scored by sand or by other abrasive in liquid.

Pump Takes Too Much Power

1. Speed too high.
2. Liquid heavier or more viscous than water.
3. Suction or discharge line obstructed.
4. Mechanical defects:
    a. Shaft bent.
    b. Rotating elements bind.
    c. Stuffing boxes too tight.

d. Misalignment due to improper connections to pipe lines, or improper installation of foundation causing spring in base.

5. Misalignment of coupling (direct-connected units).

Periodic Checks and Tests on Rotary Pumps

1. Lubricate the speed-limiting, speed-regulating, and over-speed-trip governors at least once a week.
2. Run the pump under power.
3. Trip the emergency overspeed trip by hand.
4. Check the speed-limiting or speed-regulating governor to see that it maintains unit at proper speed.
5. Lift all relief valves by hand.
6. Check operation on discharge check valves.
7. Check lubricating oil for condition and presence of water.
8. Check thrust bearings, position of pump rotors, and turbine-blade clearance.

Periodic Checks and Tests on Centrifugal And Reciprocating Pumps

1. Inspect pump and motor shaft for alignment.
2. Examine the pump shaft for cuts and ridges on the section coming in contact with the stuffing-box packing.
3. Check the impeller for wear.
4. Check pump and motor bearings for wear.
5. Determine whether or not the assembled unit is level.
6. Check pump speed and compare with that shown on pump nameplate.
7. Check suction lift to determine if total lift, including friction loss in suction pipe, is excessive. Total suction lift including friction loss should not exceed 15 feet.
8. Check suction line for possible air leakage.
9. If foot valve is used, check the maximum free opening of same to determine whether or not it is at least equal to the area of the suction pipe.
10. Check pressure gauge on pump discharge. If pressure gauge is located on top of pump casing its readings are worthless. It should be connected to pump discharge near discharge nozzle.
11. Check suction line to determine if same is correctly sloped so as not to contain air pockets.

SUMMARY

Because pumps are an integral part of any commercial refrigeration system, a brief description of their classification, construction, and operating principles has been provided. Direction of rotation, drive methods, and power requirements are other factors fully dealt with in this chapter.

Since the selection of pumps is of paramount importance, the manufacturer's recommendation should be procured and followed, but only after full details as to required capacity, speed, and all other operating requirements have been furnished. The maintenance of pumps, including tabulated data on troubleshooting, have been given in considerable detail in order to assist the maintenance man or operating engineer in the correction of any operating difficulties which may occur.

REVIEW QUESTIONS

1. Describe the function of circulating pumps in a refrigeration system.

2. How are circulating pumps classified?

3. Describe the operation of a rotary pump.

4. What is the direction of rotation in a rotary pump?

5. Describe the standard rotation of an electric motor.

6. How does a centrifugal pump differ from that of a rotary pump?

7. What is the maximum total head for a single-stage centrifugal pump?

8. Describe a reciprocating pump and its use in a refrigeration system.

9. Give the power requirements for various types of pumps.

10. Where are motor-driven centrifugal pumps used?

CHAPTER 28

Installation and Operation

A good installation is the best insurance for trouble-free operation. The refrigerating machinery should be located in a clean, dry, well lighted, and well ventilated area. Cleanliness and the absence of dampness will ensure long life to the motors and belts and will reduce the necessity of frequent painting of the exposed piping to keep down corrosion. Good ventilation will make for greater safety of the operator during maintenance operation. In the case of air-cooled units, an adequate supply of outside air must be provided and a means of exhausting the air from the room to avoid recirculation.

When an installation is composed of two or more compressors, it is desirable to install the compressors on the same or parallel center lines, both for appearance and so that the connections can be properly made. It should, however, be noted that in each case the compressors must be set to comply with the piping layout.

Leave enough room around the compressor and motor so that they can be easily serviced. Allow enough room at the oil-pump end of the compressor (approximately equal to the length of the compressor) so that the crankshaft may be removed if necessary. In locating the condenser, allow room equal to the length of the condenser for cleaning the tubes unless built-up cleaning rods are planned to be used. Be sure to protect water-cooled condensers, water lines, and other accessories from freezing damage during winter shutdown.

FOUNDATIONS

Foundations should be installed in accordance with the blueprints furnished with each machine. In the case of excavations running into quicksand, or into very marshy or soft ground, it will be necessary to make a wide footing under the foundation. In case of solid rock, a lesser foundation may be built. Foundations are preferably built of concrete in the proportion of one measure of Portland cement to three measures of sand and five measures of screened crushed stone. Allow from 36 to 48 hours to set before the machine is put in place. Be sure the sand used in making the concrete is clean and has absolutely no soil mixed with it.

Foundations which are isolated so as to absorb vibration must be carefully engineered to suit the particular job. Simply purchasing some kind of springy supports will not generally prove satisfactory. When installing a refrigerating machine on the upper floor of a building, remember that, while the structure may be amply strong to support the weight, it may by chance have a frequency of vibration in some of its members which will cause them to vibrate in resonance with any equipment in rhythmic motion. This may call for some form of stiffening, or for a foundation adapted to dampening the machine vibration until it is of a much lower intensity. A concrete base or a spread mat properly isolated will usually suffice. What is usually called noise is often really structural vibration.

If the compressor is to be located on a floor of medium or light construction, it should be supported directly over the joists or beams under the floor. When floors are weak or inadequate, supplementary floor supports must be provided. These supports should not contact the weaker floor between supports as this would tend to transmit vibrations to a larger surrounding area.

A concrete foundation used for a machine should be heavy enough to absorb operating vibrations. As a rule, the concrete slab should weigh from one to two times the total weight of the unit it supports. To insure proper isolation, locate the concrete foundation not less than 6 feet from the footings of the building walls or columns. It is recommended that vibration isolators be used in mounting all units having steel bases when these units are not mounted on concrete foundations poured in the ground. In addition to providing more quiet operation, vibration isolators will make allowance for any slight irregularities in mounting which otherwise might result in distortion of the base and drive.

Anchor Bolts and Templates

Anchor bolts and templates are usually furnished by the compressor manufacturer. Prior to pouring the concrete foundation, the template should be braced, thus preventing its movement during the pouring operation. Nail the template to the form, being sure that it is in line and square with the foundation of the driving unit. The anchor bolts should be placed through the holes in the template, as shown in Fig. 1. Place

Fig. 1. Template arrangement to provide exact placement of anchor bolts in a belt-driven compressor installation.

ferrules made of tin spouting, pipe, or very thin lumber, around the anchor bolts. These ferrules should have an inside diameter at least three times the diameter of the anchor bolt. This space is essential for moving the machine at the time of leveling and lining up before grouting. The length of the ferrules and setting of the bolt can be determined by referring to Fig. 2, and by following these directions: The distance from the top of the foundation to the top of the bolt should be equal to the thickness of the motor or compressor base casting plus the thickness of the nut and washer, plus one inch for grouting. A spacing plug should be put at the top of each ferrule to keep it concentric about the bolt. When the motor and compressor are mounted on a steel base, it is advisable to set this steel base on a solid concrete foundation, and to level and grout it to give a firm footing. Sound isolation can be used where the noise of the compressor may be objectionable.

— WOOD CASING

— TO BE FILLED WITH CEMENT WHEN MACHINE IS GROUTED

— ANCHOR PLATE

— ANCHOR NUT

— PIPE FERRULE TO SUPPORT NUT AND TO ALLOW BOLT ADJUSTMENT

— BRICK, STONE, OR HARD WOOD

Fig. 2. Anchor-bolt installation details.

Grouting

Build a frame around the top of the foundation that will hold at least 2 inches of grout. Mix two parts of clean sharp sand with one part of Portland cement, and add enough water to make a mixture that will flow freely. Wet the foundation top thoroughly and pour the grout in from all sides, working it well with a thin strip of metal to insure a solid grout over the entire base and in all the foundation bolt holes. Continue pouring and stirring until the grout rises to the top of the dam. If any low spots form after the dam is full, these should be promptly filled.

When the grout has hardened sufficiently to be self-supporting, remove the dam and trowel the grout down even with the bottom of the machine base and give it a slight pitch to the outside so that any water or oil will drain off the foundation. To make a nice smooth job, trowel a thin layer of cement on the foundation.

After the grout is reasonably hard (which usually requires four or five days), draw the foundation bolt nuts uniformly all around until rea-

562

sonably tight. Use care not to spring the base casting as this will throw the shaft out of line and cause heating of the bearings, and may even cause the crankshaft to break at some later date.

INSTALLING BELT-DRIVEN COMPRESSORS

When installing belt-driven compressors, it is necessary that the erection engineer check the foundations to see that they are built in accordance with certified blueprints and that the foundation tops are sufficiently rough to insure proper grouting. The anchor bolts should be correctly placed, preferably by using a template such as shown in Fig. 1. Set the lower half of the belt wheel in place with the hub below the keyway on the shaft, then carefully place the compressor and motor, and lower them to within one inch from the tops of their respective foundations. Level the compressor by using an accurate level on the crankshaft and on the cross-leveling pads, and level by adjusting the leveling screws. Leveling screws and iron plates are usually furnished with each compressor. If the compressor is not equipped with cross-leveling pads, it will be necessary to remove the top heads and cross level from the machined surface thus exposed. When the compressor is level, tighten the nuts on the foundation bolts slightly to prevent the compressor from shifting.

Set the key seat on the shaft, lower the top half of the belt wheel over the key, then raise the inner half and bolt the two halves of the wheel together. (Heat the bolts for this purpose.) Check to make sure the wheel is in the proper location as called for on the plans. Check the compressor to see that it has remained level, and place a jack under the shaft to relieve it of the weight of the wheel. Mount the motor pulley on its shaft with the projecting end of the hub toward the motor, and level and align it so that its belt grooves line up with the corresponding grooves in the flywheel. To do this, hold a long straightedge against the flywheel so that it touches the extreme edges of the flywheel and is in line with the centers of both the compressor and motor shafts. Make sure that the face of the motor pulley is parallel with the straightedge, and that the center of the first groove on both the motor pulley and the flywheel are an equal distance from the straightedge. When the belt grooves are in line, tighten the set screws which hold the motor pulley key in place. The compressor and motor can now be grouted. When the grout has set properly, the compressor leveling screws can be removed if desired. To apply the V-belts, slide the motor forward sufficiently to place them in the grooves without stretching.When

563

all belts are in their grooves, they should be tightened sufficiently to prevent slipping. Do not overtighten, or wear will be excessive. These belts
require no lubrication or dressing.

INSTALLING DIRECT-CONNECTED
MOTOR-DRIVEN COMPRESSORS

With this type of unit, the compressor should be placed on the foundation, leveled, and grouted. However, in this case, it is necessary to allow
the grout to set before the motor is attached to the shaft. After the grout
has set, place the stator frame and base plates in position over the anchor
bolts, with the base-plate shims in place. The rotor is usually furnished
with a split hub so that it can be easily slipped over the end of the shaft.
Sometimes, however, it is a very tight fit, in which case it will be
necessary to apply a light coating of oil and white lead to the shaft and
force the rotor in place with a jack. In the installation of a large compressor, where the rotor is to be mounted on the shaft between the outboard
bearing and the compressor, care must be used not to score the shaft.
After the rotor has been keyed to the shaft, the stator should be moved
over on the base plates until the stator laminations coincide with the
field-pole laminations when the crankshaft is moved out so that the faces
of the bearings take the thrust.

The outboard bearing should be leveled and aligned and the sole plate
grouted before the motor is mounted on the shaft. To do this, remove the
liners from between the pedestal and cap and pull the cap solid against the
shaft. When the pedestal cap is drawn down tight on the shaft, the bottom
of the box will be level providing the shaft is level. To level the pedestal
the opposite way, place a level on the machined space on the sole plate
and wedge underneath the sole plate until the pedestal is level. Also make
sure the crankshaft has remained level. The bearing-cap liners must be
replaced before the compressor is put into operation in order to give the
bearing proper running clearance.

The air gap between the stator and rotor must be adjusted so that it is
equal all around. To do this, first place the stator so that the gap is equal at
both sides. Then wedge up or down on the base plates as required. During
the process of checking the air gap, it is well to turn the rotor to several
different positions and check the gap all around for each position. The air
gap can be checked by means of a thin iron or wooden wedge, or, if
desired, a gap gauge can be obtained from the motor manufacturer. After

the motor is thus aligned, and with the stator shims still in place, bolt the stator, pedestals, and base plates securely. The base plates can then be grouted.

LEAK TESTING

When a refrigerant piping system is completed, the system should be checked very carefully for leaks. The most effective way of finding a leak in a *Freon* system is with a Halide leak detector (Fig. 6). Testing with soap and water, or with oil at the joints, will only locate the larger leaks and therefore is unsatisfactory in determining the relative tightness of a system.

Fig. 3. Lightweight portable electronic leak detector.

When testing a new system for leaks, disconnect the lines from the water valve and pressurestat and cap the connections to the system. This is done to protect the bellows from injury by the high test pressure. Charge the system with a small amount of refrigerant (a couple of pounds for systems over 10 tons and less for smaller jobs). Replace the refrigerant drum with a cylinder of nitrogen and build up the pressure to that required by local codes. *Do not use oxygen to build up pressure in the system. Serious explosions have resulted from using pure oxygen and oil under pressure.* If no local codes are in effect, a pressure of 100 pounds per square inch should be sufficient for locating any leaks. When a test pressure in excess of 150 pounds per square inch is required, the compressor should be isolated to prevent damage to the seal. Disconnect the cylinder after the maximum test pressure has been reached. Test each joint and connection for leaks. The small amount of refrigerant will act as an indicator. Relieve the pressure and repair any leaks. Reconnect the lines to the water valve and pressurestat and proceed to dehydrate the system.

DEHYDRATION AND EVACUATION

A dehydrator, as its name implies, is a device that removes moisture from the refrigerant in a system. Usually, it consists of a shell filled with a chemical moisture-absorbing agent. Its permanent installation in any field-connected installation is extremely desirable for both economy and proper operation. Its first cost is negligible compared to the service parts and labor it will save, and also the expansion valve operation is improved by the elimination of moisture.

In systems which are field connected, there is always a possibility of taking in moisture. This is especially true where connections are hastily or carelessly made. Even carefully made connections may be loosened by continued vibration or possible abuse after installation. Obviously, these leaks should be repaired as soon as they are detected. Before detection, however, considerable moisture may be absorbed by the system, especially on low-temperature installations. Damage from such sources may be avoided if a dehydrator (Fig. 7) has been installed as a permanent fixture.

Fig. 4. A typical refrigeration dehydrator filled with moisture-absorbent silica gel.

Temporary dehydrator installation as a service procedure to overcome moisture is, of course, necessary, but two things must be kept in mind. First, the condition probably would not have developed had a permanent dehydrator been installed; and second, when the temporary dehydrator is removed, the system is necessarily opened up and moisture is given a chance to re-enter and nullify all the good which had been accomplished. These points further emphasize the advisability of installing permanent dehydrators at the time of the original installation.

The dehydrating agent used is only capable of absorbing a certain amount of moisture. Should a greater quantity of moisture get into the system, the dehydrator will become saturated and inactive. It then must be replaced or refilled. Should dehydrators be opened or even not tightly sealed in the stock room, they will absorb moisture from the air, becoming saturated and inactive if put on an installation. Therefore, be sure the dehydrating agent is still active in any dehydrator which is to be installed.

It is extremely important that all refrigeration systems be absolutely free of moisture. Moisture in the system can cause sludging of the oil in the crankcase, resulting in corrosion and lack of proper lubrication. This contamination can be explained by two entirely different types of oil decomposition. The first type—reaction of the oil with oxygen in the noncondensables—usually is the first to take place. Even though the system is dry, oil breakdown products due to reaction of the oil with oxygen will start building up on the valve plate and particularly on the seat of the discharge valve at the high head temperatures with noncondensables present. In time, this can lead to wire drawing (forcing of the gas through a small leak on the discharge valve, creating very high temperatures—sometimes as high as 800°-900° F) and initiate the second type of oil breakdown. The second type involves the oil reacting directly with the refrigerant to form hydrochloric acid, hydrofluoric acid, carbon, and water. This becomes a vicious cycle, snowballing until the system has a serious wet acidic condition, oil deterioration and, with hermetic compressors, an eventual motor burnout.

If the system is properly evacuated by using a good vacuum pump which can pull down into the 50- to 100-micron range, the noncondensables are eliminated before the system is ever started. A good vacuum pump with electronic gauge is desirable and highly recommended for this procedure. Actually, 25 to 28 inches of vacuum is simply not enough to accomplish any great amount of dehydration or noncondensable removal. The gauge may be showing 50 microns at the pump, but if a long run is being pulled, it may be a completely different story at the other end of the line. The poorer the vacuum, the worse the situation becomes. Many times the situation is further complicated by using too small a connecting line between the vacuum pump and the system. This creates a high-pressure drop which greatly reduces the efficiency of the vacuum pump.

CHARGING THE SYSTEM

After the system has been completely evacuated and dehydrated (and with the compressor stopped), open the discharge and suction valves and connect the refrigerant drum to the charging valve on the system. Pay strict attention to the instruction tag attached to the drum by the manufacturer. When the connections to the drum have been made, open the valve on the drum slowly and make sure that there are no leaks between the drum and the charging valve. If no leaks are present, then open the drum valve two or three turns. Open the charging valve on the system slightly and allow the refrigerant to enter the system which has previously been pulled down to a vacuum while the evacuation was building up pressure throughout the system. Care should be exercised while charging and the drum valve should be closed immediately if any leakage is discovered in the system.

Open the main discharge valve on the manifold and turn on the water to the condenser and compressor jackets, making sure that both are getting an ample supply. If the low side is a brine or water cooler, be sure the fluid is flowing so that freezing does not occur during charging. Start the compressor and slowly open the main suction valve on the manifold. Close the valve on the liquid line from the receiver to the evaporator and make sure that the valve in the liquid line from the condenser to the receiver is open.

As soon as the valve in the liquid line from the receiver is closed, the discharge pressure will begin to build up rapidly and the suction pressure will drop correspondingly. Do not allow the suction temperature to operate below the freezing point of the cooling medium in the cooler. As the suction pressure drops, open the charging valve and the drum valve and regulate the flow from the drum to maintain the normal operating suction pressure. This will allow the refrigerant to be taken into the system from the drum and a charge to be built up in the receiver. A drop in the suction pressure when charging the system from the drum, and the frosting of the lower side of the drum near the valve end are an indication

568

that the drum is becoming empty. The drum valve should immediately be opened wide and the suction pulled down to zero to completely exhaust the drum.

As soon as the drum is empty, the charging valve should be closed and the valve on the receiver opened and regulated in conjunction with the expansion valve to give the desired suction pressure or proper amount for the low side design. The amount of refrigerant required will depend entirely upon the system and, if the initial charge proves insufficient, more may be added by connecting another drum to the charging connection, closing the valve on the receiver outlet, and allowing the refrigerant from the drum to maintain the proper suction pressure. A close watch should be kept on the liquid-level gauge on the receiver so that the proper level may be maintained. The receiver should contain enough liquid at all times to keep the outlet sealed and prevent gas from entering the liquid line. It is advisable to weigh the drums before and after charging so that the amount of refrigerant added to the system may be recorded. This information is often of value for other operating or maintenance personnel.

SUMMARY

Installation of new refrigeration equipment is normally a function of various contracting companies and thus may not necessarily be of concern to the maintenance or operating engineer. It is felt, however, that a thorough review of installation methods will be of assistance since additional units may at times have to be installed without the assistance or guidance of experienced persons.

As noted in the first part of the chapter, a good installation is the best insurance for trouble-free operation. The various details concerning this attainment have been presented, particularly that part dealing with the arrangement of the refrigerating equipment, foundations, and the methods of securing compressors and accessories to the concrete foundation.

The various operating features, such as purging, leak testing, dehydration, and charging of typical refrigeration systems have been included, as well as the necessary precautions to be observed when accomplishing the foregoing maintenance details. The importance of vacuum pumps in evacuating a system and the method to use in determining the proper vacuum ranges are discussed.

REVIEW QUESTIONS

1. What are the essential features in planning and installing refrigerating machinery?

2. What should be observed with respect to the installation of two or more compressors in one refrigeration plant?

3. Why is ventilation necessary in refrigeration plants?

4. What are the precautions to be observed when constructing the foundation?

5. What is the function of vibration insulators as installed in refrigeration plants?

6. Give the recommended ratios between the weight of concrete foundation slabs and the machinery to be supported.

7. Describe the placement of anchor bolts and the methods used for securing the motor and compressor to the foundation.

8. What is meant by grouting?

9. Describe installation methods for belt-driven compressors and the various precautions to be observed when such installations are made.

10. How do belt-driven installations differ from those of directly connected motor-driven compressor installations?

11. Why is leak testing necessary and what methods are used to leak-test a refrigeration system?

12. What is a dehydrator and what is its purpose in a refrigeration system?

13. How does moisture affect compressor operation?

CHAPTER 29

Heat Leakage Through Walls

Heat leakage is always given in Btu per hour per ° F temperature difference per square foot of exposed surface. Prior to designing a refrigeration system, an estimate must be made of the maximum probable heat loss of each room or space to be cooled. Therefore, before attempting to place even a small room cooler into service, check up on the walls and determine just what a certain sized unit will do or is supposed to do in the way of cooling and dehumidifying.

The heat losses may be divided into two groups—(a) losses through confining walls, floors, ceilings, glass, or other surfaces, and (b) infiltration losses due to leaks through cracks and crevices around doors and windows. The heat leakage through walls, floors, and ceilings can be determined by means of a formula and depends on the type and thickness of the insulating material used. The formula for heat leakage is:

$$H = KA(t_1 - t_2)$$

where,

K is the heat transfer coefficient in Btu's per square foot per hour per degree Fahrenheit,

A is the area in square feet,

$t_1 - t_2$ is the temperature gradient through the wall in degrees Fahrenheit.

Example—Calculate the heat leakage through an 8-inch brick wall having an area of 200 square feet if the inside temperature is 10° F and the outside temperature is 70° F.

Solution—If it is assumed that the heat transfer coefficient of a plain brick wall is 0.50, a substitution of values in the foregoing equation will be as follows:

$$H = 0.50 \times 200(70\text{-}10) = 6000 \text{ Btu per hour}$$

The heat leakage through floors, ceilings, and roofs may be estimated in the same manner, the **K** or heat transfer coefficient depending on the particular insulating materials used. The tables shown on the following pages cover almost every type and combination of wall, floor, and ceiling encountered in the field. Partition walls are also included, so that these may be estimated for heat leakage. Although the **K** values given in the tables do not agree entirely with the data given in various handbooks, they will give sufficiently close values for adaptation, since in most instances only approximate values can be obtained.

In making use of the data, it must be remembered that the outside area is used as a basis for estimating. Ceiling and floor construction must also be determined and, in many instances, the four walls of the room may not be of the same construction since one or more of the walls may have partitions.

SUMMARY

Because the economy and efficiency of refrigeration depends largely on the insulation properties of walls, floors, and ceilings, an effort has been made to provide information on heat leakage in various structures and the mathematical methods used to determine the amount of this leakage. Tabulated data of **K**, or heat-transfer coefficient, for various insulating materials have been furnished.

If it is recalled that one ton of refrigeration equals 12,000 Btu per hour, it is an easy matter to calculate the required size of plant equipment, since all that is necessary is to divide the combined sum of the plant leakage and cooling loads in Btu per hour by 12,000 to obtain the condensing unit capacity.

Table 1. Concrete Wall
(No Exterior Finish)

K-X+Y
X-WALL
Y-INTERIOR CONSTRUCTION

Concrete Wall
(No Exterior Finish)
Values of K in Btu/Hr/1°F/Sq. Ft.

Wall Construction (Y)	Thickness (X)				
	6″	8″	10″	12″	16″
Plain wall—no interior finish	.58	.51	.46	.41	.34
½″ plaster—direct on concrete	.52	.46	.42	.38	.32
½″ plaster on wood lath, furred	.31	.29	.26	.24	.22
¾″ plaster on metal lath, furred	.34	.32	.29	.27	.24
½″ plaster on ⅜″ plasterboard, furred	.32	.30	.27	.25	.22
½″ plaster on ½″ board insulation, furred	.21	.20	.19	.18	.17
½″ plaster on 1″ corkboard, set in ½″ cement	.16	.15	.14	.14	.13
½″ plaster on 1½″ corkboard, set in ½″ cement	.15	.14	.14	.13	.12
½″ plaster on 2″ corkboard, set in ½″ cement	.13	.12	.11	.10	.09
½″ plaster on wood lath on 2″ fur., 1⅝″ gypsum fill	.20	.19	.18	.17	.16

Table 2.   Concrete Wall (Exterior Stucco Finish)

EXTERIOR 1" STUCCO
ON WIRE MESH

X = CONCRETE
Y = INSULATION
Z = EXTERIOR STUCCO 1"
K = X + Z + Y

Concrete Wall
(Exterior Stucco Finish)
Values of K in Btu/Hr/1°F/Sq. Ft.

Wall Construction (Y)	Thickness of Concrete (X)					
	6″	8″	10″	12″	16″	18″
Plain walls—no interior finish	.54	.48	.43	.39	.33	.28
½″ plaster direct on concrete	.49	.44	.40	.36	.31	.27
½″ plaster on wood lath, furred	.31	.29	.27	.25	.23	.22
¾″ plaster on metal lath, furred	.32	.30	.28	.26	.24	.23
½″ plaster on ⅜″ plasterboard, furred	.31	.29	.27	.25	.23	.22
½″ plaster on wood lath on 2″ furring strips with 1⅝″ cellular gypsum fill	.20	.19	.18	.17	.16	.15
½″ plaster on ½″ board insulation, furred	.22	.21	.20	.19	.18	.17
1″	.17	.16	.15	.145	.140	.13
½″ plaster on 1½″ sheet cork set in cement	.145	.140	.135	.130	.12	.11
2″	.12	.118	.115	.110	.105	.100

Table 3. Concrete Wall
(Brick Veneer)

X - CONCRETE

Y - INTERIOR FINISH

K = X + 1/2' CEMENT + 4' BRICK + Y

Concrete Wall
(Brick Veneer)
Values of K in Btu/Hr/1°F/Sq. Ft.

Wall Construction (Y)	Concrete (X)			
	6"	8"	10"	12"
Plain wall—no interior finish	.39	.36	.33	.30
½" plaster direct on concrete	.37	.34	.30	.29
½" plaster on wood lath, furred	.24	.23	.21	.20
¾" plaster on metal lath, furred	.27	.25	.23	.22
½" plaster on wood lath, on 2" furring strips, with 1⅝" cellular gypsum fill	.18	.17	.16	.15
½" plaster on ⅜" plasterboard, furred	.25	.24	.22	.21
½" plaster on ½" board insulation, furred	.17	.16	.14	.12
½" plaster on 1" board insulation, furred	.15	.14	.12	.11
½" plaster on 1½" board insulation, furred	.13	.12	.11	.10
½" plaster on 2" board insulation, furred	.11	.10	.09	.08

Table 4.   Concrete Wall (4 in. Cut Stone)

½" MORTAR

K = X + Y + Z
X = CONCRETE
Y = INSULATION
Z = 4" CUT STONE

Concrete Wall
(4 in. Cut Stone)
Values of K in Btu/Hr/1°F/Sq. Ft.

Wall Construction (Y)	Thickness of Concrete (X)				
	6"	8"	10"	12"	16"
Plain walls—no interior finish	.46	.42	.38	.34	.30
½" plaster direct on concrete	.42	.38	.34	.32	.28
¾" plaster on metal lath, furred	.29	.27	.26	.24	.22
½" plaster on wood lath, furred	.28	.26	.25	.23	.21
½" plaster on ⅜" plasterboard, furred	.28	.26	.25	.23	.21
½" plaster on wood lath on 2" furring strips, filled with 1⅝" gypsum	.18	.175	.17	.16	.15
½" plaster on ½" board insulation, furred	.21	.20	.19	.18	.17
½" plaster on 1" board insulation, furred	.16	.155	.15	.14	.13
½" plaster on 1" sheet cork, set in ½" cement or mortar	.15	.145	.140	.135	.130
1½"	.138	.135	.130	.125	.12
2"	.115	.110	.108	.105	.100

Table 5. Cinder and Concrete Block Wall

K = X + Y

X - WALL

Y - INTERIOR FINISH

Cinder and Concrete Block Wall
Values of K in Btu/Hr/1°F/Sq. Ft.

Wall Construction (Y)	Thickness (X) and kind of blocks			
	Concrete		Cinder	
	8"	12"	8"	12"
Plain wall—no interior finish	.46	.34	.31	.23
½" plaster—direct on blocks	.42	.32	.29	.21
½" plaster on wood lath, furred	.28	.23	.22	.17
¾" plaster on metal lath, furred	.29	.24	.23	.17
½" plaster on ⅜" plasterboard, furred	.28	.23	.22	.17
½" plaster on ½" board insulation, furred	.21	.18	.17	.14
½" plaster on 1" board insulation, furred	.16	.14	.14	.11
½" plaster on 1½" sheet cork set in ½" cement	.13	.12	.12	.10
½" plaster on 2" sheet cork set in ½" cement	.11	.10	.10	.09
½" plaster on wood lath, on 2" fur. strips, 1⅝" Gypsum fill	.18	.16	.16	.13

Table 6.   Cinder and Concrete Block Wall (Brick Veneer)

1/2" MORTAR

4"

X-WALL & EXTERIOR
Y-INTERIOR FINISH

K   FACTORS GIVEN FOR
TOTAL WALL THICKNESS
(X + 1/2' MORTAR + 4" BRICK + Y)

Cinder and Concrete Block Wall
(Brick Veneer)
Values of K in Btu/Hr/1°F/Sq. Ft.

Wall Construction (Y)	Thickness (X) and kind of blocks			
	Concrete		Cinder	
	8"	12"	8"	12"
Plain wall—no interior finish	.33	.26	.25	.18
½" plaster—direct on blocks	.31	.25	.23	.18
½" plaster on wood lath, furred	.23	.19	.18	.15
¾" plaster on metal lath, furred	.24	.20	.19	.15
½" plaster on wood lath, on 2" fur. strips, 1⅝" gypsum fill	.16	.14	.14	.12
½" plaster on ½" board insulation, furred	.18	.16	.15	.13
½" plaster on 1" board insulation, furred	.14	.13	.12	.10
½" plaster on ⅜" plasterboard, furred	.23	.19	.18	.15
½" plaster on 1½" sheet cork, set in ½" cement mortar	.12	.11	.11	.09
½" plaster on 2" sheet cork, set in ½" cement mortar	.10	.09	.09	.08

Table 7.   Brick Veneer on Hollow Tile Wall
(Brick Veneer)

OUTSIDE

1/2" CEMENT MORTAR

K VALUES GIVEN FOR

TOTAL WALL THICKNESS AS:

X + 1/2" MORTAR + 4" BRICK + Y

X - WALL & EXTERIOR

Y - INTERIOR FINISH

Brick Veneer on Hollow Tile Wall
(Brick Veneer)
Values of K in Btu/Hr/1°F/Sq. Ft.

Wall Construction (Y)	Tile Thickness (X)				
	4"	6"	8"	10"	12"
Plain wall—no interior finish	.29	.27	.25	.22	.18
½" plaster on hollow tile	.28	.25	.24	.21	.17
½" plaster on wood lath, furred	.20	.19	.18	.17	.15
¾" plaster on metal lath, furred	.21	.20	.19	.18	.16
½" plaster on ⅜" plasterboard, furred	.20	.19	.18	.17	.15
½" plaster on wood lath, on 2" furring strips, with 1⅝" cellular gypsum fill	.14	.13	.12	.11	.10
½" plaster on ½" board insulation, furred	.16	.15	.14	.13	.12
½" plaster on 1" board insulation, furred	.14	.13	.12	.11	.10
½" plaster on 1½" board insulation, furred	.13	.12	.11	.10	.09
½" plaster on 2" board insulation, furred	.12	.11	.10	.09	.08

Table 8. Hollow Tile Wall (4 in. Cut Stone Veneer)

1/2" CEMENT

K = X + Y + Z

X = TILE
Y = INSULATION
Z = 4" CUT STONE

Hollow Tile Wall
(4 in. Cut Stone Veneer)
Values of K in Btu/Hr/1°F/Sq. Ft.

Wall Construction (Y)	Thickness of Tile (X)			
	6"	8"	10"	12"
Plain walls—no interior finish	.30	.29	.28	.22
½" plaster—direct on hollow tile	.28	.27	.26	.21
¾" plaster on metal lath, furred	.22	.21	.20	.18
½" plaster on wood lath, furred	.21	.20	.19	.17
½" plaster on ⅜" plasterboard, furred	.21	.20	.19	.17
½" plaster on ½" cork board set in ½" cement	.17	.16	.15	.14
½" plaster on 1" corkboard set in ½" cement	.14	.13	.13	.12
½" plaster on 1½" corkboard set in ½" cement	.13	.12	.11	.10
½" plaster on 2" corkboard set in ½" cement	.11	.10	.09	.08
½" plaster on wood lath, 2" furring & 1⅝" gypsum fill	.15	.14	.13	.12
½" plaster on ½" board insulation	.18	.17	.16	.15

Table 9.   Hollow Tile Wall
(Stucco Exterior)

X - TILE THICKNESS
Y - INTERIOR FINISH

K VALUES GIVEN FOR
TOTAL WALL THICKNESS
X + Y + STUCCO FINISH

EXTERIOR STUCCO

Hollow Tile Wall
(Stucco Exterior)

Values of K in Btu/Hr/1°F/Sq. Ft.

Wall Construction (Y)	Tile Thickness (X)				
	6"	8"	10"	12"	16"
Plain wall with stucco—no interior finish	.32	.30	.28	.22	.18
½" plaster—direct on hollow tile	.31	.29	.27	.21	.17
½" plaster on wood lath, furred	.22	.20	.18	.16	.12
¾" plaster on metal lath, furred	.23	.21	.19	.17	.13
½" plaster on ⅜" plasterboard, furred	.21	.20	.18	.16	.12
½" plaster on wood lath, on 2" furring strips, with 1⅝" cellular gypsum fill	.16	.15	.14	.13	.12
½" plaster on ½" board insulation, furred	.15	.14	.13	.12	.11
½" plaster on 1" board insulation, furred	.13	.12	.11	.10	.09
½" plaster on 1½" board insulation, furred	.12	.11	.10	.09	.08
½" plaster on 2" board insulation, furred	.11	.10	.09	.08	.07

Table 10. Limestone or Sandstone Wall

K = X + Y

X - WALL
Y - INTERIOR CONSTRUCTION

Limestone or Sandstone Wall

Values of K in Btu/Hr/1°F/Sq. Ft.

Wall Construction (Y)	Thickness (X)			
	8"	10"	12"	16"
Plain wall—no interior finish	.56	.50	.48	.39
½" plaster—direct on stone	.50	.45	.42	.36
¾" plaster on metal lath, furred	.33	.31	.29	.26
½" plaster on wood lath, furred	.31	.29	.28	.25
½" plaster on ⅜" plasterboard, furred	.31	.30	.28	.25
½" plaster on wood lath on 2" fur. strips, 1⅝" gypsum fill	.20	.19	.18	.17
½" plaster on ½" board insulation, furred	.23	.22	.21	.19
½" plaster on 1" board insulation, furred	.17	.16	.15	.14
½" plaster on 1½" corkboard, set in ½" cement	.14	.14	.13	.12
½" plaster on 2" corkboard, set in ½" cement	.11	.11	.10	.10
½" plaster on wood lath, on 2" fur. strips, 2" gypsum fill	.19	.18	.17	.16

Table 11.   Brick Wall
(4 in. Cut Stone)

1/2" MORTAR

Y

Z   X

4"

INTERIOR FINISH

K = X + Y
X = WALL
Y = INTERIOR
Z = 4" STONE 8-1/2" MORTAR

Brick Wall
(4 in. Cut Stone)
Values of K in Btu/Hr/1°F/Sq. Ft.

Wall Construction (Y)	Thickness (X)						
	8"	9"	12"	13"	16"	18"	24"
Plain wall—no interior finish	.33	.29	.26	.25	.22	.20	.16
½" plaster—direct on brick	.31	.28	.25	.24	.21	.19	.15
¾" plaster on metal lath, furred	.23	.22	.20	.19	.18	.17	.15
½" plaster on wood lath, furred	.22	.21	.19	.18	.17	.15	.12
½" plaster on ⅜" plasterboard, furred	.23	.22	.20	.19	.18	.17	.14
½" plaster, lath, 2" furring strips, 1⅝" gypsum fill	.16	.15	.14	.13	.12	.11	.10
½" plaster on ½" sheet cork	.18	.16	.15	.14	.13	.12	.11
½" plaster on 1" sheet cork	.14	.13	.12	.11	.10	.10	.09
½" plaster on 1½" sheet cork	.12	.12	.11	.11	.10	.10	.09
½" plaster on 2" sheet cork	.11	.10	.10	.09	.09	.08	.08

583

Table 12.   Solid Brick Wall (No Exterior Finish)

K = X+Y
X = WALL
Y = INTERIOR CONSTRUCTION

Solid Brick Wall
(No Exterior Finish)
Values of K in Btu/Hr/1°F/Sq. Ft.

Wall Construction (Y)	Thickness (X)						
	8″	9″	12″	13″	16″	18″	24″
Plain wall—no interior finish	.39	.37	.30	.29	.24	.23	.18
½″ plaster—direct on brick	.36	.35	.28	.27	.23	.20	.16
¾″ plaster on metal lath, furred	.26	.25	.23	.21	.19	.18	.15
½″ plaster on wood lath, furred	.25	.24	.21	.19	.18	.17	.14
½″ plaster on ⅜″ plasterboard, furred	.24	.23	.20	.18	.17	.16	.14
½″ plaster on ½″ board insulation, fur.	.20	.19	.18	.17	.16	.15	.13
½″ plaster on ½″ sheet cork, set in ½″ cement	.18	.17	.16	.15	.14	.13	.12
½″ plaster on 1″ sheet cork, set in ½″ cement	.15	.14	.13	.12	.11	.10	.09
½″ plaster on 1½″ sheet cork, set in ½″ cement	.14	.13	.12	.11	.10	.09	.08
½″ plaster on 2″ sheet cork, set in ½″ cement	.13	.12	.11	.10	.09	.08	.07
½″ plaster on 1½″ split furring, tiled	.28	.26	.23	.22	.20	.19	.17

Table 13. Wood Siding Clapboard Frame or Shingle Walls

PLASTER BASE SPACE 2"X4" STUD SHEATHING PLASTER WOOD SIDING ADDITIONAL INSULATION

Wood Siding Clapboard Frame or Shingle Walls
Values of K in Btu/Hr/1°F/Sq. Ft.

Type of Sheathing	Insulation between studding	Plaster Base						
		Wood lath	Metal lath	3/8" plaster board	1/2" rigid insulation	1" rigid insulation	1 1/2" sheet cork	2" sheet cork
1/2" plasterboard	None	.30	.31	.30	.22	.16	.12	.10
	Cellular gypsum, 18#	.12	.125	.12	.10	.09	.08	.07
	Flaked gypsum, 24#	.10	.108	.10	.09	.08	.07	.06
	1/2" flexible insulation	.16	.17	.16	.14	.11	.09	.08
1" wood	None	.26	.28	.26	.20	.15	.12	.10
	Cellular gypsum, 18#	.11	.118	.11	.10	.09	.08	.07
	Flaked gypsum, 24#	.10	.108	.10	.09	.08	.07	.06
	1/2" flexible insulation	.15	.16	.15	.13	.11	.09	.08
1/2" rigid insulation (board form)	None	.22	.23	.22	.17	.14	.11	.09
	Cellular gypsum, 18#	.10	.11	.10	.09	.08	.07	.06
	Flaked gypsum, 24#	.09	.10	.09	.08	.07	.06	.05
	1/2" flexible insulation	.14	.148	.14	.12	.10	.08	.07

585

Table 14. Interior Walls and Plastered Partitions (No Fill)

Values of K in Btu/Hr/1°F/Sq. Ft.

Plaster base	Single partition 1 side plastered	Double partition 2 sides plastered
Wood lath	.60	.33
Metal lath	.67	.37
3/8" plasterboard	.56	.32
1/2" board (rigid) insulation	.33	.18
1" board (rigid) insulation	.22	.12
1½" corkboard	.16	.08
2" corkboard	.12	.06

Table 15. Frame Floors and Ceilings (No Fill)

Values of K in Btu/Hr/1°F/Sq. Ft.

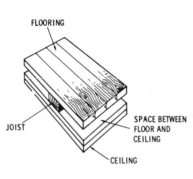

Type of ceiling with 1/2" plaster	No flooring	1" yellow pine flooring	1" yellow pine flooring and 1/2" fiberboard	1" sub oak or maple on 1" yellow pine flooring
No ceiling		.45	.27	.32
Metal lath	.66	.24	.20	.24
Wood lath	.60	.27	.19	.23
3/8" plasterboard	.59	.28	.20	.24
1/2" rigid insulation	.34	.20	.15	.17

Table 16. Masonry Partitions

BARE NO PLASTER

PLASTER

1/2"

4''

PLASTERED
ONE SIDE

PLASTERED
BOTH SIDES

1/2"

4''

Masonry Partitions

Values of K in Btu/Hr/1°F/Sq. Ft.

Wall	Bare no plaster	One side plastered	Two Sides plastered
4″ brick	.48	.46	.41
4″ hollow gypsum tile	.29	.27	.26
4″ hollow clay tile	.44	.41	.39

587

Table 17. Concrete Floors (On Ground)

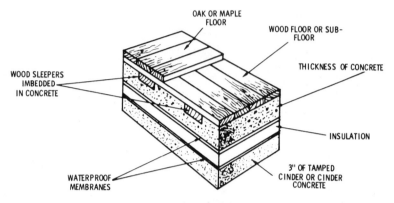

OAK OR MAPLE FLOOR

WOOD FLOOR OR SUB-FLOOR

THICKNESS OF CONCRETE

WOOD SLEEPERS IMBEDDED IN CONCRETE

INSULATION

WATERPROOF MEMBRANES

3'' OF TAMPED CINDER OR CINDER CONCRETE

Concrete Floors
(On Ground)

Values of K in Btu/Hr/1°F/Sq. Ft.

Type of insulation	Thickness of Y	Concrete thickness X	No flooring	Tile or terrazzo floor	1" yellow pine on wood sleepers	Oak or Maple on yellow pine sub floor on sleepers
None and no cinder base (Z)	0	4	1.02	.93	.51	.37
	0	5	.94	.83	.48	.36
	0	6	.86	.79	.46	.34
	0	8	.76	.70	.43	.32
None with 3" cinder base (Z)	0	4	.58	.55	.51	.31
	0	5	.55	.52	.47	.30
	0	6	.52	.49	.44	.29
	0	8	.50	.47	.40	.28
1" rigid insulation with 3" cinder base (Z)	1"	4	.22	.21	.18	.165
	1"	5	.21	.20	.175	.16
	1"	6	.20	.19	.17	.155
	1"	8	.19	.18	.16	.15
2" corkboard insulation with 3" cinder base (Z)	2"	4	.125	.12	.115	.105
	2"	5	.12	.115	.11	.10
	2"	6	.115	.11	.105	.095
	2"	8	.11	.105	.10	.09

Table 18. Concrete Floors and Ceilings

Concrete Floors and Ceilings
Values of K in Btu/Hr/1°F/Sq. Ft.

Type of ceiling and base	Thick-ness	Floor Type			
		No flooring	Tile or Terrazzo floor	1" yellow pine on wood sleepers	Oak or maple on yellow pine, sub-floor on sleeper
No ceiling	4	.60	.57	.38	.29
	6	.54	.52	.35	.28
	8	.48	.46	.33	.26
	10	.46	.44	.31	.25
½" plaster direct on concrete	4	.56	.53	.38	.30
	6	.51	.49	.35	.29
	8	.47	.44	.33	.27
	10	.42	.41	.31	.26
Suspended or furred ceiling on ⅜" plasterboard ½" plaster	4	.33	.32	.24	.20
	6	.31	.30	.23	.19
	8	.29	.28	.22	.18
	10	.28	.27	.21	.17
Suspended or furred metal lath ¾" plaster	4	.36	.34	.26	.21
	6	.34	.32	.24	.20
	8	.32	.30	.23	.19
	10	.31	.29	.22	.18
Suspended or furred on ½" rigid insulation ½" plaster	4	.22	.21	.20	.18
	6	.21	.20	.19	.17
	8	.20	.19	.18	.16
	10	.19	.18	.17	.15
1½" cork-board in ½" mortar ½" plaster	4	.16	.15	.14	.13
	6	.15	.14	.13	.12
	8	.14	.13	.12	.11
	10	.13	.12	.11	.10

589

Table 19.   Brick Veneer (Frame Walls)

**Brick Veneer
(Frame Walls)**

Values of K in Btu/Hr/1°F/Sq. Ft.

Type of sheathing	Insulation between studding	Plaster Base							
		Wood lath	Metal lath	½" fiber board	¾" plaster board	½" cork board	1" cork board	1½" cork board	2" cork board
1" wood sheathing	None	.26	.27	.20	.27	.18	.15	.12	.095
	½" fiberboard	.17	.18	.14	.16	.14	.11	.09	.08
	Flaked gypsum 24#	.095	.10	.09	.10		.08	.07	.06
	Cellular gypsum 18#	.11	.15	.10	.11		.09	.08	.07
½" plaster-board sheathing	None	.32	.34	.24	.30		.16	.13	.11
	½" fiberboard	.18	.19	.14	.17		.11	.09	.08
	Flaked gypsum 24#	.10	.11	.09	.10		.08	.07	.06
	Cellular gypsum 18#	.12	.12	.10	.12		.09	.08	.07
½" fiber-board (rigid) insulation sheathing	None	.25	.26	.19	.26		.15	.12	.10
	½" fiberboard	.16	.17	.13	.15		.10	.09	.08
	Flaked gypsum 24#	.09	.09	.08	.09		.07	.07	.06
	Cellular gypsum 18#	.10	.105	.09	.10		.08	.07	.06

Table 20. Frame Floors and Ceilings
(With Fill)

Frame Floors and Ceilings
(With Fill)
Values of K in Btu/Hr/1°F/Sq. Ft.

Type of ceiling with ½" plaster	Insulation or fill between joists	No flooring	1" yellow pine flooring	1" yellow pine flooring and ½" fiberboard	1" sub-oak, maple on 1" yellow pine flooring
Wood lath	½" flexible insulation	.22	.15	.12	.14
Wood lath	½" rigid insulation	.24	.16	.13	.15
Wood lath	Flaked 2" gypsum	.15	.12	.105	.12
Wood lath	Rock wool, 2"	.12	.095	.08	.09
Corkboard ½"	None	.155	.115	.10	.105
Corkboard ½"	None	.12	.10	.085	.095

REVIEW QUESTIONS

1. Why is heat leakage through walls, ceilings, and floors an important consideration when erecting refrigeration plants?

2. Give the formula for leakage through walls.

3. What is meant by the heat-transfer coefficient and why does this coefficient vary for different types of materials?

4. Calculate the heat leakage in Btu per hour for a plain 10-inch concrete wall (no exterior finish) having an area of 450 square feet for a temperature difference of 50° F.

5. Is the inside or outside wall area used as a basis for estimating heat leakage?

CHAPTER 30

Refrigeration Load Calculations

A refrigeration load is measured by the heat units absorbed by the refrigerant while passing through the evaporator. With respect to a cold storage plant, this load consists of the following items:

1. Heat leakage through walls, floor, and roof.
2. Heat emitted from the products to be cooled during the freezing process, usually termed *product cooling* and *freezing load.*
3. Heat due to entrance of warm air when the doors are opened.
4. Heat due to lights, motors, and people that may be in the rooms.

If it is desired to calculate the load on a locker storage plant, it is practically impossible to estimate all of the foregoing items with any assurance of accuracy. Heat leakage, however, can be estimated with fair accuracy. The heat leakage through walls, floors, and ceilings can be determined by means of a formula and depends on the insulating material used and its thickness. The formula for heat leakage is:

$$H = KA(t_1 - t_2)$$

where,

K is the heat transfer coefficient in Btu's per square foot per hour per degree F,

A is the area in square feet,

$t_1 - t_2$ is the temperature gradient through the wall, in degrees F.

CALCULATING HEAT TRANSFER COEFFICIENT

Assume that it is desired to find **K** (heat transfer coefficient) for a wall made up of 6 inches of concrete and 4 inches of cork. The outside of a concrete wall is presumably an outside wall exposed to outside air circulation. Experience has shown that, for average conditions, the heat transfer through the air film next to the wall is about 4.2 Btu per square foot per hour for each degree difference in temperature between the inside and outside. The resistance is the reciprocal of this, or $1/4.2 = 0.24$. The coefficient for concrete is about 8 Btu per square foot per hour per degree difference, for each inch of thickness; the resistance is 0.125.

The resistance for a 6-inch wall is six times greater (6×0.125, or 0.75). The coefficient for cork is about 0.31 Btu per square foot per hour per degree difference, per inch of thickness, which is a resistance of $1/0.31$ or 3.23. The resistance for a 4-inch wall is therefore $4 \times 1/0.31$ or 12.9. The inside of the wall is in contact with still air, and experience has shown that an average value of the coefficient for the film is about 1.4. The resistance then is $1/1.4$, or 0.71. The total resistance for this wall is then $0.24 + 0.75 + 12.9 + 0.71 = 14.60$, and the overall coefficient is therefore $1/14.60$, or 0.0682. It is apparent that the principal resistance is offered by the cork. By following this same procedure, the overall coefficient for any type of wall can be obtained with a fair degree of accuracy.

CALCULATING HEAT LEAKAGE

After the heat transfer coefficient is known, the heat leakage can readily be determined by the use of the formula. Assume that it is desired to compute the heat leakage of a room 25×25 ft. by 10 ft. high, with walls of 6-inch concrete and 4-inch cork. The perimeter of the room is 4×25, or 100 ft. The total wall surface is 10×100 or 1000 sq. ft. With an outside temperature of 80° F and an inside temperature of 30° F, the heat leakage through the walls according to the formula is:

$$H = 0.0682 \times 1000 \, (80\text{-}30) = 3410 \text{ Btu per hour}$$

The leakage through the floor and ceiling can be determined in a similar manner to obtain the total heat leakage of the room. Any additional refrigeration needed to cool products in the room will have to be added to the heat leakage to obtain the total load.

PRODUCT COOLING LOAD

Heat emitted from the product to be cooled may also be calculated if the amount of product per locker is known. Most authorities assume that the average locker user will place an average of 2 to 2-½ lbs. of produce per day in his storage compartment. Thus, in a chill room having a 300-locker installation, the load would be 300 × 2-½, or 750 pounds daily. The initial temperature may be as high as 95° F and the final temperature can be taken as 36° F.

Because the specific heat of various kinds of produce will vary, an average specific heat of 0.7 is generally adopted as a good value for calculations. For a 300-locker installation, the daily product cooling load would be 0.7 × 750 (95-36), or 30,975 Btu per day.

AIR-CHANGE LOADS

Heat caused by the entrance of warm air when the doors are opened is termed *air-change load*. This load is difficult to estimate with any degree of accuracy, since it is affected by such factors as usage of the room, interior volume, whether or not the room is entered through an outside door, size of the door, etc.

PRODUCT FREEZING LOAD

When meat is removed from the chill room, it may have a temperature of about 35° F. During the process of preparation for the quick freezer, it may warm up to 40° or 50° F. It is evident that this meat will have to be cooled down to 32° F before it will begin to freeze. When freezing meat, the latent heat of fusion is removed, and the average heat of fusion amounts to about 90 Btu per pound. After freezing, the meat is sub-cooled to the quick-freeze temperature, with an average specific heat of 0.4.

If it is required to calculate the amount of heat removed from a pound of meat (assuming a final temperature of the quick freeze to be −10° F), under the above conditions, we obtain

Cooling the meat to the freezing point,
  remove 0.7(45 − 28)   = 11.9 Btu
Freezing the meat (latent heat of fusion) = 90.9 Btu
Subcooling the meat to −10° F =
  0.4(28 + 10)   = 15.2 Btu
Total   117.1 Btu per lb.

Again, allowing 2.5 pounds of meat per locker, the quick freezer in a 300-locker installation should have a capacity to freeze 300 × 2.5, or 750 pounds of meat daily. In this case, the product freezing load would be 750 × 117.1, or 87,825 Btu per day, approximately.

MISCELLANEOUS LOADS

Miscellaneous loads cannot readily be calculated. They consist of such items as heat from electric motors, lights, and people who may be in the room where the freezing process is in operation. With respect to this load (often referred to as the *service load*), it is assumed that ordinary precautions in handling of doors will be observed in order to keep this part of the total load to a minimum.

It has been found from experience that the average plant can be so handled that the service load experienced on the low-temperature room will not be more than 15% to 20% of the leakage load.

SERVICE FACTOR

Considering the foregoing, it is obviously difficult to arrive at a definite total value which will assist in a determination of the refrigeration capacity required by the compressor. To obtain a relatively close approximation of the load, it has been found by experience that these indeterminable factors can be closely compensated by allowing a certain percentage of the insulation loss. This percentage factor differs with the thickness of insulation used and temperature gradient between the walls.

CAPACITY OF MACHINES

The capacity of any refrigerating compressor depends largely on the number and size of its cylinders, its speed when running, the efficiency of compression, the suction and discharge pressures, and the number of hours of operation per day. The rated capacity is always based on continuous operation through 24 hours. Capacity rating is usually based on the conditions adopted as standard by the A.S.R.E. (American Society of Refrigerating Engineers) which are 5° F and 19.6 lbs. gauge pressure for the suction, and 86° F and 154.5 lbs. for the discharge.

LOAD CALCULATION FOR A WALK-IN REFRIGERATOR

It is desired to estimate the size of the condensing unit for a walk-in refrigerator that is 8 feet wide, 8 feet high, and 10 feet long. The insulation consists of 3 inches of sheet cork. The cooling compartment is to be kept at 40° F in a 90° F room. The **K** factor for sheet cork insulation is 0.28. (See Table 1.)

Since the temperature difference is (90-40) or 50° F, and the thickness of the insulation is 3 inches, the heat leakage per hour is obtained as follows:

$$H = AK(t_1 - t_2) = \frac{448 \times 0.28 \times 50}{3} = 2090 \text{ Btu}$$

Because heat loads and equipment capacities are usually estimated on a 24-hour basis, the heat leakage load in this cooler for 24 hours will be 2090 × 24, or 50,160 Btu.

The product cooling load depends on the temperature of the products to be chilled. Assume that 200 pounds of lean beef and 100 pounds of poultry, both at room temperature, are brought into the 8 × 8 × 10 ft. walk-in refrigerator daily. Assume further that the specific heat for beef and poultry is 0.77 and 0.80, respectively. The product cooling load is therefore obtained as follows:

Beef	200 × 0.77 × 50 = 7700 Btu
Poultry	100 × 0.80 × 50 = 4000 Btu

That is, a total 24-hour product load 7700 = 4000, or 11,700 Btu.

It should be observed that when freezing ice cream, water, foods, or other products, large amounts of heat must be extracted because of the fact that the *latent heat of freezing or fusion* is very large and may add loads far in excess of the normal when merely storing these products pre-frozen.

Air-change load is sometimes termed *service load* or *usage load,* and consists of heat introduced into the refrigerator products. This load varies a great deal and is affected by so many factors that it cannot be calculated with any degree of accuracy. A convenient table supplied through the courtesy of *The American Society of Refrigeration Engineers* is used for this purpose. To use this table to determine the air-change load, it is

Table 1. Conductivity Factors for Various Insulations

Insulation	K*
Sheet cork	0.28
Glass wool	0.26
Rock wool	0.28
Sawdust	0.41
Kapok	0.24
Hair felt	0.26

*Btu/sq. ft./°F/hr/in. thickness

necessary first to calculate the inside cubic content of the walk-in refrigerator and multiply this by the *Usage Factor* from Table 2.

Assuming a total wall thickness of 6 inches, the 8 × 8 × 10 ft. refrigerator will have a net volume of 7 × 7 × 9 ft., or 441 cubic feet. This puts the walk-in unit in the 400 cubic foot class, since the total volume is well over 400 cu. ft. If it is further assumed that the walk-in unit is subjected to *normal usage* rather than heavy, keeping in mind that a temperature difference of 50° F is required, it is found that the usage factor in this particular case is 81 per 24 hours per cubic foot of interior capacity. From the foregoing, it follows that the total air-change load is 81 × 441, or 35,720 Btu per 24 hours, approximately.

Table 2. Usage Factors in Btu
Per 24 Hours Per Cubic Foot Interior Capacity

Inside Volume	Type of Service	Temperature Difference (Room temp. minus refrigerator temp.)								
		40°F.	45°F.	50°F.	55°F.	60°F.	65°F.	70°F.	75°F.	80°F.
15 cu. ft.	Normal	108	122	135	149	162	176	189	203	216
	Heavy	134	151	168	184	201	218	235	251	268
50 cu. ft.	Normal	97	109	121	133	145	157	169	182	194
	Heavy	124	140	155	171	186	202	217	233	248
100 cu. ft.	Normal	85	96	107	117	128	138	149	160	170
	Heavy	114	128	143	157	171	185	200	214	228
200 cu. ft.	Normal	74	83	93	102	111	120	130	139	148
	Heavy	104	117	130	143	156	169	182	195	208
300 cu. ft.	Normal	68	77	85	94	102	111	119	218	136
	Heavy	98	110	123	135	147	159	172	184	196
400 cu. ft.	Normal	65	73	81	89	97	105	113	122	130
	Heavy	95	107	119	130	142	154	166	178	190

The miscellaneous load consists of heat from several sources, such as that generated by electrical equipment, motors, and lamps, in addition to heat from persons during servicing operations. Therefore, this load cannot readily be calculated with any degree of accuracy. With respect to the electrical equipment, it has been found that one watt of electrical energy is equivalent to 3.415 Btu. Thus, a 25-watt lamp will generate 25 × 3.415, or 85.375 Btu. If it is assumed that three 25-watt lamps are used for lighting and are lit 12 hours per day, and that a 100-watt blower motor is running 24 hours per day, the additional load imposed on the system will be as follows:

Lamp load   75 × 3.415 × 12 = 3074 Btu

Motor load   100 × 3.415 × 24 = 8196 Btu

The total miscellaneous load will be 3074 + 8196, or 11,270 Btu.

From the foregoing calculations it follows that the total 24-hour load is obtained by adding together the heat-leakage load, the product-cooling load, the air-change load, and the miscellaneous load as follows:

Heat-leakage load.............................. 50,160 Btu

Product-cooling load .... 11,700 Btu

Air-change load .... 35,720 Btu

Miscellaneous load .... 11,270 Btu

_____

108,850 Btu

In order to compensate for the *on* and *off* periods of the condensing unit, and to maintain the same refrigerator temperature despite variations in load (from summer to winter, or in accordance with the amount of warm products chilled, or variations in opening of the doors), it is necessary to use a condensing unit with a somewhat greater capacity per 24 hours than the total heat load. It is good practice to select a condensing unit with at least a 50% greater capacity per 24 hours than the total heat load. In this manner, the unit will be of sufficient capacity if it is assumed to run only 2/3 of the time, or 16 hours out of 24. Thus, if the total load is 108,850 Btu and the condensing unit is run only 16 hours out of 24 hours, it must have an hourly capacity of 108,850/16, or approximately 6800 Btu.

At this point it is also necessary to determine the operating temperature of the evaporator. If an average temperature difference of 15° F in this 40°

F refrigerator is assumed, the average temperature of the evaporator must be 25° F. The capacity of the condensing unit also varies with the evaporator temperature (and suction pressure which is governed by the evaporator temperature). Therefore, it will be necessary to employ a condensing unit that has a capacity of 6800 Btu per hour at an evaporator temperature of 25° F, which corresponds to a *Freon-12* suction pressure of 24.5 psi.

Table 3 shows the capacity of typical air-cooled units of several motor sizes operating on 25° F evaporators and in a 90° F room. It will be observed that a 1-HP condensing unit having a capacity of 8190 Btu per hour will be required since the ¾-HP unit has a capacity of only 5820 Btu per hour. With the calculated cooling load required it should be noted that, since the chosen condensing unit is somewhat oversized, the actual operating period need be only 108,850/8190, or 13 hours approximately out of 24.

Table 3. Capacity of Typical Air-Cooled Condensing Units

Condensing Unit Horsepower	Capacity per Hour in Btu's
⅓	2460
½	4010
¾	5820
1	8190
1½	12,050
2	16,150

EVAPORATOR CAPACITY

After having selected a 1-HP condensing unit that, while in operation, refrigerates at the rate of 8190 Btu per hour, an evaporator that, at a 15° F temperature difference, will have a capacity of at least 8190 Btu per hour should be selected.

LOAD CALCULATIONS FOR A DAIRY PLANT

For a dairy handling 800 gallons of milk daily that must be cooled from a pasteurizing temperature of 145° F to 40° F, the following assumptions are made:

Specific gravity of milk—1.03

Weight of milk—8.6 pounds per gallon
Weight of water—8.34 pounds per gallon
Specific heat of milk—0.95

The equation for the heat load is:

$$H = WS(t_1 - t_2)$$

where,

W is the weight of fluid being cooled
S is the specific heat in Btu per pound per degree
t_1 is the higher temperature
t_2 is the lower temperature

Substituting the values in the equation, we obtain:

$$H = 800 \times 8.6 \times 0.95(145 - 40) = 686,280 \text{ Btu}$$

If it is assumed that ample water is available, it is entirely feasible to cool the milk with water to 80° F. Therefore, the heat to be removed by the water is:

$$H = \frac{686,280 \times 65}{105} = 424,840 \text{ Btu}$$

If the water used for cooling is allowed to rise 10° F, the amount of water necessary is:

$$\frac{424,840}{10 \times 8.34} = 5094 \text{ gallons}$$

If a 4-hour period is allowed for the cooling, the amount of water necessary per hour is 5094/4 or 1274 gallons. If a 25% loss of water is allowed, the gpm requirement is:

$$\frac{1274 \times 1.25}{60} = 26.5 \text{ gallons per minute (approximately)}$$

To cool the milk from 80° to 40° F, the amount of heat to be removed is:

$$H = 686,280 - 424,840 = 261,440 \text{ Btu}$$

The cooling substance is usually brine. The brine for this purpose should enter the cooling coil at an initial temperature of approximately 18° F and leave at about 28° F. The specific heat of brine is dependent on the amount of salt in it, but 0.86 is a reasonable value. The weight of the brine required to remove 261,440 Btu of heat can be obtained if the foregoing equation is solved for W. That is:

$$W = \frac{H}{S(t_1 - t_2)} = \frac{261,440}{0.86(28 - 18)} = 30,400 \text{ lbs.}$$

If the weight per gallon of brine is assumed to be 9.1 pounds, we obtain the number of gallons of brine required as 30,400/9.1, or approximately 3340 gallons. Again, if a 4-hour period is allowed for the cooling, the requirement of brine in gpm is 3340/4 ÷ 60, or 13.9 gallons per minute. Allowing for heat losses, a pump having a minimum capacity of 15 gpm should be provided.

If it is desired to calculate the cooling surface required for the water and brine cooler the following procedure is recommended. For that part of the cooler in which water is the cooling medium, the amount of heat to be removed per hour is 424,840/4, or 106,210 Btu, assuming a 4-hour cooling period. If 20% if added to provide for losses, the additional heat to be removed is 1.2 × 106,210, or approximately 127,450 Btu. The milk enters the cooler at 145° F and leaves the water-cooled section at 80° F.

Although the water temperature varies in different localities, and with the season of the year, for the purpose of this calculation it may be assumed that water enters the cooler at 60° F and leaves it at 70° F. If the counter-flow principal is employed, the milk enters the top of the cooler and leaves at the bottom, while the water enters the cooler at the bottom and leaves at the top. From this, it follows that the coolest milk is cooled by the coldest water and the warmest milk by the warmest water. The temperature difference at the warm end is 145 − 70 = 75° F, and at the cold end 80 − 60 = 20° F. The formula for computation of the cooling surface is:

$$H = KAT_d$$

where,

H is the heat transfer in Btu per hour,
K is the coefficient of heat transfer = 60,
A is the surface area in square feet,
T_d is the mean temperature.

If the formula is solved for A and the numerical values substituted, we obtain:

$$A = \frac{H}{KT_d} = \frac{127,450}{60\left(\frac{75 + 20}{2}\right)} = 44.7 \text{ sq. ft. (approx.)}$$

The amount of surface required for the brine cooler may be calculated in a similar manner. The heat to be removed by the brine is 261,440 Btu. Allowing 20% additional heat losses, the total is 261,440 × 1.2 = 313,700 Btu (approx.). If it is assumed that the compressor is operating 8 hours per day and that, during the cooling period, the evaporating ammonia will remove half of the 313,700 Btu and that the other half is absorbed by the brine to be removed by the evaporating ammonia later, then by applying the equation, we have:

$$H = WS(t_1 - t_2)$$

where,

H is the heat to be absorbed by the brine,
W is the weight of the brine in the system,
S is the specific heat of the brine = 0.86,
t_1 is the temperature of the brine at the end of the milk cooling period,
t_2 is the temperature of the brine at the beginning of the milk cooling period.

Allowing the brine to warm 5° F during the cooling period, then

$$t_1 - t_2 = 5$$

and

$$W = \frac{H}{S(t_1 - t_2)} = \frac{313,700}{2 \times 0.86 \times 5} = 36,500 \text{ lbs. (approx)}$$

At 9.1 pounds per gallon, this is 36,500/9.1, or about 4000 gallons of brine. On the basis of 8 hours operating time, the refrigerating equipment must remove heat at the rate of 313,700/8, or approximately 39,200 Btu per hour. In this particular case, therefore, the tonnage capacity of the compressor must be 39,200/12,000, or 3.27 tons.

The evaporating surface area in the brine tank may also be calculated. With an average brine temperature of 28° F, the temperature of the evaporating ammonia at 5° F, and an assumed coefficient of heat transfer of 15, we obtain:

$$A = \frac{H}{K \times T_d} = \frac{39,200}{15(28 - 5)} = 114 \text{ sq. ft. (approx.)}$$

It has been found by experience that it requires 2.3 lineal feet of 1-¼'' pipe to furnish one square foot of surface. Since the number of square feet in this particular case is 114, the pipe requirement is 2.3 × 114, or 262 lineal feet.

MILK STORAGE

Most dairies have a walk-in type of cooler for storage of the milk after aerating, in addition to storage of other products such as cream and butter. Immediately after aerating, the milk is usually bottled, cased, and stored. During the process of bottling, temperature of the milk usually rises two or three degrees due to contact with the warm bottles. The amount of refrigeration required for such storage varies over a rather large range, depending on such factors as insulation, amount of products to be stored, etc. Experience, however, has shown that if a certain amount of refrigeration is provided per cubic foot of room space, satisfactory conditions can be maintained. If a room having dimensions of 18 × 12 × 10 ft. (approximately 2200 cubic feet) is chosen, and the temperature is to be held at 35° F, about one ton of refrigeration should be provided. For a room 10 × 10 × 10 ft. (1000 cubic feet of space), about 0.6 ton of refrigeration should be provided.

A brine tank is often located in the milk storage room. When so located, it should be placed near the ceiling. The refrigeration stored in it during the time the machine is in operation will help to hold the room at the proper temperature during the shut-down period, and the fact that it is in an insulated room eliminates the necessity of insulating the tank itself.

If it cannot be placed near the ceiling, it should not be placed in the room at all, because it occupies valuable space, does not help to hold the room temperature very well, and renders it much more difficult to keep the room neat and sanitary.

In a dairy manufacturing plant, two compressors, each operating at a different suction pressure, are often desirable. The machine which cools the brine for freezing and cools the hardening room can be operated with an evaporating temperature of $-20°$ F, which corresponds to a pressure of 18.3 pounds absolute or 3.6 pounds gauge. The machine which serves the milk storage room, the brine tank for milk cooling, and the mix holding vat should have an evaporating temperature of about 5° F, which requires a suction pressure of 34.3 pound absolute pressure or 19.6 pound gauge. If ice is made, the ice tank should also be connected to this machine.

LOAD CALCULATIONS FOR SMALL FREEZERS

The so-called package units that are larger than the standard household refrigerator but smaller than the walk-in type freezer and storage units, and that can be used for both freezing and storing after freezing as well as for cooling products to about 35° to 40° F, and hold at that temperature, are here considered. In order to evaluate some of the engineering aspects of such a unit, the following data is given: The cabinet consists of a compartment $16'' \times 19'' \times 35''$, having a volume of about 6 cubic feet, and a larger compartment $20'' \times 22'' \times 64''$, having a volume of about 16 cubic feet. The smaller compartment is insulated with about 5 inches of a high-grade insulating material, and the larger one with about 3 inches. The small compartment is designed to be held at 5° F, and the larger compartment at 35° F. The amount of cooling surface in the low-temperature space is about 15.3 square feet and in the high-temperature compartment about 5.8 square feet.

The refrigerating unit is a two-cylinder compressor driven by a ⅓ HP motor and the refrigerant is *Freon* (F-12). The controls are set to start the compressor when the pressure in the evaporator is 12 lbs. gauge, which corresponds to about 5° F, and to stop it when the pressure is reduced to 0 lbs. gauge (atmospheric pressure), corresponding to $-22°$ F. This avoids production of a partial vacuum in the system and the attendant evils resulting from air and water-vapor leakage into the unit. The average evaporating temperature is about $-9°$ F. At this temperature, the rated capacity of the condensing unit is about 1580 Btu per hour. This is more, however, than the surface in the compartments can absorb with the temperature differences available. The result is that the compressor will continue to run until an equilibrium point or balanced condition is reached. As the following calculations will show, this equilibrium point is estimated to be at about $-20°$ F.

A reasonable heat transfer coefficient for the freezer surface is 1.65 Btu per square foot per hour per degree, and for the cooler surface about 1.75. The temperature difference is 25° in the freezer and 55° in the cooler. The heat transfer is:

$$H = AK (t_1 - t_2)$$

For the freezer,

$$H = 15.3 \times 1.65 \times 25 = 631 \text{ Btu per hour}$$

For the cooler,

$$H = 5.8 \times 1.75 \times 55 = 588 \text{ Btu per hour}$$

The total heat that the surface can absorb is the sum of these two heat transfers, 1189, or in round figures, 1200 Btu per hour. At this evaporating temperature, the rated capacity of the condensing unit is substantially the same. As long as heat is being supplied to the evaporator from the compartments at the rate of about 1200 Btu per hour, the compressor will continue to run, but if the heat supply is reduced, the evaporator temperature will begin to fall, and when it reaches the predetermined cutout point, the driving motor will be stopped automatically until the pressure has risen sufficiently to operate the starting mechanism.

The estimated heat leakage is about 160 Btu per hour, and the loss due to opening of the doors about 140 Btu per hour. These losses continue at a more or less uniform rate throughout the day and amount to about 300 Btu per hour. The necessary operating time required is

$$\frac{300}{1200} = 0.25, \text{ or } 25\%$$

The energy consumption is about 0.4 kW while the motor is running. The minimum power consumption is $0.4 \times 0.25 \times 24 = 2.4$ kW hours per day, and the maximum 4 times as much, or 9.6. The difference between

1200 and 300, or 900 Btu per hour is the refrigeration available for cooling or freezing goods. It is sufficient to cool and freeze meat or poultry at the rate of about 6 lbs. per hour or berries at the rate of about 5 lbs. per hour.

SUMMARY

This chapter deals with the detailed calculation of, and capacity requirements for, a variety of cooling applications. Thus, the heat leakage through walls, floor, and roof of the structure is first determined, after which the product cooling and freezing loads, etc., have been calculated by numerous practical examples.

Tabulated data include conductivity factors for various types of insulation, usage factors in Btu per 24 hours per cubic foot of space for various temperatures, in addition to condensing capacity when the cooling requirement is known.

Refrigeration load calculations for various structures have been made, such as for walk-in refrigerators, dairy plants, etc. In this connection it should be noted that other load problems may be calculated in a similar manner by substituting figures in the customary manner, since no two problems in the field of refrigeration will be exactly alike.

REVIEW QUESTIONS

1. Enumerate the various cooling loads that must be considered in a typical cold-storage plant in order to obtain the required machine capacity.

2. What is meant by the product cooling load in a locker-plant installation?

3. What does the capacity of a refrigerating compressor depend on?

4. Of what do the so-called miscellaneous loads consist?

5. Give the formula for computing the cooling surface area of a brine cooler.

6. How does the calculation for small freezers differ from that of cold-storage plants?

Glossary

Absolute Pressure—The sum, at any particular time, of the gauge pressure and the atmospheric pressure. Thus, for example, if the gauge reads 164.5 lbs. per sq. in. (psi) the absolute pressure will be 164.5 + 14.7, or 179.2 lbs per sq. in.

Absolute Temperature—The temperature of a substance measured above absolute zero.

Absolute Zero—The temperature at which the molecular motion of a substance theoretically ceased. This temperature is equal to $-459.6°$ on the Fahrenheit scale and $-273.1°$ on the Centigrade scale. At this temperature, the volume of an ideal gas maintained at a constant pressure becomes zero.

Absorber—A device for absorbing a refrigerant. A low-side element in an absorption system.

Absorption Machine—A refrigeration machine in which the gas involved in an evaporator is taken up by an absorber.

Accumulator—A shell placed in a suction line for separating the liquid entrained in the suction gas.

609

Acrolein—A warning agent often used with methyl chloride to call attention to the escape of refrigerant. The material has a compelling, pungent odor and causes irritation of the throat and eyes. Acrolein reacts with sulfur dioxide to form a sludge.

Activated Aluminum—A form of aluminum oxide (Al_2O_3) which absorbs moisture readily and is used as a drying agent.

Adiabatic—Referring to a change in gas conditions where no heat is added or removed except in the form of work.

Adiabatic Process—Any thermodynamic process taking place in a closed system without the addition or removal of heat.

Absorption—The adhesion of the molecules of gases or dissolved substances to the surface of solid bodies, resulting in a concentration of the gas or solution at the place of contact.

Aeration—A term generally employed with reference to air circulation or ventilation. In milk cooling, it refers to a method where the milk flow over refrigerated surfaces is exposed to the atmosphere.

Agitation—A condition in which a device causes circulation in a tank containing fluid.

Air Conditioning—The simultaneous control of all, or at least the first three, of the following factors affecting the physical and chemical conditions of the atmosphere within a structure: temperature, humidity, motion, distribution, dust, bacteria, odors, toxic gases, and ionization, most of which affect human health or comfort.

Air-Conditioning Unit—A piece of equipment designed as a specific air-treating combination, consisting of a means for ventilation, air circulation, air cleaning, and heat transfer, with a control means for maintaining temperature and humidity within prescribed limits.

Air Infiltration—The in-leakage of air through cracks, crevices, doors, windows, or other openings, caused by wind pressure or temperature differences.

Air Washer—An enclosure in which air is forced through a spray of water in order to cleanse, humidify, or dehumidify the air.

Ambient Temperature—The temperature of the medium surrounding an object. In a domestic or commercial system having an air-cooled condenser, it is the temperature of the air entering this condenser.

Ammonia Machine—An abbreviation for a compression refrigerating machine using ammonia as a refrigerant. Similarly, *Freon* machine, sulfur dioxide machine, etc.

Analyzer—A device used in the high side of an absorption system for increasing the concentration of vapor entering the rectifier or condenser.

Anemometer—An instrument for measuring the velocity of air in motion.

Antifreeze Liquid—A substance added to the refrigerant to prevent formation of ice crystals at the expansion valve. Antifreeze agents in general do not prevent corrosion due to moisture. The use of liquid should be a temporary measure where large quantities of water are involved, unless a drier is used to reduce the moisture content. Ice crystals may form when moisture is present below the corrosion limits and, in such instances, a suitable noncorrosive antifreeze liquid is often of value. Materials such as alcohol are corrosive and, if used, should be allowed to remain in the machine for a limited time only.

Atmospheric Condenser—A condenser operated with water which is exposed to the atmosphere.

Atmospheric Pressure—The pressure exerted by the atmosphere in all directions as indicated by a barometer. Standard atmospheric pressure is considered to be 14.695 lbs. per sq. in. (usually written 14.7 psi), which is equivalent to 29.92 inches of mercury.

Atomize—To reduce to a fine spray.

Automatic Expansion Valve—A pressure-actuated device which regulates the flow of refrigerant from the liquid line into the evaporator to maintain a constant evaporator pressure.

Automatic Refrigerating System—A system which regulates itself to maintain a definite set of conditions by means of automatic controls and valves usually responsive to temperature or pressure.

Baffle—A partition used to divert the flow of a fluid.

Balanced Pressure—The same pressure in a system or container that exists outside the system or container.

Barometer—An instrument for measuring atmospheric pressure.

Bleeder—A pipe sometimes attached to a condenser to lead off liquid refrigerant parallel to the main flow.

Boiler—A closed vessel in which steam is generated or in which water is heated.

Boiling Point—The temperature at which a liquid is vaporized upon the addition of heat, dependent upon the refrigerant and the absolute pressure at the surface of the liquid and vapor.

Brine—Any liquid cooled by the refrigerating system and used for the transmission of heat.

Brine System of Cooling—A system whereby brine, cooled by a refrigerating system, is circulated through pipes to the point where the refrigeration is needed.

British Thermal Unit (Btu)—The amount of heat required to raise the temperature of one pound of water one degree Fahrenheit. It is also the measure of the amount of heat removed in cooling one pound of water one degree Fahrenheit and is so used as a measure of refrigerating effect.

Butane—A flammable, hydrocarbon refrigerant used to a limited extent in small units.

Calcium Chloride—A chemical having the formula $CaCl_2$ which, in granular form, is used as a drier. This material is soluable in water, and in the presence of large quantities of moisture may dissolve and plug up the drier unit or even pass into the system beyond the drier.

Calcium Sulfate—A solid chemical, of the formula $CaSO_4$, which may be used as a drying agent.

Calibration—The process of dividing and numbering the scale of an instrument; also of correcting and determining the error of an existing scale.

Calorie—The heat per unit of weight required to raise the temperature of water one degree Centigrade. Thus, a gram calorie is the amount of heat required to raise the temperature of one gram of water one degree Centigrade.

Capacity—In a refrigerating machine, it is the heat absorbing capacity per unit time, usually measured in tons or Btu/hr.

Capacity, Container—The ability of a container to hold a material. The quantity of material which may safely be contained in a container.

Capacity, Refrigerating—The ability of a refrigerating system, or part thereof, to remove heat. Expressed as a rate of heat removal, it is usually measured in Btu/hr., or tons/24 hrs.

Carbon Dioxide Ice—Compressed solid CO_2. Dry ice.

Carbon Tetrachloride—A liquid having the formula CCl_4 (also known as carbona) which is a nonflammable solvent used for removing grease and oil and loosening sludges.

Centigrade—A thermometric system in which the freezing point of water is called $0°$ and its boiling point $100°$ at normal barometric pressure.

Centipoise—A unit of viscosity used in figuring pressure drop, etc.

Centrifugal Machine—A compressor employing centrifugal force for compression.

Centrifuge—A device for separating liquids of different densities by centrifugal action.

Change of State—A change from one state to another, as from a liquid to a solid, from a liquid to a gas, etc.

Charge—The amount of refrigerant in a system.

Charging—Putting in a charge.

Coefficient of Expansion—The fractional increase in length or volume of a material per degree rise in temperature.

Coil—Any cooling element made of pipe or tubing.

Cold Storage—A trade or process of preserving perishables on a large scale by refrigeration.

Comfort Chart—A psychrometric chart. Strictly, a chart showing effective temperatures.

Compound Compressor—A compressor in which compression is accomplished by stages in two or more cylinders.

Compression System—A refrigerating system in which the pressure imposing element is mechanically operated.

Compressor—That part of a mechanical refrigerating system which receives the refrigerant vapor at low pressure and compresses it into a smaller volume at higher pressure.

Condenser—A heat transfer device which receives high-pressure vapor at temperatures above that of the cooling mediums, such as air or water, to which the condenser passes latent heat from the refrigerant, causing the refrigerant vapor to liquefy.

Condensing—The process of giving up latent heat of vaporization in order to liquefy a vapor.

Conduction, Thermal—Passage of heat from one point to another by transmission of molecular energy from particle to particle through a conductor.

Conductivity, Thermal—The ability of a material to pass heat from one point to another, generally expressed in terms of Btu per hour per square foot of material per inch of thickness per degree temperature difference.

Conductor, Electrical—A material which will pass an electric current as part of an electrical system.

Constant-Pressure Valve—A valve of the throttling type, responsive to pressure, located in the suction line of an evaporator to maintain a desired constant pressure in the evaporator higher than the main suction line pressure.

Constant-Temperature Valve—A valve of the throttling type, responsive to the temperature of a thermostatic bulb. This valve is located in the suction line of an evaporator to reduce the refrigerating effect on the coil to just maintain a desired minimum temperature.

Control—Any device for the regulation of a machine in normal operation, either manual or automatic. If automatic, the implication is that it is responsive to temperature, pressure, liquid level, or time.

Control, Low-Pressure—An electric switch, responsive to pressure, connected into the low-pressure part of a refrigerating system. Usually closes at high pressure and opens at low pressure.

Control, Temperature—An electric switch responsive to the temperature of a thermostatic bulb or element.

Cooling Unit—A specific air-treating combination consisting of a means for air circulation and cooling within prescribed temperature limits.

Cooling Water—Water used for condensation of refrigerant. Condenser water.

Convection—Passage of heat from one point to another by means of a gravity fluid circulation due to changes in density resulting from absorbing and giving up heat.

Copper Plating—Formation of a film of copper, usually on compressor walls, pistons, or discharge valves.

Counterflow—In the heat exchange between two fluids, the opposite direction of flow, the coldest portion of one meeting the coldest portion of the other.

Critical Pressure—The vapor pressure corresponding to the critical temperature.

Critical Temperature—The temperature above which a vapor cannot be liquefied, regardless of pressure.

Critical Velocity—The velocity above which fluid flow is turbulent.

Crohydrate—A eutectic brine mixture of water and any salt mixed in proportions to give the lowest freezing temperature.

Cycle of Refrigeration—A complete course of operation of a refrigerant back to the starting point, measured in thermodynamic terms. Also used in general for any repeated process for any system.

Defrosting—Removal of accumulated ice from the cooling unit.

Defrosting Cycle—A cycle which permits the cooling unit to defrost during an off period.

Degree Day—A unit based upon temperature difference and time used to specify the nominal heating load in winter. For one day, there exists as many degree days as there are degrees F difference in temperature between the average outside air temperature, taken over a 24-hour period, and a temperature of 65° F.

Dehumidify—To remove water vapor from the atmosphere. To remove water or liquid from stored goods.

Dehydrator—A device used to remove moisture from the refrigerant.

Density—The mass or weight per unit of volume.

Dew Point (of air)—The temperature at which a specified sample of air, with no moisture added or removed, is completely saturated. The temperature at which the air, on being cooled, gives up moisture or dew.

Differential (of a control)—The difference between the cut-in and cut-out temperature or pressure.

Direct Expansion—A system in which the evaporator is located in the material or space refrigerated or in the air-circulating passages communicating with such space.

Displacement, Actual—The volume of gas at the compressor inlet actually moved in a given time.

Displacement, Theoretical—The total volume displaced by all the pistons of a compressor for every stroke during a definite interval. Usually measured in cubic feet per minute.

Domestic Refrigerator—A refrigerator for home use.

Drier—Synonymous with dehydrator.

Dry-Bulb Temperature—The actual temperature of air, as opposed to wet-bulb temperature.

Dry-Type Evaporator—An evaporator of the continuous-tube type where the refrigerant from a pressure-reducing device is fed into one end and the suction line connected to the outlet end.

Ebullator—A device inserted in flooded evaporator tubes to prevent the evaporator from becoming oil bound.

Efficiency, Mechanical—The ratio of the output of a machine to the input in equivalent units.

Efficiency, Volumetric—The ratio of the volume of gas actually pumped by a compressor or pump to the theoretical displacement of the compressor.

Ejector—A device which utilizes static pressure to build up a high fluid velocity in a restricted area to obtain a lower static pressure at that point so that fluid from another source may be drawn in.

Emulsification—Formation of an emulsion, i.e., a mixture of small droplets of two or more liquids which do not dissolve in each other.

Entropy—In thermodynamics, the base of a heat diagram, the area of which is heat units and the altitude of which is absolute temperature.

Eutectic Solution—A solution of such concentration as to have a constant freezing point at the lowest freezing temperature for the solution.

Evaporation—The change of state from liquid to vapor.

Evaporative Condenser—A refrigerant condenser utilizing the evaporation of water by air at the condenser surface as a means of dissipating heat.

Evaporative Cooling—The process of cooling by means of the evaporation of water in air.

Evaporator—A device in which the refrigerant evaporates while absorbing heat.

Expansion Valve, Automatic—A device which regulates the flow of refrigerant from the liquid line into the evaporator to maintain a constant evaporator pressure.

Expansion Valve, Thermostatic—A device to regulate the flow of refrigerant into an evaporator so as to maintain an evaporation temperature in a definite relationship to the temperature of a thermostatic bulb.

Extended Surface—The evaporator or condenser surface which is not a primary surface. Fins or other surfaces which transmit heat from or to a primary surface which is part of the refrigerant container.

External Equalizer—In a thermostatic expansion valve, a tube connection from the chamber containing the evaporation pressure-actuated element of the valve to the outlet of the evaporator coil. A device to compensate for excessive pressure drop through the coil.

Fahrenheit—A thermometric system in which 32° denotes the freezing point of water and 212° the boiling point under normal pressure.

Filter—A device to remove solid material from a fluid by a straining action.

Flammability—The ability of a material to burn.

Flare Fitting—A type of soft-tube connector which involves the flaring of the tube to provide a mechanical seal.

Flash Gas—The gas resulting from the instantaneous evaporation of the refrigerant in a pressure-reducing device to cool the refrigerant to the evaporation temperature obtained at the reduced pressure.

Float Valve—Valve actuated by a float immersed in a liquid container.

Flooded System—One in which the refrigerant enters into a header from a pressure-reducing valve and in which the evaporator maintains a liquid level. Opposed to dry evaporator.

Fluid—A gas or liquid.

Foaming—Formation of a foam or froth of oil refrigerant due to rapid boiling out of the refrigerant dissolved in the oil when the pressure is suddenly reduced. This occurs when the compressor operates, and, if large quantities of refrigerant have been dissolved, large quantities of oil may ''boil'' out and be carried through the refrigerant lines.

Freeze Up—Failure of a refrigeration unit to operate normally due to formation of ice at the expansion valve. The valve may be frozen closed or open, causing improper refrigeration in either case.

Freezing Point—The temperature at which a liquid will solidify upon the removal of heat.

Freon-12—The common name for dichlorodifluoromethane (CCl_2F_2).

Frost Back—The flooding of liquid from an evaporator into the suction line, accompanied by frost formation on the suction line in most cases.

Fusible Plug—A safety plug used in vessels containing refrigerant. The plug is designed to melt at high temperatures (usually about 165° F) to prevent excessive pressure from bursting the vessel.

Gas—The vapor state of a material.

Gauge—An instrument for measuring pressure or liquid level.

Glycerol—This material has been suggested as a lubricant, but has found little favor because of its tendency to rapidly absorb moisture. A suitable oil is more satisfactory

Head Pressure—The pressure in the compressor head. Discharge pressure.

Heat—Basic form of energy which may be partially converted into other forms and into which all other forms may be entirely converted.

Heat Content—See Total heat.

Hermetically-Sealed Unit—A refrigerating unit having no mechanical connections and containing no shaft seal.

High-Pressure Cutout—A control device connected into the high-pressure part of a refrigerating system to stop the machine when the pressure becomes excessive.

High Side—That part of the refrigerating system containing the high-pressure refrigerant. Also used to refer to the condensing unit, consisting of the motor, compressor, condenser, and receiver mounted on a single base.

High-Side Float Valve—A float valve which floats in high-pressure liquid. Opens on an increase in liquid level.

Hold Over—In an evaporator, the ability to stay cold after heat removal from the evaporator stops.

Horsepower—A unit of power. Work done at the rate of 33,000 foot pounds per minute or 550 foot pounds per second.

Humidifier—A device to add moisture to the air.

Humidistat—A control device sensitive to relative humidity.

Humidity, Absolute—The definite amount of water contained in a definite quantity of air. Usually measured in grains of water per pound or per cubic foot of air.

Humidity, Relative—The ratio of the water-vapor pressure of air compared to the vapor pressure it would have if saturated at its dry-bulb temperature. Very nearly the ratio of the amount of moisture contained in air compared to what it could hold at the existing temperature.

Hydrocarbons—A series of chemicals of similar chemical nature ranging from methane (the main constituent of natural gas) through butane, octane, etc., to heavy lubricating oils. All are more or less flammable. Butane and isobutane have been used to a limited extent as refrigerants.

Hydrochloric Acid, Hydrogen Chloride—An acid commonly used in soldering operations. It is formed when appreciable quantities of moisture are present in *Freon-12* or methyl chloride, or when these refrigerants come in contact with a flame or hot object. The acid vapor is pungent and warns of its presence.

Hydrofluoric Acid, Hydrogen Fluoride—This acid is formed when *Freon-12* passes through a flame. Although the vapor is pungent at moderate concentrations, lower, harmful concentrations may be undetected.

Hydrolysis—Reaction of a material, such as *Freon-12* or methyl chloride, with water. Acid materials in general are formed.

Hydrostatic Pressure—The pressure due to liquid in a container which contains no gas space.

Ice-Melting Equivalent—The amount of heat (144 Btu) absorbed by one pound of ice at 32° F in liquefying to water at 32° F.

Indirect System of Cooling—See **Brine System of Cooling.**

Infiltration—The leakage of air into a building or space.

Insulation (heat)—Material of low heat conductivity.

Irritant Refrigerant—Any refrigerant which has an irritating effect on the eyes, nose, throat, or lungs.

Isobutane—A hydrocarbon refrigerant used to a limited extent. It is flammable.

Kilowatt—Unit of electrical power equal to 1000 watts or 1.34 horse-power, approximately.

Lag of Temperature Control—The delay in action of a temperature-responsive element due to the time required for the temperature of the element to reach the surrounding temperature.

Latent Heat—The quantity of heat which may be added to a substance during a change of state without causing a temperature change.

Latent Heat of Evaporation—The quantity of heat required to change one pound of liquid into a vapor with no change in temperature. Reversible.

Leak Detector—A device used to detect refrigerant leaks in a refrigerating system.

Liquid—The state of a material in which the top surface in a vessel will become horizontal. Distinguished from solid or vapor forms.

Liquid Line—The tube or pipe which carries the refrigerant liquid from the condenser or receiver of a refrigerating system to a pressure-reducing device.

Liquid Receiver—That part of the condensing unit that stores the liquid refrigerant.

Load—The required rate of heat removal.

Low-Pressure Control—An electric switch and pressure-responsive element connected into the suction side of a refrigerating unit to control the operation of the system.

Low Side—That part of a refrigerating system which normally operates under low pressure as opposed to the high side. Also used to refer to the evaporator.

Low-Side Float Valve—A valve operated by the low-pressure liquid. Opens at a low level and closes at a high level.

Manometer—A U-shaped tube for measuring pressure differences.

Mechanical Efficiency—The ratio of work done by a machine to the work done on it or energy used by it.

Methanol—See **Methyl Alcohol.**

Methyl Alcohol—This material, also called wood alcohol, has a formula of CH_3OH. It has been used as an antifreeze, but is corrosive.

Methyl Chloride—A refrigerant having the chemical formula CH_3Cl.

Micron Measurement—Actually, a micron is a unit of measure in the metric system and is equivalent to 1/1000 millimeter. There are 25.4 millimeters to an inch. Therefore, one micron is equal to 0.00004 inch. One hundred microns are equivalent to 0.004 inch, or 29.996 inches of vacuum. Vacuum of less than 29.0 inches of mercury are not satisfactory for evacuating compressors.

Mollier Chart—A graphical representation of thermal properties of fluids, with total heat and entropy as co-ordinates.

Motor—A device for transforming electrical energy into mechanical energy.

Non-condensibles—Foreign gases mixed with a refrigerant and which cannot be condensed into the liquid form at the temperatures and pressures at which the refrigerant condenses.

Oil Trap—A device to separate oil from the high-pressure vapor from the compressor. Usually contains a float valve to return the oil to the compressor crankcase.

Over-Run—In ice-cream freezing, the ratio of the volume of ice cream to the volume of the mix used.

Ozone—The O_3 form of oxygen. Used in air conditioning as an odor eliminator. Harmful in high concentrations.

Packing—The stuffing around a shaft to prevent fluid leakage between the shaft and parts around the shaft.

Packless Valve—A valve which does not use packing to prevent leaks around the valve stem. Flexible material is usually used to seal against leaks and still permit valve movement.

Performance Factor—The ratio of the heat moved by a refrigerating system to the heat equivalent of the energy used. Varies with conditions.

Phosphorus Pentoxide—An efficient drier material which becomes gummy reacting with moisture, and hence is not used alone as a drying agent.

Pour Point (of oils)—The temperature below which the oil surface will not change when the oil container is tilted.

Power—The rate of doing work, measured in horsepower, watts, kilowatts, etc.

Power Factor (of an electrical device)—The ratio of watts to volt amperes in an alternating current circuit.

Pressure—The force exerted per unit of area.

Pressure Drop—Loss in pressure, as from one end of a refrigerant line to the other, due to friction, static head, etc.

Psychrometric Chart—A chart used to determine the specific volume, heat content, dew point, relative humidity, absolute humidity, and wet-bulb and dry-bulb temperatures, knowing any two independent items of those mentioned.

Purging—The act of blowing out refrigerant gas from a refrigerant-containing vessel, usually for the purpose of removing noncondensibles.

Pyrometer—An instrument for the measurement of high temperatures.

Radiation—The passage of heat from one object to another without warming the space between. The heat is passed by wave motion similar to light.

Refrigerant—The medium of heat transfer in a refrigerating system which picks up heat by evaporating at a low temperature and gives up heat by condensing at a higher temperature.

Refrigerating System—A combination of parts in which a refrigerant is circulated for the purpose of extracting heat.

Relative Humidity—The ratio of the water-vapor pressure of air compared to the vapor pressure it would have if saturated at its dry-bulb temperature. Very nearly the ratio of the amount of moisture contained in air compared to what it could hold at the existing temperature.

Relief Valve—A valve designed to open at excessively high pressures to allow the refrigerant to escape.

Resistance (electrical)—The opposition to electric current flow, measured in ohms.

Resistance (thermal)—The reciprocal of thermal conductivity.

Room Cooler—A cooling element for a room. In air conditioning, a device for conditioning small volumes of air for comfort.

Rotary Compressor—A compressor in which compression is attained in a cylinder by rotation of a semiradial member.

Running Time—Usually indicates percent of time a refrigerant compressor operates.

Saponifiable—Capable of reacting with an alkali (i.e., lye) to form a soap. Fats may be saponified.

Saturated Vapor—Vapor not superheated, but of 100% quality; i.e., containing no unvaporized liquid.

Seal Shaft—A mechanical system of parts for preventing gas leakage between a rotating shaft and a stationary crankcase.

Sealed Units—See **Hermetically Sealed Units.**

Sensible Heat—Heat which raises temperature.

Shell and Tube—Pertaining to heat exchangers in which a coil of tubing or pipe is contained in a shell or container. The pipe is provided with openings to allow the passage of a fluid through it, while the shell is also provided with an inlet and outlet for a fluid flow.

Silica Gel—A drier material having the formula SiO_2.

Sludge—A decomposition product formed in a refrigerant due to impurities in the oil or due to moisture. Sludges may be gummy or hard.

Soda Lime—A material used for removing moisture. Not recommended for refrigeration use.

Solenoid Valve—A valve opened by the magnetic effect of an electric current through a solenoid coil.

Solid—The state of matter in which a force can be exerted in a downward direction only when not confined. As distinguished from fluids.

Solubility—The ability of one material to enter into solution with another.

Solution—The homogeneous liquid misture of two or more materials.

Specific Gravity—The weight of a volume of a material compared to the weight of the same volume of water.

Specific Heat—The quantity of heat required to raise the temperature of a definite mass of a material a definite amount compared to that required to raise the temperature of the same mass of water the same amount. May be expressed as Btu lb./°F.

Specific Volume—The volume of a definite weight of a material. Usually expressed in terms of cubic feet per pound. The reciprocal of density.

Standard Air—Air weighing 0.7488 pound per cubic foot, which is air at 68° F dry bulb and 50% relative humidity at a barometric pressure of 29.92 in. of mercury, or approximately dry air at 70° F at the same pressure.

Static Pressure—Pressure against the walls of tubes, pipes or ducts.

Subcooled—Cooled below the condensing temperature corresponding to the existing pressure.

Sublimation—The change from a solid to a vapor state without an intermediate liquid state.

Suction Line—The tube or pipe which carries the refrigerant vapor from the evaporator to the compressor inlet.

Suction Pressure—Pressure on the suction side of the compressor.

Sulfur Dioxide—A refrigerant having the formula SO_2. Forms sulfuric acid when mixed with water.

Superheat—The heat contained in a vapor above its heat content at the boiling point at the existing pressure.

Superheated Gas—A gas whose temperature is higher than the evaporation temperature at the existing pressure.

Superheat Setting of Thermostatic Expansion Valve—The depression of the evaporation temperature at the valve feed pressure below the temperature of the thermostatic bulb.

Sweating—Condensation of moisture from the air on surfaces below the dewpoint temperature.

Temperature—Heat level or pressure. The thermal state of a body with respect to its ability to pick up heat from or pass heat to another body.

Temperature Control Lag—The delay in action of a temperature-responsive element due to the time required for the element to reach the surrounding temperature.

Thermal Conductivity—The ability of a material to conduct heat from one point to another. Indicated in terms of Btu/hr./sq. ft./in. thickness/°F.

Thermocouple—A device consisting of two electrical conductors having two junctions, one at a point whose temperature is to be measured, and the other at a known temperature. The temperature between the two junctions is determined by the material characteristics and the electrical potential set up.

Thermodynamic—The science of the mechanics of heat.

Thermometer—A device for indicating temperature.

Thermostat—A temperature-actuated switch.

Thermostatic Expansion Valve—A device to regulate the flow of refrigerant into an evaporator so as to maintain an evaporation temperature in a definite relationship to the temperature of a thermostatic bulb.

Ton of Refrigeration Capacity—Refrigeration equivalent to the melting of one ton of ice per 24 hours. 288,000 Btu/day, 12,000 Btu/hr., or 200 Btu/minute.

Total Heat—The total heat added to a refrigerant above an arbitrary starting point to bring it to a given set of conditions (usually expressed in Btu/lb.). For instance, in a superheated gas, the combined heat added to the liquid necessary to raise its temperature from an arbitrary starting point to the evaporation temperature to complete evaporation, and to raise the temperature to the final temperature where the gas is superheated.

Total Pressure—In fluid flow, the sum of static pressure and velocity pressure.

Toxicity—The characteristic of a material to intoxicate or poison.

Turbulent Flow—Fluid flow in which the fluid moves transversely as well as in the direction of the tube or pipe axis, as opposed to streamline or viscous flow.

Unit System—A system which can be removed from the user's premises without disconnecting any refrigerant-containing parts, water connections, or fixed electrical connections.

Unloader—A device in a compressor for equalizing high-side and low-side pressures when the compressor stops, and for a brief period after it starts, so as to decrease the starting load on the motor.

Vacuum—A pressure below atmospheric, usually measured in inches of mercury below atmospheric pressure.

Vapor Pressure—The pressure existing at the liquid and vapor surface of a refrigerant. Depends upon temperature.

Viscosity—The property of fluid to resist flow or change of shape.

Wax—A material which may separate when oil/refrigerant mixtures are cooled. Wax may plug the expansion valve and reduce heat transfer of the coil.

Wet Compression—A system of refrigeration in which some liquid refrigerant is mixed with vapor entering the compressor so as to cause the discharge vapors from the compressor to tend to be saturated rather than superheated.

Xylene—A flammable solvent, similar to kerosene, used for dissolving or loosening sludges, and for cleaning compressors and lines.

Zero, Absolute, of Pressure—The pressure existing in a vessel which is entirely empty. The lowest possible pressure. Perfect vacuum.

Zero, Absolute, of Temperature—The temperature at which a body has no heat in it ($-459.6°$ F, or $-273.1°$C).

Appendix

Temperature Requirements for Coolers and Refrigerators

Name of Product	Approximate Cooler Temperature	Approximate Operating Refrigerant Suction Temperature	Approximate Temperature Difference in Ordering Refrigerating Coils
Chee.e, Eggs, Butter	Serving Cooler, 50°to 60° Storage Cooler, 31° to 36°	30° 15°	15° to 20° 15°
Poultry and Meats (Fresh)	Cooler, 34° to 38° Cases, 38° to 42°	20° 20° to 25°	12° to 15°
Poultry and Meats (Frozen)	−5° to +5°	−15° to −5°	10°
Vegetables, Fruits	35° to 45°	25° to 30°	10°
Cream and Milk	Can Water Coiling, 36° to 38° Cooler Storage, 36° to 40°	15° to 20° 20°	15° to 20°
Beer Storage	36° to 45°	20°	15° to 20°
Fur Storage	Shock Temperature, 10° to 15° Max. Temperature, 38°	0° 15°	10° 20° to 25°
Ice Cream, Hardening	−20°	−30°	5° to 10°
Ice Cream, Storing	−5° to −10°	−15° to −20°	10°

Courtesy Dunham-Bush, Inc.

Solutions of Calcium Chloride (CaCl₂)

Percent of Pure Salt by Weight	Freezing Point, ° F	Ammonia Gauge Pressure, Lbs. at Freezing Point	Specific Gravity † 60° F †	Baume Density 60° F	Specific Heat Btu per Lb. ° F at 60° F	Weight in Lbs. per Cu. Ft.	
						CaCl,	Brine
0	32.0	47.6	1.000	0.0	1.000	0.00	62.40
5	29.0	43.8	1.044	6.1	0.9246	3.26	65.15
6	28.0	42.6	1.050	7.0	.9143	3.93	65.52
7	27.0	41.4	1.060	8.2	.8984	4.63	66.14
8	25.5	39.0	1.069	9.3	.8842	5.34	66.70
9	24.0	37.9	1.078	10.4	.8699	6.05	67.27

Solutions of Calcium Chloride (CaCl₂) (Cont'd)

Percent of Pure Salt by Weight	Freezing Point, ° F	Ammonia Gauge Pressure, Lbs. at Freezing Point	Specific Gravity † 60° F † 39° F	Baume Density 60° F	Specific Heat Btu per Lb. ° F at 60° F	Weight in Lbs. per Cu. Ft.	
						CaCl₂	Brine
10	23.0	36.8	1.087	11.6	.8556	6.78	67.83
11	21.5	35.0	1.096	12.6	.8429	7.52	68.33
12	19.0	32.5	1.105	13.8	.8284	8.27	68.95
13	17.0	30.4	1.114	14.8	.8166	9.04	69.51
14	14.5	28.0	1.124	15.9	.8043	9.81	70.08
15	12.5	26.0	1.133	16.9	.793	10.6	70.64
16	9.5	23.4	1.143	18.0	.7798	11.4	71.26
17	6.5	20.8	1.152	19.1	.7672	12.22	71.89
18	3.0	18.0	1.162	20.2	.7566	13.05	72.51
19	0.0	15.7	1.172	21.3	.746	13.9	73.13
20	— 3.0	13.6	1.182	22.1	.7375	14.73	73.63
21	— 5.5	11.9	1.192	23.0	.729	15.58	74.19
22	—10.5	8.8	1.202	24.4	.7168	16.50	75.0
23	—15.5	5.9	1.212	25.5	.7076	17.4	75.63
24	—20.5	3.4	1.223	26.4	.6979	13.32	76.32
25	—25.0	1.3	1.233	27.4	.6899	19.24	76.94
26	—30.0	1.6*	1.244	28.3	.682	20.17	77.56
27	—36.0	6.1*	1.254	29.3	.6735	21.13	78.25
28	—43.5	10.6*	1.265	30.4	.6657	22.1	78.94
29	—53.0	15.7*	1.276	31.4	.6584	23.09	79.62
29.5	—58.0	17.8*	1.280	31.7	.6557	23.56	79.87

†Specific Gravity and Weights at 60° F referred to water at 39° F.
*Inches Vacuum.
Note: To obtain the weight of commercial salt, divide the values given for anhydrous salt in the table by the percent purity of the commercial calcium salt. A calcium brine freezing at 0° and requiring 13.9 pounds of "pure" salt per cubic foot will require .73 into 13.9, or 19 pounds of 73% salt. To obtain the weights of water for other than anhydrous salt, subtract the weight of that salt, as just determined, from the weight of the brine. (73.13—19 equals 54.13).

Courtesy Frick Company, Inc.

Solutions of Sodium Chloride (NaCl)
(common salt)

Percent of Pure Salt by Weight	Freezing Point, ° F	Ammonia Gauge Pressure, Lbs. at Freezing Point	Specific Gravity † 60° F † 39° F	Baume Density 60° F	Specific Heat Btu	per Lb. per Lb. per per Lb. per ° F at 60° F	Weight in Lbs. per Cu. Ft.	
							NaCl	Brine
0	32.0	47.6	1.000	0.0	0.0	1.000	0.000	62.4
5	27.0	41.4	1.035	5.1	18.2	0.938	3.230	64.6
6	25.5	39.6	1.043	6.1	22.5	0.927	3.906	65.1
7	24.0	37.9	1.050	7.0	26.0	0.917	4.585	65.5
8	23.2	37.0	1.057	8.0	29.6	0.907	5.280	66.0
9	21.8	35.6	1.065	9.0	33.5	0.897	9.985	66.5
10	20.4	34.1	1.072	10.1	37.2	0.888	6.690	66.9
11	18.5	32.0	1.080	10.8	41.1	0.879	7.414	67.4
12	17.2	30.6	1.087	11.8	44.8	0.870	8.136	67.8

Solutions of Sodium Chloride (NaCl) (Cont'd)
(common salt)

Percent of Pure Salt by Weight	Freezing Point, ° F	Ammonia Gauge Pressure, Lbs. at Freezing Point	Specific Gravity † 60° F † 39° F	Baume Density 60° F	Specific Heat 60° F Btu	per Lb. per Lb. per per Lb. per ° F at 60° F	Weight in Lbs. per Cu. Ft.	
							NaCl	Brine
13	15.5	28.9	1.095	12.7	48.7	0.862	8.879	68.3
14	13.9	27.4	1.103	13.6	52.6	0.854	9.632	68.8
15	12.0	25.6	1.111	14.5	56.8	0.847	10.395	69.3
16	10.2	24.0	1.118	15.4	60.0	0.840	11.168	69.8
17	8.2	22.3	1.126	16.3	64.0	0.833	11.951	70.3
18	6.1	20.5	1.134	17.2	68.0	0.826	12.744	70.8
19	4.0	18.8	1.142	18.1	71.7	0.819	13.547	71.3
20	1.8	17.1	1.150	19.0	75.2	0.813	14.360	71.8
21	— 0.8	15.1	1.158	19.9	79.1	0.807	15.183	72.3
22	— 3.0	13.6	1.166	20.8	82.8	0.802	16.016	72.8
23	— 6.0x	11.6	1.175	21.7	86.8	0.796	16.854	73.3
24	3.8	18.6	1.183	22.5	90.2	0.791	17.712	73.8
25	16.1	29.5	1.191	23.4	94.0	0.786	18.575	74.3
26	32.0	47.6	1.200			....	....	......

†Specific Gravity and Weights at 59° referred to water at 39° F.
Note: Weight given for pure NaCl salt can be used for commercial salt with a theoretical error of about 1%.

x Eutectic Point: addition of more salt raises the freezing point.

Courtesy Frick Company, Inc.

*Suction Line Sizes for R-12
(O.D. IN INCHES)

Evaporator Capacity Net	40°				20°				0°				−20°				−40°			
	Total Equivalent Length of Suction Line (Feet)																		
	25	50	100	150	25	50	100	150	25	50	100	150	25	50	100	150	25	50	100	150
3000	⅜	½	½	⅝	½	½	½	⅝	⅝	⅝	¾	¾	⅝	¾	⅞	⅞	¾	⅞	1⅛	1⅛
6000	½	½	⅝	¾	½	⅝	¾	¾	¾	⅞	⅞	1⅛	⅞	1⅛	1⅛	1⅛	1⅛	1⅛	1⅜	1⅜
9000	⅝	⅝	¾	¾	⅝	¾	⅞	⅞	⅞	1⅛	1⅛	1⅛	1⅛	1⅛	1⅜	1⅜	1⅛	1⅜	1⅜	1⅝
12,000	⅝	¾	¾	⅞	¾	¾	⅞	1⅛	⅞	1⅛	1⅛	1⅜	1⅛	1⅛	1⅜	1⅜	1⅜	1⅜	1⅝	2⅛
18,000	¾	¾	⅞	1⅛	¾	⅞	1⅛	1⅛	1⅛	1⅛	1⅜	1⅜	1⅛	1⅜	1⅝	1⅝	1⅜	1⅝	2⅛	2⅛
24,000	⅞	⅞	1⅛	1⅛	⅞	1⅛	1⅛	1⅜	1⅛	1⅜	1⅝	1⅝	1⅜	1⅝	1⅝	2⅛	1⅝	2⅛	2⅛	2⅝
30,000	⅞	1⅛	1⅛	1⅛	1⅛	1⅛	1⅛	1⅜	1⅜	1⅝	1⅝	2⅛	1⅜	1⅝	2⅛	2⅛	2⅛	2⅛	2⅝	2⅝
36,000	⅞	1⅛	1⅛	1⅜	1⅛	1⅜	1⅜	1⅜	1⅜	1⅝	2⅛	2⅛	1⅝	2⅛	2⅛	2⅛	2⅛	2⅛	2⅝	2⅝
42,000	1⅛	1⅛	1⅜	1⅜	1⅛	1⅜	1⅜	1⅝	1⅝	1⅝	2⅛	2⅛	1⅝	2⅛	2⅛	2⅛	2⅛	2⅝	2⅝	2⅝
48,000	1⅛	1⅛	1⅜	1⅜	1⅛	1⅜	1⅝	1⅝	1⅝	1⅝	2⅛	2⅛	1⅝	2⅛	2⅝	2⅝	2⅛	2⅝	2⅝	3⅛
54,000	1⅛	1⅛	1⅜	1⅜	1⅜	1⅜	1⅝	1⅝	1⅝	2⅛	2⅛	2⅛	2⅛	2⅛	2⅝	2⅝	2⅛	2⅝	2⅝	3⅛
60,000	1⅛	1⅜	1⅜	1⅝	1⅜	1⅜	1⅝	1⅝	1⅝	2⅛	2⅛	2⅝	2⅛	2⅛	2⅝	2⅝	2⅝	2⅝	3⅛	3⅛
72,000	1⅛	1⅜	1⅝	1⅝	1⅜	1⅝	1⅝	2⅛	2⅛	2⅛	2⅝	2⅝	2⅛	2⅝	2⅝	2⅝	2⅝	3⅛	3⅛	3⅛
90,000	1⅜	1⅜	1⅝	2⅛	1⅜	1⅝	2⅛	2⅛	2⅛	2⅝	2⅝	2⅝	2⅝	2⅝	3⅛	3⅛	2⅝	3⅛	3⅛	3⅝
120,000	1⅜	1⅝	2⅛	2⅛	1⅝	2⅛	2⅛	2⅛	2⅛	2⅝	2⅝	3⅛	2⅝	3⅛	3⅛	3⅛	3⅛	3⅛	3⅝	4⅛
150,000	1⅝	1⅝	2⅛	2⅛	1⅝	2⅛	2⅛	2⅝	2⅝	2⅝	3⅛	3⅛	2⅝	3⅛	3⅛	3⅝	3⅛	4⅛	4⅛	5⅛
180,000	1⅝	2⅛	2⅛	2⅝	2⅛	2⅛	2⅝	2⅝	2⅝	2⅝	3⅛	3⅛	2⅝	3⅛	3⅝	3⅝	3⅛	4⅛	4⅛	5⅛
210,000	1⅝	2⅛	2⅛	2⅝	2⅛	2⅛	2⅝	2⅝	2⅝	3⅛	3⅛	3⅛	3⅛	3⅛	3⅝	4⅛	3⅝	4⅛	5⅛	5⅛
240,000	2⅛	2⅛	2⅝	2⅝	2⅛	2⅝	2⅝	2⅝	2⅝	3⅛	3⅛	3⅝	3⅛	3⅝	4⅛	4⅛	3⅝	4⅛	5⅛	5⅛
300,000	2⅛	2⅛	2⅝	2⅝	2⅛	2⅝	3⅛	3⅛	3⅛	3⅛	3⅝	4⅛	3⅛	3⅝	4⅛	5⅛	4⅛	5⅛	5⅛	6⅛

* Type L Copper Tubing

Courtesy Dunham-Bush, Inc.

*Suction Line Sizes for R-22
(O.D. IN INCHES)

Net Evaporator Capacity	Suction Temperature																			
	40°				20°				0°				−20°				−40°			
	Total Equivalent Length of Suction Line (Feet)																			
	25	50	100	150	25	50	100	150	25	50	100	150	25	50	100	150	25	50	100	150
3,000	3/8	3/8	3/8	1/2	3/8	3/8	1/2	1/2	1/2	1/2	5/8	5/8	1/2	5/8	5/8	3/4	5/8	3/4	7/8	7/8
6,000	1/2	1/2	1/2	5/8	1/2	1/2	5/8	5/8	5/8	5/8	3/4	7/8	5/8	3/4	7/8	7/8	3/4	7/8	1 1/8	1 1/8
9,000	1/2	1/2	5/8	5/8	1/2	5/8	5/8	3/4	3/4	3/4	7/8	7/8	3/4	7/8	1 1/8	1 1/8	7/8	1 1/8	1 1/8	1 3/8
12,000	5/8	5/8	5/8	3/4	5/8	5/8	3/4	3/4	3/4	7/8	7/8	1 1/8	3/4	7/8	1 1/8	1 1/8	1 1/8	1 1/8	1 3/8	1 3/8
18,000	5/8	3/4	3/4	7/8	3/4	3/4	7/8	7/8	7/8	7/8	1 1/8	1 1/8	7/8	1 1/8	1 1/8	1 3/8	1 1/8	1 3/8	1 5/8	1 5/8
24,000	5/8	3/4	7/8	7/8	3/4	7/8	1 1/8	1 1/8	7/8	1 1/8	1 1/8	1 3/8	1 1/8	1 1/8	1 3/8	1 3/8	1 3/8	1 3/8	1 5/8	1 5/8
30,000	3/4	7/8	7/8	1 1/8	7/8	1 1/8	1 1/8	1 1/8	1 1/8	1 1/8	1 1/8	1 3/8	1 1/8	1 1/8	1 3/8	1 3/8	1 3/8	1 5/8	1 5/8	2 1/8
36,000	3/4	7/8	1 1/8	1 1/8	7/8	1 1/8	1 1/8	1 1/8	1 1/8	1 1/8	1 3/8	1 5/8	1 1/8	1 3/8	1 5/8	1 5/8	1 5/8	1 5/8	2 1/8	2 1/8
42,000	7/8	1 1/8	1 1/8	1 1/8	1 1/8	1 1/8	1 3/8	1 3/8	1 1/8	1 3/8	1 3/8	1 5/8	1 3/8	1 3/8	1 5/8	2 1/8	1 5/8	2 1/8	2 1/8	2 1/8
48,000	7/8	1 1/8	1 1/8	1 1/8	1 1/8	1 1/8	1 3/8	1 3/8	1 3/8	1 3/8	1 3/8	1 5/8	1 3/8	1 3/8	1 5/8	2 1/8	1 5/8	2 1/8	2 1/8	2 1/8
54,000	7/8	1 1/8	1 1/8	1 3/8	1 1/8	1 1/8	1 3/8	1 3/8	1 3/8	1 3/8	1 5/8	1 5/8	1 5/8	1 5/8	2 1/8	2 1/8	1 5/8	2 1/8	2 1/8	2 5/8
60,000	7/8	1 1/8	1 1/8	1 3/8	1 1/8	1 1/8	1 3/8	1 3/8	1 5/8	1 5/8	1 5/8	2 1/8	1 5/8	1 5/8	2 1/8	2 1/8	1 5/8	2 1/8	2 1/8	2 5/8
72,000	1 1/8	1 1/8	1 3/8	1 3/8	1 1/8	1 3/8	1 5/8	1 5/8	1 3/8	1 5/8	2 1/8	2 1/8	1 5/8	2 1/8	2 1/8	2 1/8	2 1/8	2 1/8	2 5/8	2 5/8
90,000	1 1/8	1 1/8	1 3/8	1 5/8	1 3/8	1 3/8	1 5/8	1 5/8	1 5/8	1 5/8	2 1/8	2 1/8	2 1/8	2 1/8	2 1/8	2 5/8	2 1/8	2 5/8	2 5/8	3 1/8
120,000	1 1/8	1 3/8	1 5/8	1 5/8	1 3/8	1 5/8	2 1/8	2 1/8	1 5/8	2 1/8	2 1/8	2 5/8	2 1/8	2 1/8	2 5/8	2 5/8	2 1/8	2 5/8	3 1/8	3 1/8
150,000	1 3/8	1 3/8	1 5/8	2 1/8	1 3/8	1 5/8	2 1/8	2 1/8	1 5/8	2 1/8	2 1/8	2 5/8	2 1/8	2 5/8	2 5/8	2 5/8	2 5/8	2 5/8	3 1/8	3 1/8
180,000	1 3/8	1 5/8	2 1/8	2 1/8	1 5/8	2 1/8	2 1/8	2 1/8	2 1/8	2 1/8	2 5/8	2 5/8	2 5/8	2 5/8	2 5/8	3 1/8	2 5/8	3 1/8	3 5/8	3 5/8
210,000	1 3/8	1 5/8	2 1/8	2 1/8	1 5/8	2 1/8	2 1/8	2 5/8	2 1/8	2 1/8	2 5/8	2 5/8	2 5/8	2 5/8	3 1/8	3 1/8	2 5/8	3 1/8	3 5/8	3 5/8
240,000	1 5/8	1 5/8	2 1/8	2 1/8	1 5/8	2 1/8	2 1/8	2 5/8	2 1/8	2 5/8	2 5/8	2 5/8	2 5/8	2 5/8	3 1/8	3 1/8	3 1/8	3 1/8	3 5/8	4 1/8
300,000	1 5/8	2 1/8	2 1/8	2 5/8	2 1/8	2 1/8	2 5/8	2 5/8	2 1/8	2 5/8	3 1/8	3 1/8	2 5/8	3 1/8	3 1/8	3 5/8	3 1/8	3 1/8	4 1/8	5 1/8

*Type L Copper Tubing

Courtesy Dunham-Bush, Inc.

Suction Line Sizes for R-502
(O.D. IN INCHES)

Net Evaporator Capacity	40°				20°				0°				−20°				−40°			
	Total Equivalent Length of Suction Line (Feet)																			
	25	50	100	150	25	50	100	150	25	50	100	150	25	50	100	150	25	50	100	150
3,000	3/8	3/8	3/8	1/2	3/8	3/8	1/2	1/2	1/2	1/2	5/8	5/8	1/2	5/8	5/8	3/4	5/8	3/4	7/8	7/8
6,000	3/8	1/2	1/2	5/8	1/2	1/2	5/8	5/8	5/8	5/8	3/4	7/8	5/8	3/4	7/8	7/8	3/4	7/8	1 1/8	1 1/8
9,000	1/2	1/2	5/8	5/8	1/2	5/8	5/8	3/4	3/4	3/4	7/8	7/8	3/4	7/8	1 1/8	1 1/8	7/8	1 1/8	1 1/8	1 3/8
12,000	1/2	5/8	3/4	3/4	5/8	5/8	3/4	3/4	3/4	7/8	7/8	1 1/8	7/8	7/8	1 1/8	1 1/8	1 1/8	1 3/8	1 3/8	1 3/8
18,000	5/8	3/4	3/4	7/8	3/4	3/4	7/8	7/8	7/8	7/8	1 1/8	1 1/8	7/8	1 1/8	1 1/8	1 3/8	1 1/8	1 3/8	1 5/8	1 5/8
24,000	3/4	3/4	7/8	7/8	3/4	7/8	1 1/8	1 1/8	7/8	1 1/8	1 1/8	1 3/8	1 1/8	1 1/8	1 3/8	1 3/8	1 3/8	1 3/8	1 5/8	1 5/8
30,000	3/4	7/8	1 1/8	1 1/8	7/8	7/8	1 1/8	1 1/8	1 1/8	1 1/8	1 3/8	1 3/8	1 1/8	1 1/8	1 3/8	1 5/8	1 3/8	1 5/8	1 5/8	2 1/8
36,000	3/4	7/8	1 1/8	1 1/8	7/8	1 1/8	1 1/8	1 1/8	1 1/8	1 1/8	1 3/8	1 3/8	1 1/8	1 3/8	1 5/8	1 5/8	1 3/8	1 5/8	2 1/8	2 1/8
42,000	7/8	7/8	1 1/8	1 3/8	7/8	1 1/8	1 1/8	1 3/8	1 1/8	1 3/8	1 3/8	1 5/8	1 3/8	1 3/8	1 5/8	2 1/8	1 5/8	1 5/8	2 1/8	2 1/8
48,000	7/8	1 1/8	1 1/8	1 3/8	1 1/8	1 1/8	1 3/8	1 3/8	1 3/8	1 3/8	1 5/8	1 5/8	1 3/8	1 5/8	1 5/8	2 1/8	1 5/8	2 1/8	2 1/8	2 1/8
54,000	7/8	1 1/8	1 1/8	1 3/8	7/8	1 1/8	1 3/8	1 3/8	1 3/8	1 3/8	1 5/8	1 5/8	1 5/8	1 5/8	2 1/8	2 1/8	1 5/8	2 1/8	2 1/8	2 5/8
60,000	7/8	1 1/8	1 1/8	1 3/8	1 1/8	1 3/8	1 3/8	1 3/8	1 3/8	1 3/8	1 5/8	2 1/8	1 5/8	1 5/8	2 1/8	2 1/8	1 5/8	2 1/8	2 1/8	2 5/8
72,000	1 1/8	1 1/8	1 3/8	1 3/8	1 1/8	1 3/8	1 3/8	1 5/8	1 3/8	1 5/8	1 5/8	2 1/8	1 5/8	1 5/8	2 1/8	2 1/8	2 1/8	2 1/8	2 5/8	2 5/8
90,000	1 1/8	1 3/8	1 3/8	1 5/8	1 1/8	1 3/8	1 5/8	1 5/8	1 5/8	1 5/8	2 1/8	2 1/8	1 5/8	2 1/8	2 1/8	2 5/8	2 1/8	2 5/8	3 1/8	3 1/8
120,000	1 1/8	1 5/8	1 5/8	1 5/8	1 3/8	1 5/8	1 5/8	1 5/8	1 5/8	2 1/8	2 1/8	2 1/8	2 1/8	2 1/8	2 5/8	2 5/8	2 1/8	2 5/8	3 1/8	3 1/8
150,000	1 3/8	1 3/8	1 5/8	2 1/8	1 3/8	1 5/8	2 1/8	2 1/8	1 5/8	2 1/8	2 1/8	2 5/8	2 1/8	2 1/8	2 5/8	2 5/8	2 5/8	3 1/8	3 1/8	3 1/8
180,000	1 3/8	1 5/8	2 1/8	2 1/8	1 5/8	2 1/8	2 1/8	2 1/8	2 1/8	2 1/8	2 5/8	2 5/8	2 1/8	2 5/8	2 5/8	3 1/8	2 5/8	3 1/8	3 5/8	3 5/8
210,000	1 3/8	1 5/8	2 1/8	2 1/8	1 5/8	2 1/8	2 1/8	2 5/8	2 1/8	2 1/8	2 5/8	2 5/8	2 5/8	2 5/8	3 1/8	3 1/8	3 1/8	3 1/8	3 5/8	3 5/8
240,000	1 5/8	1 5/8	2 1/8	2 1/8	1 5/8	2 1/8	2 1/8	2 5/8	2 1/8	2 5/8	2 5/8	3 1/8	2 5/8	3 1/8	3 1/8	3 1/8	3 1/8	3 5/8	3 5/8	4 1/8
300,000	1 5/8	1 5/8	2 1/8	2 5/8	2 1/8	2 1/8	2 5/8	2 5/8	2 5/8	2 5/8	3 1/8	3 1/8	2 5/8	3 1/8	3 1/8	3 5/8	3 1/8	3 5/8	4 1/8	5 1/8

*Type L Copper Tubing

Courtesy Dunham-Bush, Inc.

Mechanical, Electrical, and Heat Equivalents

Unit	Equivalent Value in Other Units	
1 kilowatt hour	1000	watt hours
	1.34	horsepower hours
	2,654,200	foot-pounds
	3,600,000	joules
	3412	heat units
	367,000	kilogram meters
	0.235	pound carbon oxidized with perfect efficiency
	3.53	pounds water evaporated from and at 212° Fahrenheit
	22.75	pounds water raised from 62° to 212° Fahrenheit
1 horsepower hour	0.746	kilowatt hour
	1,980,000	foot-pounds
	2545	heat units
	273,740	kilogram meters
	0.175	pound carbon oxidized with perfect efficiency
	2.64	pounds water evaporated from and at 212° Fahrenheit
	17.0	pounds water raised from 62° to 212° Fahrenheit
1 kilowatt	1000	watts
	1.34	horsepower
	2,654,200	foot-pounds per hour
	44,240	foot-pounds per minute
	737.3	foot-pounds per second
	3412	heat units per hour
	56.9	heat units per minute
	0.948	heat unit per second
	0.2275	pound carbon oxidized per hour
	3.53	pounds water evaporated per hour from and at 212° Fahrenheit
1 horsepower	746	watts
	0.746	kilowatt
	33,000	foot-pounds per minute
	550	foot-pounds per second
	2545	heat units per hour
	42.4	heat units per minute
	0.707	heat unit per second
	0.175	pound carbon oxidized per hour
	2.64	pounds water evaporated per hour from and at 212° Fahrenheit

Courtesy Calcium Chloride Institute

Properties of Standard Ammonia

(U. S. Bureau of Standards)

Temp. °F	Pressure in Pounds per Sq. In.		Volume of Vapor Cu. ft./lb.	Density of Vapor Pound per Cu. ft.	Heat Content		Latent Heat Btu per lb.	Entropy	
	Absolute	Gauge			Liquid Btu per lb.	Vapor Btu per lb.		Liquid Btu per °F	Vapor Btu per lb. per °F
−40	10.41	* 8.7	24.86	0.04022	0.0	597.6	597.6	0.0000	1.4242
−35	12.05	* 5.4	21.68	.04613	5.3	599.5	594.2	.0126	1.4120
−30	13.90	* 1.6	18.97	.05271	10.7	601.4	590.7	.0250	1.4001
−25	15.98	1.3	16.66	.06003	16.0	603.2	587.2	.0374	1.3886
−20	18.30	3.6	14.68	.06813	21.4	605.0	583.6	.0497	1.3774
−18	19.30	4.6	13.97	.07161	23.5	605.7	582.2	.0545	1.3729
−16	20.34	5.6	13.29	.07522	25.6	606.4	580.8	.0594	1.3686
−14	21.43	6.7	12.66	.07898	27.8	607.1	579.3	.0642	1.3643
−12	22.56	7.9	12.06	.08289	30.0	607.8	577.8	.0690	1.3600
−10	23.74	9.0	11.50	.08695	32.1	608.5	576.4	.0738	1.3558
− 9	24.35	9.7	11.23	0.08904	33.2	608.8	575.6	0.0762	1.3537
− 8	24.97	10.3	10.97	.09117	34.3	609.2	574.9	.0786	1.3516
− 7	25.61	10.9	10.71	.09334	35.4	609.5	574.1	.0809	1.3495
− 6	26.26	11.6	10.47	.09555	36.4	609.8	573.4	.0833	1.3474
− 5	26.92	12.2	10.23	.09780	37.5	610.1	572.6	.0857	1.3454
− 4	27.59	12.9	9.991	.1001	38.6	610.5	571.9	.0880	1.3433
− 3	28.28	13.6	9.763	.1024	39.7	610.8	571.1	.0904	1.3413
− 2	28.98	14.3	9.541	.1048	40.7	611.1	570.4	.0928	1.3393
− 1	29.69	15.0	9.326	.1072	41.8	611.4	569.6	.0951	1.3372
0	30.42	15.7	9.116	.1097	42.9	611.8	568.9	.0975	1.3352
1	31.16	16.5	8.912	0.1122	44.0	612.1	568.1	0.0998	1.3332
2	31.92	17.2	8.714	.1148	45.1	612.4	567.3	.1022	1.3312
3	32.69	18.0	8.521	.1174	46.2	612.7	566.5	.1045	1.3292
4	33.47	18.8	8.333	.1200	47.2	613.0	565.8	.1069	1.3273
5	34.27	19.6	8.150	.1227	48.3	613.3	565.0	.1092	1.3253
6	35.09	20.4	7.971	.1254	49.4	613.6	564.2	.1115	1.3234
7	35.92	21.2	7.798	.1282	50.5	613.9	563.4	.1138	1.3214
8	36.77	22.1	7.629	.1311	51.6	614.3	562.7	.1162	1.3195
9	37.63	22.9	7.464	.1340	52.7	614.6	561.9	.1185	1.3176
10	38.51	23.8	7.304	.1369	53.8	614.9	561.1	.1208	1.3157
12	40.31	25.6	6.996	0.1429	56.0	615.5	559.5	0.1254	1.3118
14	42.18	27.5	6.703	.1492	58.2	616.1	557.9	.1300	1.3081
16	44.12	29.4	6.425	.1556	60.3	616.6	556.3	.1346	1.3043
18	46.13	31.4	6.161	.1623	62.5	617.2	554.7	.1392	1.3006
20	48.21	33.5	5.910	.1692	64.7	617.8	553.1	.1437	1.2969
22	50.36	35.7	5.671	.1763	66.9	618.3	551.4	.1483	1.2933
24	52.59	37.9	5.443	.1837	69.1	618.9	549.8	.1528	1.2897
26	54.90	40.2	5.227	.1913	71.3	619.4	548.1	.1573	1.2861
30	59.74	45.0	4.825	.2073	75.7	620.5	544.8	.1663	1.2790
35	66.26	51.6	4.373	.2287	81.2	621.7	540.5	.1775	1.2704

Properties of Standard Ammonia (Cont'd)

(U. S. Bureau of Standards)

Temp. °F	Pressure in Pounds per Sq. In.		Volume of Vapor Cu. ft./lb.	Density of Vapor Pound per Cu. ft.	Heat Content		Latent Heat Btu per lb.	Entropy	
	Absolute	Gauge			Liquid Btu per lb.	Vapor Btu per lb.		Liquid Btu per lb. per °F	Vapor Btu per lb. per °F
40	73.32	58.6	3.971	0.2518	86.8	623.0	536.2	0.1885	1.2618
45	80.96	66.3	3.614	.2767	92.3	624.1	531.8	.1996	1.2535
50	89.19	74.5	3.294	.3036	97,9	625.2	527.3	.2105	1.2453
55	98.06	83.4	3.008	.3325	103.5	626.3	522.8	.2214	1.2373
60	107.6	92.9	2.751	.3635	109.2	627.3	518.1	.2322	1.2294
65	117.8	103.1	2.520	.3968	114.8	628.2	513.4	.2430	1.2216
70	128.8	114.1	2.312	.4325	120.5	629.1	508.6	.2537	1.2140
75	140.5	125.8	2.125	.4707	126.2	629.9	503.7	.2643	1.2065
78	147.9	133.2	2.021	.4949	129.7	630.4	500.7	.2706	1.2020
80	153.0	138.3	1.955	.5115	132.0	630.7	498.7	.2749	1.1991
82	158.3	143.6	1.892	0.5287	134.3	631.0	496.7	0.2791	1.1962
84	163.7	149.0	1.831	.5462	136.6	631.3	494.7	.3833	1.1933
86	169.2	154.5	1.772	.5643	138.9	631.5	492.6	.2875	1.1904
88	174.8	160.1	1.716	.5828	141.2	631.8	490.6	.2917	1.1875
90	180.6	165.9	1.661	.6019	143.5	632.0	488.5	.2958	1.1846
92	186.6	171.9	1.609	.6214	145.8	632.2	486.4	.3000	1.1818
94	192.7	178.0	1.559	.6415	148.2	632.5	484.3	.3041	1.1789
96	198.9	184.2	1.510	.6620	150.5	632.6	482.1	.3083	1.1761
98	205.3	190.6	1.464	.6832	152.9	632.9	480.0	.3125	1.1733
100	211.9	197.2	1.419	0.7048	155.2	633.0	477.8	0.3166	1.1705
105	228.9	214.2	1.313	.7615	161.1	633.4	472.3	.3269	1.1635
110	247.0	232.3	1.217	.8219	167.0	633.7	466.7	.3372	1.1566
115	266.2	251.5	1.128	.8862	173.0	633.9	460.9	.3474	1.1497
120	286.4	271.7	1.047	.9549	179.0	634.0	455.0	.3576	1.1427

*Inches of mercury below one standard atmosphere (29.92 in.).

Courtesy Calcium Chloride Institute

*Hot Gas Line Sizes

Net Evaporator Capacity	Total Equivalent Length of Hot Gas Line (Feet)							
	R12				R22 & R502			
	25	50	100	150	25	50	100	150
3000	1/2	1/2	5/8	5/8	3/8	3/8	1/2	1/2
4500	1/2	5/8	5/8	3/4	1/2	1/2	1/2	5/8
6500	5/8	5/8	3/4	3/4	1/2	1/2	5/8	5/8
8500	5/8	3/4	7/8	7/8	1/2	5/8	3/4	3/4
12,000	3/4	3/4	7/8	1 1/8	5/8	5/8	3/4	7/8
18,000	3/4	7/8	1 1/8	1 1/8	5/8	3/4	7/8	1 1/8
24,000	7/8	1 1/8	1 1/8	1 3/8	3/4	7/8	1 1/8	1 1/8
30,000	1 1/8	1 1/8	1 3/8	1 3/8	3/4	7/8	1 1/8	1 1/8
36,000	1 1/8	1 1/8	1 3/8	1 3/8	7/8	1 1/8	1 1/8	1 1/8
42,000	1 1/8	1 3/8	1 3/8	1 5/8	7/8	1 1/8	1 1/8	1 3/8
48,000	1 1/8	1 3/8	1 3/8	1 5/8	1 1/8	1 1/8	1 3/8	1 3/8
54,000	1 1/8	1 3/8	1 5/8	1 5/8	1 1/8	1 1/8	1 3/8	1 3/8
60,000	1 1/8	1 3/8	1 5/8	1 5/8	1 1/8	1 1/8	1 3/8	1 3/8
72,000	1 3/8	1 5/8	1 5/8	2 1/8	1 1/8	1 3/8	1 3/8	1 5/8
90,000	1 3/8	1 5/8	2 1/8	2 1/8	1 1/8	1 3/8	1 5/8	1 5/8
120,000	1 5/8	2 1/8	2 1/8	2 1/8	1 3/8	1 5/8	1 5/8	2 1/8
150,000	1 5/8	2 1/8	2 1/8	2 5/8	1 3/8	1 5/8	2 1/8	2 1/8
180,000	2 1/8	2 1/8	2 5/8	2 5/8	1 5/8	1 5/8	2 1/8	2 1/8

*Type L Copper Tubing
For Suction Temperatures Less Than — 20°F, The Next Larger Line
Size Must Be Used.

Courtesy Dunham-Bush, Inc.

**Liquid Lines Sizes

Net Evaporator Capacity	Total Equivalent Length of Liquid Line (Feet)							
	R12 & R502				R22			
	25	50	100*	150	25	50	100*	150
3000	1/4	1/4	1/4	3/8	1/4	1/4	1/4	1/4
4500	1/4	1/4	3/8	3/8	1/4	1/4	3/8	3/8
6500	1/4	3/8	3/8	3/8	1/4	1/4	3/8	3/8
8500	3/8	3/8	3/8	1/2	1/4	3/8	3/8	3/8
12,000	3/8	3/8	1/2	1/2	3/8	3/8	3/8	1/2
18,000	3/8	1/2	1/2	1/2	3/8	3/8	1/2	1/2
24,000	1/2	1/2	1/2	5/8	3/8	1/2	1/2	1/2
30,000	1/2	1/2	5/8	5/8	3/8	1/2	1/2	5/8
36,000	1/2	5/8	5/8	5/8	1/2	1/2	5/8	5/8
42,000	1/2	5/8	5/8	3/4	1/2	1/2	5/8	5/8
48,000	1/2	5/8	3/4	3/4	1/2	5/8	5/8	3/4
54,000	5/8	5/8	3/4	3/4	1/2	5/8	5/8	3/4
60,000	5/8	5/8	3/4	3/4	1/2	5/8	3/4	3/4
72,000	5/8	3/4	7/8	7/8	5/8	5/8	3/4	3/4
90,000	3/4	3/4	7/8	7/8	5/8	3/4	7/8	7/8
120,000	3/4	7/8	1 1/8	1 1/8	3/4	7/8	7/8	1 1/8
150,000	7/8	7/8	1 1/8	1 1/8	3/4	7/8	1 1/8	1 1/8
180,000	7/8	7/8	1 1/8	1 1/8	7/8	7/8	1 1/8	1 1/8
210,000	1 1/8	1 1/8	1 3/8	1 3/8	7/8	7/8	1 1/8	1 1/8
240,000	1 1/8	1 1/8	1 3/8	1 3/8	7/8	1 1/8	1 3/8	1 3/8
300,000	1 1/8	1 1/8	1 5/8	1 5/8	1 1/8	1 1/8	1 3/8	1 3/8

*Line Sizes Indicated Are Suitable For Condenser-to-Receiver Applications
**Type L Copper Tubing

Courtesy Dunham-Bush, Inc.

*Discharge Line Sizes

Net Evaporator Capacity	Total Equivalent Length of Hot Gas Line (Feet)							
	R12				R22 & R502			
	25	50	100	150	25	50	100	150
3000	3/8	3/8	1/2	1/2	3/8	3/8	3/8	3/8
4500	3/8	1/2	1/2	1/2	3/8	3/8	1/2	1/2
6500	1/2	1/2	5/8	5/8	3/8	1/2	1/2	1/2
8500	1/2	1/2	5/8	5/8	3/8	1/2	1/2	5/8
12,000	1/2	5/8	3/4	3/4	1/2	1/2	5/8	5/8
18,000	5/8	3/4	3/4	7/8	1/2	5/8	3/4	3/4
24,000	5/8	3/4	7/8	1 1/8	5/8	5/8	3/4	7/8
30,000	3/4	7/8	1 1/8	1 1/8	5/8	3/4	7/8	7/8
36,000	3/4	7/8	1 1/3	1 1/8	5/8	3/4	7/8	7/8
42,000	7/8	7/8	1 1/8	1 1/8	3/4	3/4	7/8	1 1/8
48,000	7/8	1 1/8	1 1/8	1 3/8	3/4	7/8	1 1/8	1 1/8
54,000	7/8	1 1/8	1 1/8	1 3/8	3/4	7/8	1 1/8	1 1/8
60,000	7/8	1 1/8	1 3/8	1 3/8	3/4	7/8	1 1/8	1 1/8
72,000	1 1/8	1 1/8	1 3/8	1 3/8	7/8	1 1/8	1 1/8	1 1/8
90,000	1 1/8	1 3/8	1 3/8	1 5/8	7/8	1 1/8	1 1/8	1 3/8
120,000	1 1/8	1 3/8	1 5/8	1 5/8	1 1/8	1 1/8	1 3/8	1 3/8
150,000	1 3/8	1 3/8	1 5/8	2 1/8	1 1/8	1 3/8	1 3/8	1 5/8
180,000	1 3/8	1 5/8	2 1/8	2 1/8	1 1/8	1 3/8	1 5/8	1 5/8
210,000	1 3/8	1 5/8	2 1/8	2 1/8	1 3/8	1 3/8	1 5/8	1 5/8
240,000	1 5/8	1 5/8	2 1/8	2 1/8	1 3/8	1 5/8	1 5/8	2 1/8
300,000	1 5/8	2 1/8	2 1/8	2 5/8	1 3/8	1 5/8	2 1/8	2 1/8

*Type L Copper Tubing

Courtesy Dunham-Bush, Inc.

Index

The Audel® Mail Order Bookstore

Here's an opportunity to order the valuable books you may have missed before and to build your own personal, comprehensive library of Audel books. You can choose from an extensive selection of technical guides and reference books. They will provide access to the same sources the experts use, put all the answers at your fingertips, and give you the know-how to complete even the most complicated building or repairing job, in the same professional way.

Each volume:

- ● **Fully illustrated**
- ● **Packed with up-to-date facts and figures**
- ● **Completely indexed for easy reference**

APPLIANCES

REFRIGERATION: HOME AND COMMERCIAL

Covers the whole realm of refrigeration equipment from fractional-horsepower water coolers, through domestic refrigerators to multi-ton commercial installations. 656 pages; 5½ x 8¼; hardbound. **Cat. No. 23286**

AIR CONDITIONING: HOME AND COMMERCIAL

A concise collection of basic information, tables, and charts for those interested in understanding, troubleshooting, and repairing home air conditioners and commercial installations. 464 pages; 5½ x 8¼; hardbound. **Cat. No. 23288**

HOME APPLIANCE SERVICING, 3rd Edition

A practical book for electric & gas servicemen, mechanics & dealers. Covers the principles, servicing, and repairing of home appliances. 592 pages; 5¼ x 8¼; hardbound. **Cat. No 23214**

REFRIGERATION AND AIR CONDITIONING LIBRARY—2 Vols.

Cat. No. 23305

OIL BURNERS, 3rd Edition

Provides complete information on all types of oil burners and associated equipment. Discusses burners—blowers—ignition transformers—electrodes—nozzles—fuel pumps—filters—Controls. Installation and maintenance are stressed. 320 pages; 5½ x 8¼; hardbound. **Cat. No. 23277**

See price list for cost.

All prices are subject to change without notice.

Use the order coupon on the back page of this book.

AUTOMOTIVE

AUTO BODY REPAIR FOR THE DO-IT-YOURSELFER

Shows how to use touch-up paint; repair chips, scratches, and dents; remove and prevent rust; care for glass, doors, locks, lids, and vinyl tops; and clean and repair upholstery. 96 pages; 8½ x 11; softcover. **Cat. No. 23238**

AUTOMOBILE REPAIR GUIDE, 4th Edition

A practical reference for auto mechanics, servicemen, trainees, and owners. Explains theory, construction, and servicing of modern domestic motorcars. 800 pages; 5½ x 8¼; hardbound. **Cat. No. 23291**

CAN-DO TUNE-UP™ SERIES

Each book in this series comes with an audio tape cassette. Together they provide an organized set of instructions that will show you and talk you through the maintenance and tune-up procedures designed for your particular car. All books are softcover.

AMERICAN MOTORS CORPORATION CARS

(The 1964 thru 1974 cars covered include: Matador. Rambler. Gremlin. and AMC Jeep (Willys).). 112 pages; 5½ x 8½; softcover. **Cat. No. 23843**
Cat. No. 23851 Without Cassette

CHRYSLER CORPORATION CARS

(The 1964 thru 1974 cars covered include: Chrysler, Dodge, and Plymouth.) 112 pages; 5½ x 8½; softcover. **Cat. No. 23825**
Cat. No. 23846 Without Cassette

FORD MOTOR COMPANY CARS

(The 1954 thru 1974 cars covered include: Ford, Lincoln, and Mercury.) 112 pages; 5½ x 8½; softcover. **Cat. No. 23827**
Cat. No. 23848 Without Cassette

GENERAL MOTORS CORPORATION CARS

(The 1964 thru 1974 cars covered include: Buick, Cadillac, Chevrolet, Oldsmobile and Pontiac.) 112 pages; 5½ x 8½; softcover. **Cat. No. 23824**
Cat. No. 23845 Without Cassette

PINTO AND VEGA CARS,

1971 thru 1974. 112 pages· 5½ x 8½; softcover. **Cat. No. 23831**
Cat. No. 23849 Without Cassette

TOYOTA AND DATSUN CARS.

1964 thru 1974. 112 pages; 5½ x 8½; softcover. **Cat. No. 23835**
Cat. No. 23850 Without Cassette

VOLKSWAGEN CARS

(The 1964 thru 1974 cars covered include: Beetle. Super Beetle. and Karmann Ghia.) 96 pages; 5½ x 8½; softcover. **Cat. No. 23826**
Cat. No. 23847 Without Cassette

AUTOMOTIVE AIR CONDITIONING

You can easily perform most all service procedures you've been paying for in the past. This book covers the systems built by the major manufacturers, even after-market installations. Contents: introduction—refrigerant—tools—air conditioning circuit—general service procedures—electrical systems—the cooling system—system diagnosis—electrical diagnosis—troubleshooting. 232 pages; 5½ x 8½; softcover. **Cat. No. 23318**

See price list for cost.

All prices are subject to change without notice.

Use the order coupon on the back page of this book.

DIESEL ENGINE MANUAL, 3rd Edition

A practical guide covering the theory, operation, and maintenance of modern diesel engines. Explains diesel principles—valves—timing—fuel pumps—pistons and rings—cylinders—lubrication—cooling system—fuel oil and more. 480 pages; 5½ x 8¼; hardbound. **Cat. No. 23199**

GAS ENGINE MANUAL, 2nd Edition

A completely practical book covering the construction, operation, and repair of all types of modern gas engines. 400 pages; 5½ x 8¼; hardbound. **Cat. No. 23245**

BUILDING AND MAINTENANCE

ANSWERS ON BLUEPRINT READING, 3rd Edition

Covers all types of blueprint reading for mechanics and builders. This book reveals the secret language of blueprints, step-by-step in easy stages. 312 pages; 5½ x 8¼; hardbound. **Cat. No. 23283**

BUILDING MAINTENANCE, 2nd Edition

Covers all the practical aspects of building maintenance. Painting and decorating; plumbing and pipe fitting; carpentry; heating maintenance; custodial practices and more. (A book for building owners, managers and maintenance personnel.) 384 pages; 5½ x 8¼; hardbound. **Cat. No. 23278**

COMPLETE BUILDING CONSTRUCTION

At last—a *one-volume* instruction manual to show you how to construct a frame or brick building from the footings to the ridge. Build your own garage, tool shed, other outbuilding—even your own house or place of business. Building construction tells you how to lay out the building and excavation lines on the lot; how to make concrete forms and pour the footings and foundation; how to make concrete slabs, walks, and driveways; how to lay concrete block, brick and tile; how to build your own fireplace and chimney. It's one of the newest Audel books, clearly written by experts in each field and ready to help you every step of the way. 800 pages; 5½ x 8¼; hardbound. **Cat. No. 23323**

GARDENING & LANDSCAPING

A comprehensive guide for homeowners and for industrial, municipal, and estate groundskeepers. Gives information on proper care of annual and perennial flowers; various house plants; greenhouse design and construction; insect and rodent controls; and more. 384 pages; 5½ x 8¼; hardbound. **Cat. No. 23229**

CARPENTERS & BUILDERS LIBRARY, 4th Edition (4 Vols.)

A practical, illustrated trade assistant on modern construction for carpenters, builders, and all woodworkers. Explains in practical, concise language and illustrations all the principles, advances, and shortcuts based on modern practice. How to calculate various jobs. **Cat. No. 23244**

      Vol. 1—Tools, steel square, saw filing, joinery cabinets. 384 pages; 5½ x 8¼; hardbound. **Cat. No. 23240**

      Vol. 2—Mathematics, plans, specifications, estimates. 304 pages; 5½ x 8¼; hardbound. **Cat. No. 23241**

      Vol. 3—House and roof framing, laying out foundations. 304 pages; 5½ x 8¼; hardbound. **Cat. No. 23242**

      Vol. 4—Doors, windows, stairs, millwork, painting. 368 pages; 5½ x 8¼; hardbound. **Cat. No. 23243**

See price list for cost.

All prices are subject to change without notice.

Use the order coupon on the back page of this book.

CARPENTRY AND BUILDING

Answers to the problems encountered In today's building trades. The actual questions asked of an architect by carpenters and builders are answered in this book. 448 pages; 5½ x 8¼; hardbound. **Cat. No.** 23142

WOOD STOVE HANDBOOK

The wood stove handbook shows how wood burned in a modern wood stove offers an immediate, practical, low-cost method of full-time or part-time home heating. The book points out that wood is plentiful, low in cost (sometimes free), and nonpolluting, especially when burned in one of the newer and more efficient stoves. In this book, you will learn about the nature of heat and its control, what happens inside and outside a stove, how to have a safe and efficient chimney, and how to install a modern wood burning stove. You will also learn about the different types of firewood and how to get it, cut it, split it, and store it. 128 pages; 8½ x 11; softcover. **Cat. No.** 23319

HEATING, VENTILATING, AND AIR CONDITIONING LIBRARY (3 Vols.)

This three-volume set covers all types of furnaces, ductwork, air conditioners, heat pumps, radiant heaters, and water heaters, including swimming-pool heating systems. **Cat. No.** 23227

Volume 1

Partial Contents: Heating Fundamentals . . . Insulation Principles . . . Heating Fuels . . . Electric Heating System . . . Furnace Fundamentals . . . Gas-Fired Furnaces . . . Oil-Fired Furnaces . . . Coal-Fired Furnaces . . . Electric Furnaces. **Cat. No.** 23248

Volume 2

Partial Contents: Oil Burners . . . Gas Burners . . . Thermostats and Humidistats . . . Gas and Oil Controls . . . Pipes, Pipe Fitting, and Piping Details . . . Valves and Valve Installations. 560 pages; 5½ x 8¼; hardbound.   **Cat. No.** 23249

Volume 3

Partial Contents: Radiant Heating . . . Radiators, Convectors, and Unit Heaters . . . Stoves, Fireplaces, and Chimneys . . . Water Heaters and Other Appliances . . . Central Air Conditioning Systems . . . Humidifiers and Dehumidifiers.   544 pages; 5½ x 8¼; hardbound. **Cat. No.** 23250

HOME MAINTENANCE AND REPAIR: Walls, Ceilings, and Floors

Easy-to-follow instructions for sprucing up and repairing the walls, ceiling, and floors of your home. Covers nail pops, plaster repair, painting, paneling, ceiling and bathroom tile, and sound control. 80 pages; 8½ x 11; softcover. **Cat. No.** 23281

HOME PLUMBING HANDBOOK , 2nd Edition

A complete guide to home plumbing repair and installation. 200 pages; 8½ x 11; softcover. **Cat. No.** 23321

MASONS AND BUILDERS LIBRARY—2 Vols.

A practical, illustrated trade assistant on modern construction for bricklayers, stone-masons, cement workers, plasterers, and tile setters. Explains all the principles, advances, and shortcuts based on modern practice—including how to figure and calculate various jobs. **Cat. No.** 23185

Vol. 1—Concrete, Block, Tile, Terrazzo. 368 pages; 5½ x 8¼; hardbound. **Cat. No.** 23182

Vol. 2—Bricklaying, Plastering, Rock Masonry, Clay Tile. 384 pages; 5½ x 8¼; hardbound. **Cat. No.** 23183

See price list for cost.

All prices are subject to change without notice.

Use the order coupon on the back page of this book.

PLUMBERS AND PIPE FITTERS LIBRARY—3 Vols.

A practical, illustrated trade assistant and reference for master plumbers, journeymen and apprentice pipe fitters, gas fitters and helpers, builders, contractors, and engineers. Explains in simple language, illustrations, diagrams, charts, graphs, and pictures, the principles of modern plumbing and pipe-fitting practices. **Cat. No. 23255**

Vol. 1—Materials, tools, roughing-in. 320 pages; 5½ x 8¼; hardbound. **Cat. No. 23256**

Vol. 2—Welding, heating, air-conditioning. 384 pages; 5½ x 8¼; hardbound. **Cat. No. 23257**

Vol. 3—Water supply, drainage, calculations. 272 pages; 5½ x 8¼; hardbound. **Cat. No. 23258**

PLUMBERS HANDBOOK

A pocket manual providing reference material for plumbers and/or pipe fitters. General information sections contain data on cast-iron fittings, copper drainage fittings, plastic pipe, and repair of fixtures. 288 pages; 4 x 6; softcover. **Cat. No. 23339**

QUESTIONS AND ANSWERS FOR PLUMBERS EXAMINATIONS,
2nd Edition

Answers plumbers' questions about types of fixtures to use, size of pipe to install, design of systems, size and location of septic tank systems, and procedures used in installing material. 256 pages; 5½ x 8¼; softcover. **Cat. No. 23285**

TREE CARE MANUAL

The conscientious gardener's guide to healthy, beautiful trees. Covers planting, grafting, fertilizing, pruning, and spraying. Tells how to cope with insects, plant diseases, and environmental damage. 224 pages; 8½ x 11; softcover. **Cat. No. 23280**

UPHOLSTERING

Upholstering is explained for the average householder and apprentice upholsterer. From repairing and reglueing of the bare frame, to the final sewing or tacking, for antiques and most modern pieces, this book covers it all. 400 pages; 5½ x 8¼; hardbound. **Cat. No. 23189**

WOOD FURNITURE: Finishing, Refinishing, Repairing

Presents the fundamentals of furniture repair for both veneer and solid wood. Gives complete instructions on refinishing procedures, which includes stripping the old finish, sanding, selecting the finish and using wood fillers. 352 pages; 5½ x 8¼; hardbound. **Cat. No. 23216**

ELECTRICITY/ELECTRONICS

ELECTRICAL LIBRARY

If you are a student of electricity or a practicing electrician, here is a very important and helpful library you should consider owning. You can learn the basics of electricity, study electric motors and wiring diagrams, learn how to interpret the NEC, and prepare for the electrician's examination by using these books. **Cat. No. 23359**

Electric Motors, 3rd Edition. 528 pages; 5½ x 8¼; hardbound. **Cat. No. 23264**

Guide to the 1978 National Electrical Code. 672 pages; 5½ x 8¼; hardbound. **Cat. No. 23308**

House Wiring, 4th Edition. 256 pages; 5½ x 8¼; hardbound. **Cat. No. 23315**

Practical Electricity, 3rd Edition. 496 pages; 5½ x 8¼; hardbound. **Cat. No. 23218**

Questions and Answers for Electricians Examinations, 6th Edition. 288 pages; 5½ x 8¼; hardbound. **Cat. No. 23307**

ELECTRICAL COURSE FOR APPRENTICES AND JOURNEYMEN

A study course for apprentice or journeymen electricians. Covers electrical theory and its applications. 448 pages; 5½ x 8¼; hardbound. **Cat. No. 23209**

See price list for cost.
All prices are subject to change without notice.
Use the order coupon on the back page of this book.

RADIOMANS GUIDE, 4th Edition

Contains the latest information on radio and electronics from the basics through transistors. 480 pages; 5½ x 8¼; hardbound. **Cat. No. 23259**

TELEVISION SERVICE MANUAL, 4th Edition

Provides the practical information necessary for accurate diagnosis and repair of both black-and-white and color television receivers. 512 pages; 5½ x 8¼; hardbound. **Cat. No. 23247**

ENGINEERS/MECHANICS/ MACHINISTS

MACHINISTS LIBRARY, 2nd Edition

Covers modern machine-shop practice. Tells how to set up and operate lathes, screw and milling machines, shapers, drill presses, and all other machine tools. A complete reference library. **Cat. No. 23300**

Vol. 1—Basic Machine Shop. 352 pages; 5½ x 8¼; hardbound. **Cat. No. 23301**

Vol. 2—Machine Shop. 480 pages; 5½ x 8¼; hardbound. **Cat. No. 23302**

Vol. 3—Toolmakers Handy Book. 400 pages; 5½ x 8¼; hardbound. **Cat. No. 23303**

MECHANICAL TRADES POCKET MANUAL

Provides practical reference material for mechanical tradesmen. This handbook covers methods, tools, equipment, procedures, and much more. 256 pages; 4 x 6; softcover. **Cat. No. 23215**

MILLWRIGHTS AND MECHANICS GUIDE, 2nd Edition

Practical information on plant installation, operation, and maintenance for millwrights, mechanics, maintenance men, erectors, riggers, foremen, inspectors, and superintendents. 960 pages; 5½ x 8¼; hardbound. **Cat. No. 23201**

POWER PLANT ENGINEERS GUIDE, 2nd Edition

The complete steam or diesel power-plant engineer's library. 816 pages; 5½ x 8¼; hardbound. **Cat. No. 23220**

QUESTIONS AND ANSWERS FOR ENGINEERS AND FIREMANS EXAMINATIONS, 3RD EDITION

Presents both legitimate and "catch" questions with answers that may appear on examinations for engineers and firemans licenses for stationary, marine, and combustion engines. 496 pages; 5½ x 8¼; hardbound. **Cat. No. 23327**

WELDERS GUIDE, 2nd Edition

This new edition is a practical and concise manual on the theory, practical operation, and maintenance of all welding machines. Fully covers both electric and oxy-gas welding. 928 pages; 5½ x 8¼; hardbound. **Cat. No. 23202**

WELDER/FITTERS GUIDE

Provides basic training and instruction for those wishing to become welder/fitters. Step-by-step learning sequences are presented from learning about basic tools and aids used in weldment assembly, through simple work practices, to actual fabrication of weldments. 160 pages· 8½ x 11; softcover; **Cat. No. 23325**

See price list for cost.

All prices are subject to change without notice.

Use the order coupon on the back page of this book.

FLUID POWER

PNEUMATICS AND HYDRAULICS, 3rd Edition

Fully discusses installation, operation, and maintenance of both HYDRAULIC AND PNEUMATIC (air) devices. 496 pages; 5½ x 8¼; hardbound. **Cat. No. 23237**

PUMPS, 3rd Edition

A detailed book on all types of pumps from the old-fashioned kitchen variety to the most modern types. Covers construction, application, installation, and troubleshooting. 480 pages; 5½ x 8¼; hardbound. **Cat. No. 23292**

HYDRAULICS FOR OFF-THE-ROAD EQUIPMENT

Everything you need to know from basic hydraulics to troubleshooting hydraulic systems on off-the-road equipment. Heavy-equipment operators, farmers, fork-lift owners and operators, mechanics—all need this practical, fully illustrated manual. 272 pages; 5½ x 8¼; hardbound. **Cat. No. 23306**

HOBBY

COMPLETE COURSE IN STAINED GLASS

Written by an outstanding artist in the field of stained glass, this book is dedicated to all who love the beauty of the art. Ten complete lessons describe the required materials, how to obtain them, and explicit directions for making several stained glass projects. 80 pages; 8½ x 11; softbound. **Cat. No. 23287**

BUILD YOUR OWN AUDEL DO-IT-YOURSELF LIBRARY AT HOME!

Use the handy order coupon today to gain the valuable information you need in all the areas that once required a repairman. Save money and have fun while you learn to service your own air conditioner, automobile, and plumbing. Do your own professional carpentry, masonry, and wood furniture refinishing and repair. Build your own security systems. Find out how to repair your TV or Hi-Fi. Learn landscaping, upholstery, electronics and much, much more.

See price list for cost.
All prices are subject to change without notice.
Use the order coupon on the back page of this book.

PRICE LIST

	Price		Price
23142	$10.95	23264	$10.95
23182	9.95	23277	9.95
23183	9.95	23278	9.95
23185	17.95	23280	8.95
23189	9.95	23281	6.95
23199	10.95	23283	9.95
23201	14.95	23284	21.95
23202	14.95	23285	8.95
23209	10.95	23286	12.95
23214	12.95	23287	6.95
23215	8.95	23288	10.95
23216	9.95	23291	14.95
23218	10.95	23292	10.95
23227	32.95	23300	29.95
23229	9.95	23301	10.95
23237	10.95	23302	10.95
23238	6.95	23303	10.95
23240	9.95	23305	21.95
23241	9.95	23306	8.95
23242	9.95	23307	8.95
23243	9.95	23308	12.95
23244	35.95	23315	8.95
23245	9.95	23318	7.95
23247	11.95	23319	7.95
23248	11.95	23321	7.95
23249	11.95	23323	19.95
23250	11.95	23325	7.95
23255	26.95	23327	10.95
23256	9.95	23329	15.95
23257	9.95	23339	8.95
23258	9.95	23359	47.95
23259	11.95		

USE ORDER COUPON ON THE BACK PAGE OF THIS BOOK
ALL PRICES SUBJECT TO CHANGE WITHOUT NOTICE